TWENTIETH CENTURY PHYSICS

Claude Garrod
Professor of Physics
University of California

Faculty Publishing, Inc.
Davis, California

To my wife
Nadja

Published by

Faculty Publishing, Inc.
1421 Tulane Drive, Davis, CA 95616

ISBN 0-915141-00-0

Printed in the United States of America

PREFACE

The physics of the Twentieth Century has a distinct intellectual flavor or style that sets it off from most earlier scientific work. On one's first exposure to it, it is likely to seem abstract and unnatural in comparison to older parts of physics such as mechanics and optics.

There is a simple explanation for the change in the content and style of physical theory that occurred at about 1900. Previous to that time most of physics grew out of people's ordinary experiences. The books of Galileo make it clear that the science of mechanics originated with problems in building construction and simple machinery. Optics developed once lenses had been invented and it was necessary to predict their properties. In fact almost all of earlier physics could be described as a sharpening and deepening of one's common sense knowledge of the world. Of course, in some cases, notions that had seemed like simple common sense were in fact simply wrong and had to be eliminated. For example, the rather obvious generalization from everyday experience that is embodied in the Aristotelian law that an object will come to rest unless a force is continuously applied to keep it moving was a serious obstacle to progress in mechanics. In fact, it is still difficult to eliminate from most people's unconscious thinking about mechanical problems. But, with time and effort, one can revise and refine one's physical intuition to incorporate Newtonian ideas in place of Arostotelian ones. Once that has been done, Newtonian mechanics seems to provide a beautifully clear picture of how the world works.

During the Nineteenth Century, because of the steady improvement in scientific technology, it became possible to extend a person's powers of observation beyond anything that could be considered ordinary experience. Both the subtlety and the range of human observation was increased tremendously by means of scientific instruments. One could now make precise measurements involving atoms and electrons, whose sizes, in comparison with any ordinary things, were incomprehensibly small. One could, by a combination of measurement and analysis, determine what would be the effects of traveling at speeds approaching the speed of light; a speed completely removed from anything ever experienced by a human being.

What was discovered is that, far outside the range of experience in which it was formed, 'common sense' is a very unreliable guide. Atoms and galaxies are not constrained by the limits of human imagination. They are free to do things that we are completely unable to picture. Fortunately, they are constrained by the rules of logic and mathematics, and therefore they are amenable to mathematical and scientific analysis.

With a great deal of work and inventiveness, it has been possible to construct laws that can be extended far outside our normal range of experience. Those laws allow us to make accurate predictions in those extended regions, but even with the power to make predictions, we still cannot form adequate 'pictures' in our minds of the phenomena we are analyzing. For example, relativity theory shows that the basic geometrical structure that must be used in analyzing the world is a four-dimensional geometry. Now it is possible, and not particularly difficult, to learn how to mathematically analyze four-dimensional objects, but it is not possible to form pictures of them. In both of the major fields of Twentieth Century Physics, namely relativity and quantum mechanics, we have to learn not to be misled by our imperfect visualizations of the things we are analyzing. For most of us, who have a deep feeling that we only really know something when we can form a clear picture of it, it is somewhat difficult to find our way by rules and logic rather than by pictures. It is difficult, but it is also exciting, to discover that with careful thought we can understand things that are fundamentally different from anything in our common experience; to discover that our minds can guide us where our eyes cannot.

In writing a textbook for a small course on the large subject of modern physics I have had to make many compromises. In each field, such as atomic physics, nuclear physics, or statistical mechanics, I have tried to strike a reasonable balance between breadth and depth. I have had to decide what are the most important basic principles in that field and then find ways to present them in sufficient detail for the student to comprehend them. I have then presented applications of those basic principles to as many specific problems as I felt was reasonable and useful. When one considers the fact that the instructor who uses this text will have about one week in which to complete each one of these fields it becomes clear why I have had to leave out more things than I put in.

Each chapter ends with a Summary, that repeats the most important ideas of the chapter in abbreviated form. The summary is followed by a set of problems. The difficulty of each problem is indicated by a system of stars, ranging from none for an easy problem to three stars for a very challenging one.

Before I close this preface I must express my deepest thanks to certain people whose assistance and support was invaluable in producing this book. First to my daughter Susanna, who did most of the drawings and helped in many other ways with the manuscript. Second to my wife Nadja, whose understanding and encouragement made its completion possible. I would also like to thank Professor Winston Ko for the cover illustration.

CONTENTS

CHAPTER ONE - THE THEORY OF RELATIVITY

1.1	Galilean Relativity	1
1.2	The Effect of Electromagnetic Theory on the Relativity Principle	4
1.3	Attempts to Measure the Velocity of the Ether	6
1.4	Einstein's Theory of Relativity	12
1.5	The Lorentz Transformation	14
1.6	Time Dilation	20
1.7	Length Contraction	21
1.8	The Geometry of Spacetime	22
1.9	The Relativistic Addition of Velocity Law	24
1.10	Three-Dimensional Velocity Addition	26
1.11	The Doppler Effect	28
	Summary	31
	Problems	33

CHAPTER TWO - RELATIVISTIC DYNAMICS AND GENERAL RELATIVITY

2.1	Momentum Conservation	36
2.2	Relativistic Velocity	37
2.3	Relativistic Momentum	39
2.4	Conservation of Relativistic Momentum	40
2.5	"Does the Inertia of a Body Depend Upon Its Energy Content?"	44
2.6	Particles of Zero Mass	48
2.7	The Equivalence Principle	48
2.8	General Relativity	51
2.9	The Moving Clock Problem	53
2.10	The Spacetime Metric	55
2.11	The Pound-Rebka Experiment	57
2.12	The Curvature of Space	58
2.13	Black Holes	59
	Summary	61
	Problems	63

CHAPTER THREE - THE ORIGINS OF QUANTUM PHYSICS

3.1	Classical Physics	66
3.2	Cavity Radiation	67
3.3	The Photoelectric Effect	73
3.4	The Compton Effect	74
3.5	Alpha-Particle Scattering	78
3.6	Energy Quantization in Atoms	79
3.7	The Franck-Hertz Experiment	81
3.8	The Hydrogen Atom	82
3.9	De Broglie Waves	85
3.10	The Davisson-Germer Experiment	89
	Summary	92
	Problems	94

CHAPTER FOUR - THE WAVE EQUATION

4.1	The Diffraction of Light and Electrons	97
4.2	The Relationship Between Quantum and Classical Physics	100
4.3	The Schroedinger Equation	102
4.4	The Physical Meaning of the Wave Function	104
4.5	Standing Waves and Discrete Energy Values	106
4.6	Normalization of the Wave Function	109
4.7	The Heisenberg Uncertainty Relation	110
4.8	The Schroedinger Equation With a Potential	113
4.9	Quantum States of a Particle in a Potential	114
	Summary	125
	Problems	127

CHAPTER FIVE - SOLVING THE SCHROEDINGER EQUATION

5.1	Stationary States	130
5.2	The Step Potential	131
5.3	Particle Conservation	133
5.4	Barrier Penetration	133
5.5	The Quantized Bead on a Circular Wire	138
5.6	Quantization of Angular Momentum	139
5.7	The Harmonic Oscillator	141
5.8	The Wave Packet Solution	146
	Summary	148
	Problems	151

CHAPTER SIX - STATES, OPERATORS, AND MEASUREMENTS

6.1	The Need for More Detail	154
6.2	States	154
6.3	Operators	156
6.4	Operator Eigenvalues	156
6.5	The Momentum Operator	157
6.6	Measurements	158
6.7	The Position Operator	159
6.8	Other Observables	160
6.9	The Calculation of Probabilities	161
6.10	Setting Up a Quantum State	163
6.11	The Standard Quantum Mechanical Experiment	165
6.12	Electron Scattering by Nuclei	167
6.13	Commutation of Operators	169
6.14	The Fundamental Commutator Theorem	170
	Summary	171
	Problems	173

CHAPTER SEVEN - THE STRUCTURE OF ATOMS AND MOLECULES

7.1	The Hydrogen Atom	175
7.2	Positive and Negative Energy Solutions	178
7.3	The Atomic Quantum Length	179
7.4	Radial Solutions of the Schroedinger Equation	180
7.5	Nonradial Solutions of the Schroedinger Equation	183
7.6	The Detailed Wave Functions	186
7.7	The Zeeman Effect	189
7.8	Electron Spin	192
7.9	The Exclusion Principle	193
7.10	The Periodic Table of the Elements	194
7.11	The Self-Consistent Field Approximation	196
7.12	Lengths of the Periods	197
7.13	Molecular Structure	198
7.14	The Hydrogen Molecular Ion	199

7.15	Molecular Vibration	201
7.16	Molecular Rotation	204
7.17	Selection Rules for Molecular Transitions	205
7.18	The Vibrational-Rotational Spectrum of Diatomic Molecules	206
7.19	Covalent Bonding in Carbon Compounds	207
7.20	Other Tetrahedral Compounds	211
	Summary	213
	Problems	216
	CHAPTER EIGHT - NUCLEAR PHYSICS	
8.1	The Nuclear Force	218
8.2	Nuclear Fluid	220
8.3	Binding Energies of Nuclei	222
8.4	The Law of Radioactive Decay	228
8.5	Nuclear Line Widths	230
8.6	The Mossbauer Effect	231
8.8	Nuclear Shells	237
8.9	Nuclear Fission and Fusion	238
8.10	Fusion	239
8.11	Magnetic Confinement	241
8.12	Inertial Confinement	245
8.13	Nuclear Fission	245
8.14	Neutron Induced Fission	250
8.15	The Controlled Fission Reactor	251
8.16	The Yukawa Theory of the Nuclear Force	251
	Summary	255
	Problems	257
	CHAPTER NINE - HIGH ENERGY PHYSICS	
9.1	Introduction	259
9.2	Particle Production	260
9.3	Colliding Beam Experiments	267
9.4	Particle Detectors	270
9.5	The Structure of Matter	274
9.6	The Fundamental Fermions	275
9.7	The Fundamental Bosons	276
9.8	Quantum Chromodynamics	276
9.9	Quark Structure of the Hadrons	278
9.10	Interactions in QCD	280
9.11	Strong, Electromagnetic, and Weak Decays	285
9.12	Quantum Field Theory	290
9.13	Three-Dimensional Waves	292
9.14	The Particle Interpretation	293
	Summary	295
	Problems	299
	CHAPTER TEN - THE QUANTUM THEORY OF SOLIDS	
10.1	Amorphous and Crystalline Solids	301
10.2	Chemical Bonding in Solids	302
10.3	The Crystal Lattice	305
10.4	Crystal Symmetry Types	306
10.5	The Classical Theory of Lattice Vibrations	311
10.6	Lattice Vibrations in 2- and 3-Dimensional Crystals	314
10.7	Polarization of Vibrational Waves	315
10.8	The Dispersion Relation	316
10.9	Quantized Vibrational Waves	317
10.10	The Density of Normal Modes	319
10.11	Crystal Momentum	322
10.12	Phonons	323

10.13	The Phonon Gas	324
10.14	Lattice Energy and Heat Capacity	325
10.15	Electrons in Solids	327
10.16	Energy Band Structure in 1-Dimensional Crystals	328
10.17	Band Structure in 2 and 3 Dimensions	329
10.18	Insulators and Conductors	330
10.19	Semiconductors	332
10.20	Electrical Conduction by Holes	333
10.21	P-Type and N-Type Semiconductors	337
10.22	Doped Germanium Semiconductors	338
10.23	PN Junctions	340
10.24	Transistors	341
	Summary	343
	Problems	347

CHAPTER ELEVEN - QUANTUM STATISTICAL MECHANICS

11.1	What is Quantum Statistical Mechanics?	349
11.2	Entropy	350
11.3	The Approach to Equilibrium of a Quantum System	352
11.4	The Gibbs Distribution	357
11.5	The Heat Capacity of a Quantum Mechanical System	360
11.6	Vibrational Specific Heats	361
11.7	The Heat Capacity of a Solid	362
11.8	The Derivation of Planck's Formula	364
11.9	The Rotational Heat Capacity of a Diatomic Gas	365
11.10	The Gibbs Distribution for an Open System	367
11.11	The Ideal Quantum Gas	368
11.12	The Ideal Fermi-Dirac Gas	369
11.13	The Ideal Bose-Einstein Gas	371
11.14	The Absorption and Emission of Radiation	372
11.15	The Laser	374
	Summary	376
	Problems	379
	Index	381

CHAPTER ONE

THE THEORY OF RELATIVITY

1.1. GALILEAN RELATIVITY

Although, in the popular mind, Einstein's theory of relativity has become the very symbol of something deep and difficult, the truth is that the principle of relativity is a technical name for a fact about our experience that seems so trivial and natural that few of us ever notice it unless it is pointed out to us. It is common knowledge that one experiences no difficulty in walking in a train, bus, or plane that is moving smoothly at constant velocity. Sitting in a train traveling at fifty miles per hour, one can toss up a ball and confidently expect it to behave in the same way it does for a person sitting on a bench in the station. In fact, in a smoothly moving train with the window shades down, there is no way at all of detecting the motion of the train by means of mechanical experiments. This is an example of the *Galilean principle of relativity.* It was described in a very lively way by Galileo in his book entitled *Dialogue on the Great World Systems.*

> *Salviatus*: As a final proof that all the aforementioned experiments would yield null results, let me now describe a simple way to carry them out. With some friends, shut yourself up in the cargo hold of a large ship. Take with you some gnats, flies, and other winged insects. Also set up a big tub of water containing a few fish and hang up an inverted bottle which slowly drips into another narrow-mouth bottle below it. While the ship is stationary observe how the insects fly with equal facility and speed in all directions, how the fish swim indifferently towards all sides of the tub, and how the drops fall straight down into the bottle below. And if you throw an object to your friend, you need throw it no harder in one direction than another, provided the distances be equal. You will also find that you can jump as far in one direction as in any other. Of course, no one doubts that, on a still ship, all these things will take place as described. Having observed them, now let the ship sail with whatever speed you like but smoothly, not tossing or lurching this way and that. You will not find the smallest change in any of the phenomena we've discussed. And therefore you will not be able to determine by means of them whether the ship be moving or stationary. The underlying cause of this remarkable effect is that the ship's motion is shared by all the things contained in it; including the air in the hold....

Sagredus: Although I've never actually carried out those experiments while I was at sea, I'm completely sure that they would all come out exactly as you've described. For I recall a hundred occasions when, being in my cabin I wondered whether the ship was moving or stationary. In fact, sometimes I felt sure it was moving forward when in reality it was moving backward.

This is the earliest, and yet one of the clearest, descriptions of the principle of relativity. The fact that it was written in 1638 shows that the idea was widely known during the period when the fundamental laws of motion were under development.

There is no doubt that Newton used the Galilean principle of relativity in formulating his basic laws of mechanics. Once those laws are postulated, the principle that uniform motion, such as takes place on a train, is undetectable by mechanical experiments may then be derived as a mathematical theorem from those laws. More exactly, that theorem states that Newton's laws of motion are valid without modification if all position measurements are made relative to an object (e.g. the train) that is moving at constant velocity without rotation. Since a ball tossed in the air by someone on the train platform moves on a trajectory that satisfies Newton's equations of motion, the principle of relativity states that a ball tossed into the air by a person on the moving train should move on the same trajectory *relative to the person in the train.* This principle is not a new law of mechanics to be added to the usual three laws of mechanics. Rather it is a direct consequence of the mathematical form of those laws and the law of addition of velocities. To see this, let us consider the motion of a ball tossed into the air by the train passenger. We assume that the passenger is an experimental physicist with a meager travel allowance who must therefore do her traveling on flatcars (see Figure 1.1). By means of a large carpenter's square she has set up a coordinate system on the flatcar with respect to which she measures the velocity of the moving ball. The velocity she observes at time t we shall call $\mathbf{v}(t)$. Her

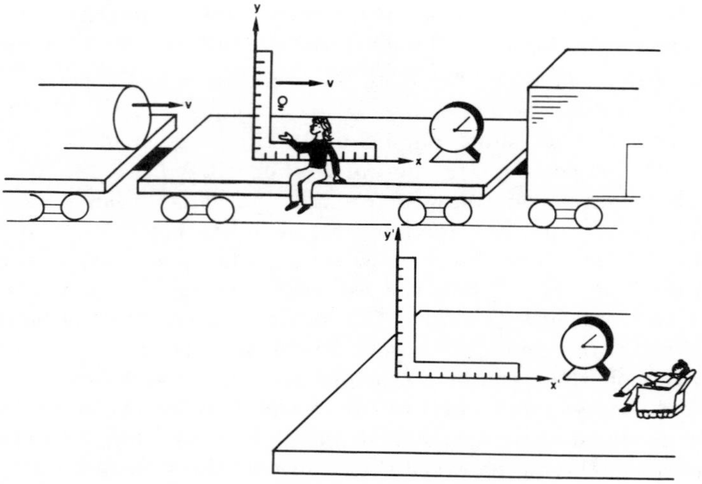

FIGURE 1.1 The motion of a ball being recorded by observers in two different reference frames.

more sedentary counterpart on the station platform has also set up a coordinate system in which he measures the velocity of the same ball. He naturally observes a different velocity which we shall call $\mathbf{v}'(t)$. The relationship between $\mathbf{v}'(t)$ and $\mathbf{v}(t)$ is given by the velocity addition formula,

$$\mathbf{v}'(t) = \mathbf{v}(t) + \mathbf{V}, \tag{1.1}$$

where $\mathbf{V}$ is the constant velocity of the train with respect to the platform. The gravitational force on the ball is $m\mathbf{g}$ where $\mathbf{g}$ is the gravitational field vector. Newton's law of motion tells us that

$$\frac{d\mathbf{v}'}{dt} = \mathbf{g} \tag{1.2}$$

If we differentiate both sides of Equation 1.1 and use the fact that $d\mathbf{V}/dt = 0$ (i.e. the velocity of the train is constant), we get

$$\frac{d\mathbf{v}'}{dt} = \frac{d\mathbf{v}}{dt}$$

Thus

$$\frac{d\mathbf{v}}{dt} = \mathbf{g}$$

which shows that the ball has exactly the same acceleration when its position is measured by the uniformly moving observer as it has when its position is measured by the observer at rest on the platform. Newton himself was quite familiar with the fact that his laws of mechanics satisfied a relativity principle. It is clear that, if the laws of mechanics did not satisfy such a relativity principle, then their application to motion on the surface of the earth would be extremely complicated. If one assumes that the laws of mechanics are valid in the frame of reference of the sun but not in any system of coordinates moving at constant velocity with respect to the sun, then, since a piece of the earth's surface carries out a complicated high speed motion, it would be necessary to know your velocity through space at a given instant in order to predict the motion of any mechanical objects near you. The local rules of mechanics, which must be used in such simple operations as walking, would change from hour to hour as the combination of our rotational velocity around the earth's axis and the velocity of the earth in its orbit around the sun added in different ways to give different total velocities with respect to the sun. Actually the laws of Newtonian mechanics are not exactly valid if position and velocity measurements are made with respect to the surface of the earth. The surface of the earth does not move at constant velocity. It accelerates, and unlike constant velocity motion, accelerated motion does have observable effects. Fortunately, although the velocity of the earth's surface is very large by ordinary standards, its acceleration is very small and can therefore usually be ignored. When it cannot be ignored such as in calculating satellite orbits, things really do get very complicated.

> **QUESTION**: When we use a coordinate system attached to the surface of the earth, the most important acceleration effect is due to the rotation of the earth about its axis. At the equator, what is the magnitude of the centripetal acceleration and how does it compare with the gravitational acceleration, $g = 9.8\,\mathrm{m/s^2}$?

ANSWER:

$$a = \omega^2 r$$

where $\omega = 2\pi$ radians/day $= 7.27\times10^{-5}$ rad/s and $r = 6.3\times10^{6}$ m.

Therefore $a = 3.33\times10^{-2}\,m/s^2$ and $a/g = 3.4\times10^{-3}$. The rotation of the earth affects the motion of a projectile at a level of about one part in three hundred. The effect is small but easily noticed with accurate measurements.

1.2. THE EFFECT OF ELECTROMAGNETIC THEORY ON THE GALILEAN RELATIVITY PRINCIPLE.

When Maxwell developed the mathematical theory of electromagnetic wave propagation it seemed as though the principle of relativity would have to be abandoned. It was perfectly obvious to Maxwell and to all of his contemporaries that a wave was some kind of disturbance that traveled through a medium. Sound waves were pressure variations that traveled through air or other materials, water waves were height variations that traveled through water, and electromagnetic waves must be variations in some other quantity associated with the electric and magnetic fields that traveled through some yet undescribed medium. The medium or material that carried electromagnetic waves came to be called the ether. Maxwell and others tried without much success to construct theoretical models of a mechanical medium that would propagate waves in accordance with Maxwell's equations for electromagnetic wave propagation. It is not worthwhile to discuss these attempts in any detail. They are mentioned only to bring out the fact that it was universally expected among nineteenth century physicists that one would see light waves propagate at the same speed in all directions only if one were at rest with respect to the ether. A coordinate system at rest with respect to the ether was called an absolute rest frame.

The Galilean principle of relativity says that constant velocity motion through empty space cannot be detected by means of mechanical experiments. A more general relativity principle, incorporating electromagnetic theory would state that constant velocity motion could not be detected by means of any type of experiments. By considering the situation shown in Figure 1.2 it seems clear that

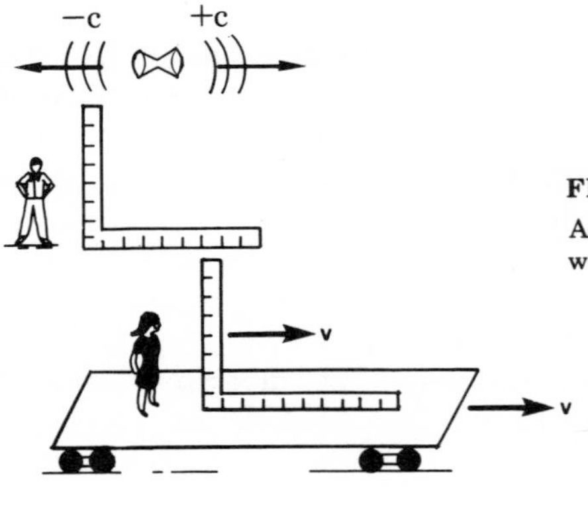

FIGURE 1.2

As observed by the woman the two light signals would propagate with speeds c+v and c−v.

no such relativity principle could be valid. If the man is assumed to be in an absolute rest frame so that the two light signals are traveling in opposite directions with speed c as measured by him, then, by the velocity addition law, they must be traveling at speeds $c + v$ and $c - v$ with respect to the woman. Thus she could easily determine her velocity through the absolute rest frame in which Maxwell's equations hold. Considering the fact that electromagnetic theory seemed not to satisfy any principle of relativity, certain aspects of the theory appeared very puzzling or strangely coincidental. These puzzling aspects of the theory are exemplified by the situation shown in Figure 1.3. Two people approach one another with constant relative speed v. The woman holds a bar magnet in her hand with the North pole pointed at the man. The man holds a large loop of wire in his hand. In the loop of wire is an ammeter. Both see the ammeter

FIGURE 1.3
Both observers can explain the current generated in the wire loop, but their explanations differ.

register a pulse of current in the loop as the magnet approaches and enters it. From the point of view of the man the generation of the current would be explained as follows.

Man's Explanation.

The loop is stationary. The electrons in the loop are also initially stationary and would therefore experience no force due to the magnetic field of the magnet. Since the magnet is moving, the magnetic field in the vicinity of the loop is changing. Changing magnetic fields create electric fields. The electric field created is in such a direction as to drive a current through the loop of just the observed value.

From the point of view of the woman the observed current would be explained as follows.

Woman's Explanation.

The magnet is stationary and therefore there is no electric field anywhere. The electrons in the loop are moving, along with the loop, toward the magnet. A charged particle moving in a magnetic field experiences a force given by

$$\mathbf{F} = q\mathbf{v}\times\mathbf{B}.$$

That force is in the direction of the loop and therefore causes the electrons to move around the loop. Again a detailed calculation of the current that should be caused by the magnetic force gives a result that agrees with the ammeter reading.

Each person has no difficulty in predicting the current in the loop by applying Maxwell's equations in his own frame of reference. This seemed to be a very odd coincidence for a theory that was supposed to be applicable only in the ether rest frame.

1.3. ATTEMPTS TO MEASURE THE VELOCITY OF THE ETHER.

In order to clarify the details of the electromagnetic theory of light a number of experiments were devised whose aims were to detect the effects of ether motion on light propagation. We shall describe three of the most important ones.

The Fizeau Experiment: In 1851, before the publication of Maxwell's treatise on electricity and magnetism and the widespread identification of light as electromagnetic waves, the French physicist Armand Hippolyte Fizeau carried out the following experiment to measure the velocity of light propagation through a moving medium. The medium was water. The basic apparatus is shown in Figure 1.4. By means of a half-silvered mirror* a monochromatic light beam from a mercury lamp is split into two beams. Each beam passes through a different tube of water. The light beams are then brought together again and produce an interference pattern consisting of light and dark fringes on the screen. A bright fringe passes through any point P for which the time taken by a wave to traverse the path ABCP differs from the time to traverse the path AB'C'P by an integral number of periods. The time it takes for the light to traverse one of the paths obviously depends on the speed of the light as it goes through the corresponding tube. If the speed of the light is in any way affected by the water flow, then the time taken along the two paths will be shifted in opposite directions whenever the flow rate of the water is changed. A point of the screen that was occupied by a bright line for zero flow rate might be occupied by a dark line for some other flow rate. By looking at the shift in the interference pattern on the screen one can determine the speed of light through the moving water.

When the experiment was carried out by Fizeau, he found that the speed of light in the two tubes was

$$V = \frac{c}{n} \pm (1 - \frac{1}{n^2})\, v_{water} \qquad (1.3)$$

where c is the speed of light in vacuum and n is the refractive index of water. This is the formula one would expect to see if the ether were being partially dragged along with the moving water at a speed

$$v_{ether} = (1 - \frac{1}{n^2})\, v_{water}.$$

This result, which was so puzzling during the nineteenth century, will later be seen to follow very naturally from the relativity theory.

---

*An ordinary mirror is made by coating the back surface of a sheet of glass with an uncolored metal such as silver or aluminum. A "half- silvered" mirror is made in the same way except for the fact that the metallic coating is so light (it is only a fine spray of metal) that half the light is transmitted and only half reflected.

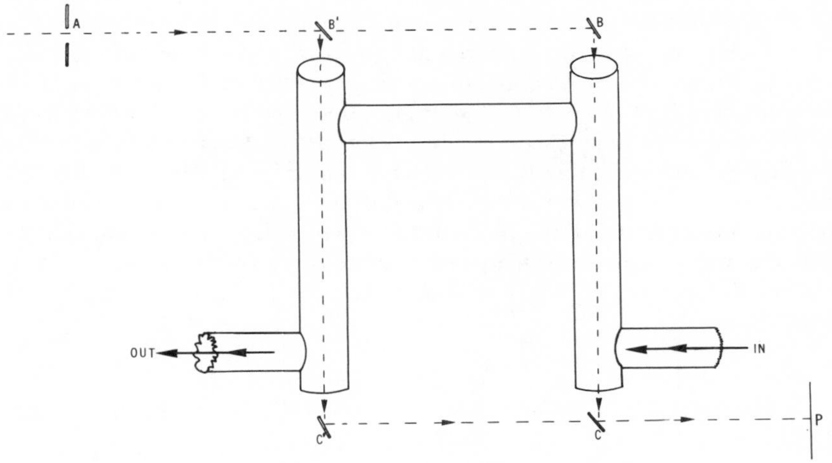

FIGURE 1.4 The Fizeau experiment. Water flows in from the right. The mirrors at B′ and C are half-silvered.

The Trouton-Noble Experiment: In order to grasp the theory behind this experiment we consider two metal spheres connected by a rigid nonconducting rod. The spheres carry equal and opposite electric charges (See Figure 1.5). The object moves through the ether (i.e. through the absolute rest frame) at constant velocity without rotation. According to Maxwell's equations the electric field produced by a positive charge moving through space at constant velocity points directly away from the charge in all directions. It is not exactly the Coulomb field because it is somewhat weaker in directions along the line of motion than it is in directions perpendicular to the line of motion. It can be described by a set of straight field lines that move along with the charge like a rigid body. (See Figure 1.6) From a consideration of this field pattern it is clear that the mutual attractive electrical forces on the oppositely charged spheres will cancel so as to produce zero net force and zero net torque on the whole object. Now let us show

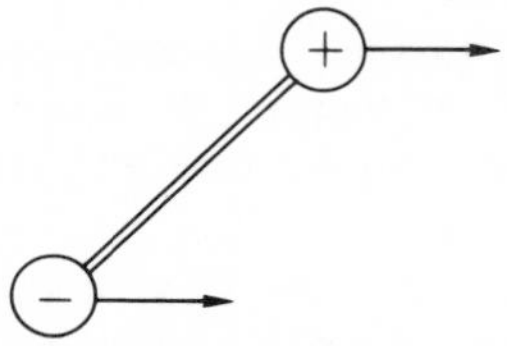

FIGURE 1.5
A rod with two oppositely charged spheres attached, moves through the ether.

FIGURE 1.6
The electric field of a moving charge.

that the same is not true for the magnetic forces. The combination of the magnetic forces on the two spheres produces a torque on the object. The magnetic field lines of a charge moving at constant velocity are circles with their centers along the line of motion and their planes perpendicular to the velocity vector. The magnetic field of the moving positive charge produces a force on the negative charge perpendicular to the velocity vector, as shown in Figure 1.7. The magnetic force on the positive charge is equal and opposite, thus producing no net force on the object. But the magnetic forces do produce a net torque which tends to line up the rod in a direction perpendicular to the velocity vector.

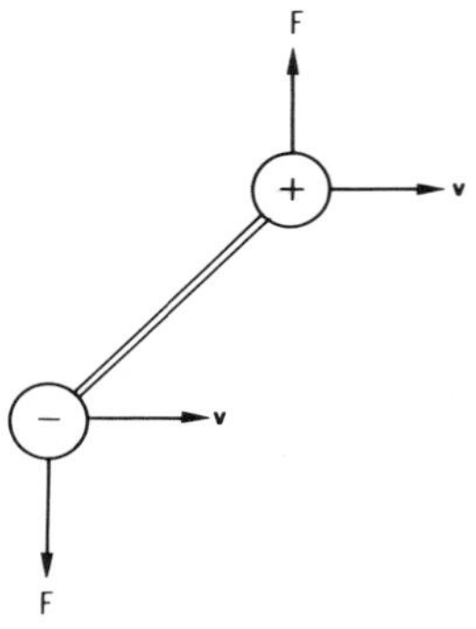

FIGURE 1.7
The magnetic field of the moving positive charge creates a force on the moving negative charge and vice versa.

If we suspend such an object on a thin thread attached at its center of mass, we could expect it to turn so as to align itself perpendicular to the velocity vector of the earth moving through the ether. This device seems to be capable of detecting our absolute velocity through space, where *space* means that coordinate frame in which Maxwell's equations are valid. Such an experiment was carried out by Trouton and Noble using a charged parallel plate capacitor suspended on a fine thread. They found no such effect although the sensitivity of their device was more than ample to detect the torque that would be produced by a velocity of the size of the earth's orbital velocity around the sun.

The Michelson-Morley Experiment: The most sensitive of all the experiments done to detect the motion of the earth through the ether was that carried out by the American physicists A. A. Michelson and E. W. Morley in 1887. Essentially the same experiment, but with modifications to improve the sensitivity, was repeated a number of times in subsequent years. We shall first consider a drastically simplified model of the experimental apparatus and then a more realistic model. (See Figure 1.8.)

(a) The aim of the experiment is to compare the times it takes wave fronts to travel from the source to the two mirrors and back to the source. We shall assume that the whole apparatus is rigidly constructed and is moving, through the stationary ether, to the right with a velocity v that is less than the velocity of light c.

(b) With respect to the apparatus the light wave travels at a speed c−v when moving toward the right-hand mirror and with a speed c+v when returning to the source (Figure 1.9). The total time taken for the trip back and forth in the direction parallel to the velocity of the apparatus is

$$T_{para} = \frac{L}{c-v} + \frac{L}{c+v} = \frac{2cL}{c^2 - v^2}$$

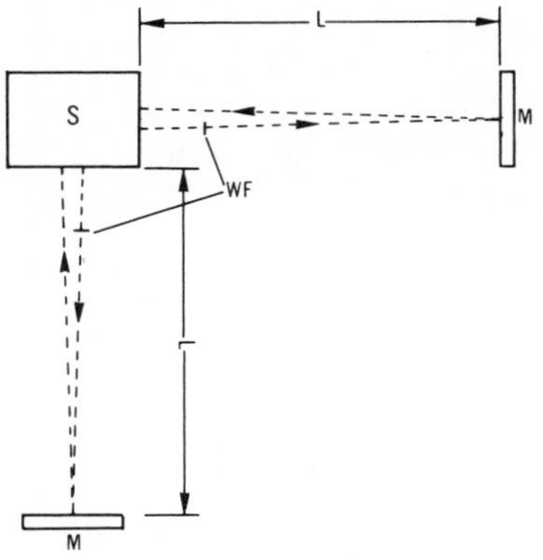

FIGURE 1.8

A simplified model of the Michelson-Morley experiment. S is a source of wave fronts (WF) that travel in perpendicular directions to the mirrors (M) and back.

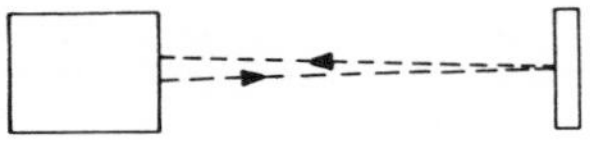

FIGURE 1.9

The light wave moving back and forth in the parallel arm of the Michelson-Morley experiment.

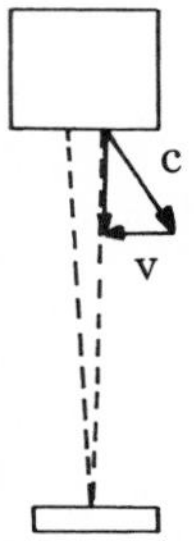

FIGURE 1.10

The light wave moving back and forth in the perpendicular arm.

(c) If we now look at the wave front that travels toward the mirror that is in a direction perpendicular to the motion of the apparatus through the ether, we see that we must determine the velocity of the wave front with respect to the apparatus by using the vector form of the velocity addition law. We know that the velocity of the ether with respect to the apparatus is to the left with a magnitude v. We know that the magnitude of the velocity of the wave front with respect to the ether is just the velocity of light c. We also know that the resultant velocity of the wave front with respect to the apparatus must be in the direction toward the mirror. This tells us that the magnitude of the velocity of the wave front with respect to the apparatus is $(c^2 - v^2)^{1/2}$. (Apply the Pythagorean Theorem to the vector diagram in Figure 1.10.) The magnitude of the velocity on the return path is the same, as can be seen by drawing a similar vector diagram at the mirror. Thus the total time taken for the round trip by the wave front is

$$T_{perp} = \frac{2L}{\sqrt{c^2 - v^2}}$$

Unless $v = 0$, the times T_{para} and T_{perp} are different. Two wave fronts which start from the source simultaneously will not return to the center of the apparatus at the same time. Given L and c, the difference in arrival times for the two wave fronts should allow one to calculate v. Thus one could determine the velocity of the apparatus through the ether.

The experiment we have described is not a practical one due to the extremely high speed of light waves. However, by an interference method similar to that used in the Fizeau experiment, Michelson and Morley were able to measure the tiny time differences involved. In the actual Michelson-Morley experiment a half-silvered mirror is placed at the intersection of a T attached to a heavy marble slab. It is set at a 45 degree angle as shown in Figure 1.11. The marble

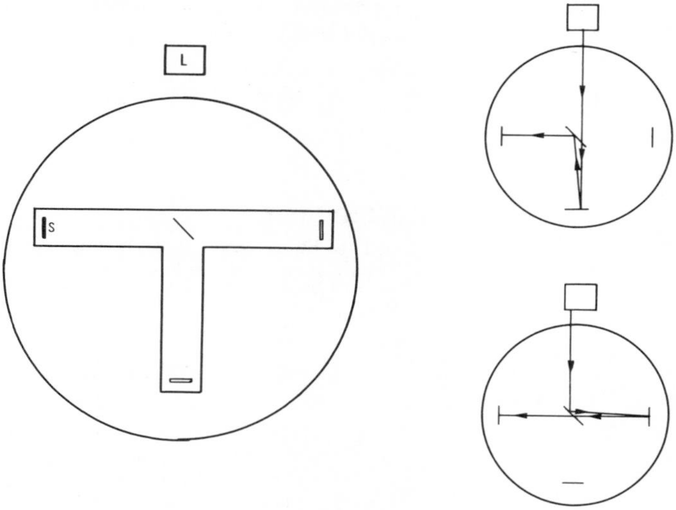

FIGURE 1.11 The Michelson-Morley apparatus, showing the two possible paths from the light source (L) to the screen (S).

slab floats in a pool of mercury so that it may be slowly turned by applying extremely weak forces. Two fully coated mirrors and a screen on which to view an interference pattern are set up at the three ends of the T. A monochromatic light beam from an external source strikes the half-silvered mirror where it is broken into a transmitted and a reflected beam. Those beams travel in perpendicular directions to two ends of the T where they are reflected back toward the half-silvered mirror. Half of each beam is then sent to the screen; the other half being sent back toward the source and absorbed. On the screen the two beams create an interference pattern. The smallest change in the relative phase of the two beams causes noticeable shifts in the interference pattern. Such a change in relative phase could be caused by a slight change in the time it takes one of the beams to travel to the fully coated mirror and back. The interference pattern was observed with different orientations of the T. This was done at various times of day and at different parts of the year so as to eliminate any possibility that the apparatus was accidentally at rest in the ether at the time of observation. No significant shift in the interference pattern was detected. The Michelson-Morley experiment and all similar experiments failed to detect any motion of the earth relative to the ether. A number of attempts were made to explain the null result of the experiments. Some of the important ones and the reasons for their failure are:

1. The earth and other bodies drag the ether with them as they move.

   This explanation is untenable because such a flow of the light conducting medium would cause a deflection of light traveling to the earth from outer space. It would be detectable as a constant wandering of the apparent positions of the stars. It would also contradict the Fizeau experiment by predicting that the propagation velocities in the two tubes should be $c/n \pm v_{water}$.

2. The propagation velocity of light depends upon the motion of the source just the way the velocity of a bullet depends upon the motion of the gun. That is, light travels at constant velocity with respect to its source only, not with respect to the ether.

   This explanation is untenable on two counts. Firstly, Maxwell's equations, which have been verified in great detail, predict that the velocity of light propagation is independent of the motion of the source. Secondly, if light from a receding source traveled to earth more slowly than light from an approaching source, there would be a detectable change in the apparent motion of astronomical objects such as the satellites of Venus and Jupiter that are constantly changing their velocities relative to the earth.

3. Solid bodies are distorted by motion through the ether. In particular they shrink in the direction of the motion by a factor $(1-v^2/c^2)^{1/2}$. That is, a square of side L would become a rectangle of sides L and $L(1 - v^2/c^2)^{1/2}$. (See Figure 1.12).

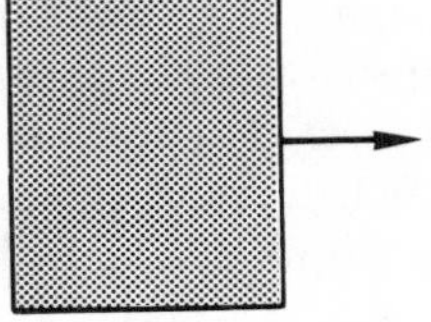

FIGURE 1.12
The Lorentz-Fitzgerald contraction.

   This would mean that the arm of the Michelson-Morley apparatus that was in the direction of the motion, would shrink by just the right amount to cancel the effect that had been expected. This contraction, called the Lorentz-Fitzgerald contraction, was analyzed and explained in great detail by Lorentz. If one assumed that electromagnetic forces between the particles of a body were what made a rigid body rigid, then one could show that just such a contraction could be expected so that the explanation was not quite as unconvincing as it appears at first sight. One of the difficulties with the Lorentz-Fitzgerald "explanation" was that it could not explain the null results of other experiments, such as the Trouton-Noble experiment, which would have detected the motion of the earth through the ether in spite of the contraction. In actuality Lorentz' explanation was simply superseded by the publication, a short time later, of Einstein's first paper on relativity which explained the null result of all such experiments and at the same time predicted a number of startling new effects that were subsequently confirmed. Incidentally, a length contraction equal to the Lorentz-Fitzgerald contraction is a consequence of the relativity theory.

1.4. EINSTEIN'S THEORY OF RELATIVITY

These experiments seem to show that the relativity principle is valid not only for mechanics but for electromagnetic theory as well. One might reasonably ask why people who were familiar with the experiments did not immediately recognize the fact that the equations of electromagnetic theory were valid in any constant velocity frame of reference. [A constant velocity frame is usually called an *inertial frame*. More exactly, an inertial frame is any system of coordinates in which isolated particles travel at constant velocities.] The reason why physicists were very reluctant to accept the principle of relativity for electromagnetic theory was that it seemed to lead to clear logical contradictions. To illustrate this let us consider the following imaginary experiment.

A man stands in the center of a long dark train. At each end of the train are light-sensitive devices that will cause a firecracker to pop as soon as they detect any light. We imagine ourselves riding on the train with the man. He lights a match. Now, if the train were stationary the light would travel in both directions with the speed, c, and cause the firecrackers to pop simultaneously. But, if unaccelerated motion is undetectable, then *to the observers on the train* (even if it is moving with respect to the surface of the earth) the light will still travel at the same speed in both directions and cause the firecrackers to pop simultaneously. Now let us suppose we are on the ground as the train moves by and we view the same sequence of events. We see the man light the match. We also see* the light move in both directions at constant velocity *with respect to us on the ground.* But we see the back end of the train move toward the light while the front end of the train moves away from the light. Therefore we would see that the firecracker in the rear of the train popped before the firecracker in the front of the train! This experiment is illustrated from both points of view in Figures 1.13 and 1.14.

We seem to have shown that any attempt to combine Maxwell's electromagnetic theory with a principle of relativity leads to internal logical inconsistencies. The resulting theory predicts that the firecrackers pop simultaneously and that the firecrackers do not pop simultaneously. Before we conclude that this proves that the theory is absurd, we must note that there is no contradiction at all in saying that an object is both moving and not moving. If we consider an object on the train, then it may be stationary according to all observers on the train and moving according to all observers on the ground. By

*The word "see" here should not be taken literally. One cannot see light propagate nor could the human eye resolve a sequence of events which took place with such rapidity as the events described here.

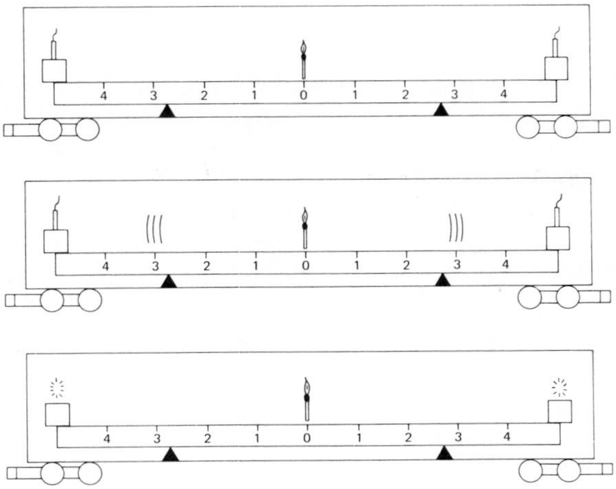

FIGURE 1.13 The experiment as viewed by a person on the train. If light travels at speed c in both directions within this frame of reference then the firecrackers pop simultaneously.

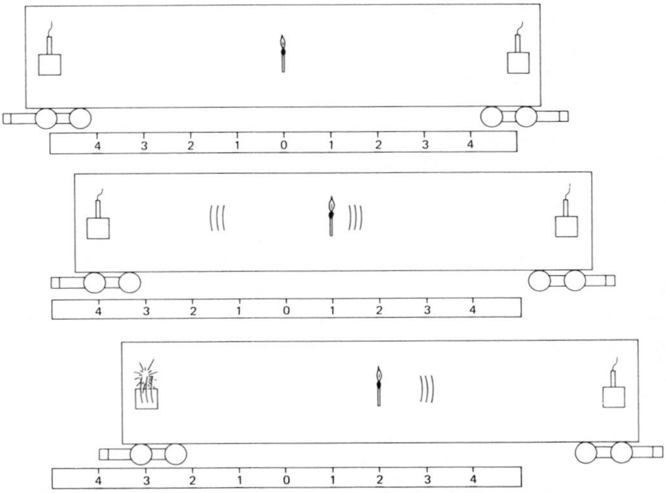

FIGURE 1.14 The experiment as viewed by a person on the ground. If light travels at speed c in both directions within this frame of reference then the back firecracker pops before the front one.

making a careful analysis of the experimental procedures we use to verify simultaneity, Einstein realized that two events happening at different places in space might be simultaneous and not simultaneous in the same sense as an object might be moving and not moving. They would be simultaneous to one set of observers and not simultaneous to another set of observers. This idea is not as outrageous as it sounds at first. Suppose we look up into the night sky and see explosions occur simultaneously on two distant stars. Were those two explosions actually simultaneous? If we were completely naive we would simply say that if we saw them occur at the same time, then they must have happened at the same time. Being more sophisticated than that, we realize that it may take many years for the light from a star to reach our eyes. Thus, if one star is twice as far from us as the other, events on the two stars that appear simultaneous on earth may have been separated by years of time. In order to determine simultaneity we must first know the distances to the stars and then take into account the time required for light signals to travel from the stars to our eyes using the fact that light travels at velocity c. Being still more sophisticated we might realize that if exactly the same thing is done by another intelligent being on another planet that is moving with respect to ours, then the conclusion he comes to regarding the simultaneity of the two explosions may be different from ours. In other words, he may assign time values to events in the universe that are different from the time values we assign to them. (We mean more here than the trivial difference of using a different zero of the time scale.) This may be true even though he uses the same basic rules for assigning time values to events that we do. Furthermore, the laws of physics may have exactly the same form when expressed in terms of his space and time values as they do when expressed in terms of ours.

1.5. THE LORENTZ TRANSFORMATION

It is clear that the only way we shall be able to settle the question of whether the principle of relativity is self-contradictory or not is by a very careful logical analysis. While we make the analysis we shall have to be on our guard against making assumptions based upon "common sense" since our common sense notions have grown out of our everyday experiences and none of us has had any experience with velocities that are comparable to the velocity of light. We have suggested that things that are simultaneous for one observer may not be so for another. We have also mentioned the possibility that rigid bodies shrink when they move. Certainly a good place to start our analysis is by establishing the relationship between the position and time of a particular event as measured by one observer and the position and time of the same event as measured by another observer who is moving with respect to the first. In order to do this we shall make certain definite assumptions. We shall try to be very explicit in our reasoning.

FIGURE 1.15 Mr. Unprime and his subordinates. Collectively they make up one inertial frame.

We picture a Chief Observer (call him Mr. Unprime) sitting on an infinitely long meter stick. He is situated at the zero of the meter stick and he holds a clock. By choosing a peculiar unit of time, namely $(1/3\times10^8)$ sec., he has arranged that the velocity of light is equal to one. We call that extremely short time unit a subsecond (abbreviated ss). We temporarily introduce this artificial time unit only in order to prevent the reader from being distracted from following the central logic of the analysis by the necessity of having to work through the more complicated algebra that occurs with ordinary units. However, at the end of this section it will be quite easy to convert our results to the form they take in ordinary units. Mr. Unprime has employed an infinite army of Junior Observers who are distributed at small intervals along the stick to take measurements of the time and place of events which do not occur close to the zero of his meter stick. Each of the Junior Observers also has a clock. In order to synchronize the clocks, Mr. Unprime sends a light pulse in both directions each unit of time. The nameless Junior Observers are told that these light pulses are separated by exactly one subsecond in time. When the first pulse arrives at the Junior Observer who is at the Seventeen Meter mark, he sets his clock to 17 subseconds (since he knows that it took the light that long to reach him). He also checks that his clock is running at the same rate as the Chief's. Mr. Unprime and his army of Junior Observers are aware of the fact that close by there is another endless meter stick passing them at the speed of v m/ss. Sitting at the zero of the other meter stick is another Chief Observer, Ms Prime, who has also synchronized the clock she carries with her own army of Junior Observers on her meter stick. When the two Chief Observers pass, they both notice that their clocks read exactly zero. Whenever an event occurs (i.e., a firecracker pops, an object passes, etc.) the nearest of Mr. Unprime's observers records the time and place of the event. We will call his measurements t and x. The nearest of Ms Prime's observers also records her measurements of the time and place of the event. We shall call those t' and x' Our problem is now to calculate the relationship between the measurements taken by Ms Prime's observers and Mr. Unprime's observers. We make the following assumptions:

1. Any isolated object (an object with no force applied to it) moves without acceleration according to both observers. That is: both frames are inertial frames.
2. Light pulses propagate at the velocity of plus or minus one meter per subsecond according to both observers.
3. The origins of the meter sticks coincide when both clocks read zero.
4. The equation for the position of Ms Prime, namely $x' = 0$, is transformed into $x = vt$, while the equation for the position of Mr. Unprime, namely $x = 0$, is transformed into $x' = -vt'$. That is, Ms Prime travels with velocity v through the unprimed frame while Mr. Unprime travels with velocity $-v$ through the primed frame.

Assumptions #1 and #2 express the principle of relativity as it applies to two of the simplest laws of mechanics and electromagnetic theory respectively. Assumptions #3 and #4 are of a much more trivial nature. Assumption #3 simply guarantees that both observers use a mutually convenient zero for their time scales. Assumption #4 simply expresses the fact that their relative speed is v meters/subsecond.

Before we proceed with the formal derivation let us reiterate exactly what we aim to do. Consider an event that happens at position x' and time t' according to Ms Prime's observers. This same event can be observed by Mr. Unprime's observers also. They determine that it happened at position x and time t. We shall see that, from the four assumptions given, we can determine the relationship between the pair of numbers (x', t') and the pair of numbers (x, t). That is, if we know the space and time coordinates that Ms Prime assigns to an event, we shall have formulas that allow us to calculate the space and time coordinates that Mr. Unprime assigns to the same event. They will be of the form

$$x = f(x', t') \text{ and } t = g(x', t') \tag{1.4}$$

where $f(x', t')$ and $g(x', t')$ are two functions we shall determine. First we make use of Assumption #1. The equation that describes the motion of any unaccelerated object in Ms Prime's frame is of the form

$$x' = u't' + x'_o \tag{1.5}$$

where u' is the velocity of the object and x'_o is the location of the object at time $t' = 0$. Thus any event that happens at the location of the object must have space and time coordinates that satisfy Equation 1.5. We can picture the object as a brick moving freely down the x' axis with no forces on it. In order to keep our attention focused on events rather than objects, we imagine that on the brick is an ant with hiccups. Each hiccup is an event. If we plot the (x', t') coordinates of these events, we get a series of points that all lie on the straight line given by Equation 1.5.

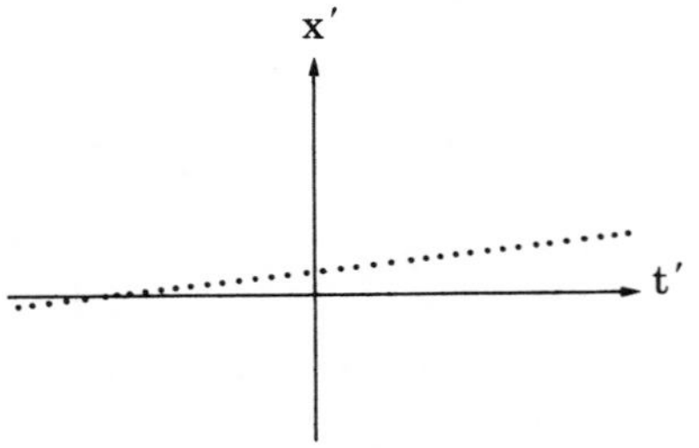

FIGURE 1.16
The times and places of the ant's hiccups according to Ms Prime.

Now the unprimed observers are also observing the ant's distress. If they plot the space and time coordinates which they assign to the hiccups, they must also obtain points on a straight line. This is so because they also see the ant moving along on an isolated object and isolated objects move without acceleration in both frames. (In general the speed of the brick and its initial position are not the same in the two frames.) Thus the transformation given by Equation 1.4

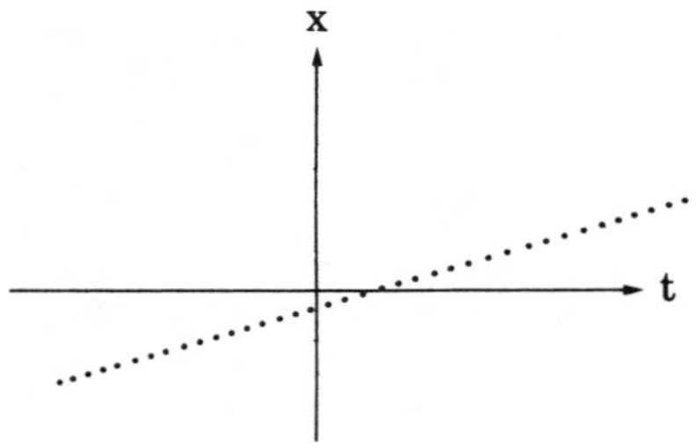

FIGURE 1.17
The times and places of the ant's hiccups according to Mr. Unprime.

must have the property that when the points on any straight line in the (x', t') variables are plotted in the (x,t) plane, they must still lie on a straight line. A transformation with this property is called a *linear transformation.* It can be shown that a linear transformation is always of the form

$$x = Ax' + Bt' + C \tag{1.6}$$

and

$$t = Dt' + Ex' + F \tag{1.7}$$

We must still determine the constants A, B, C, D, E, and F. An event that happens right at the place and time at which the two Chief Observers pass one another has space and time coordinates $x' = 0$ and $t' = 0$ according to Ms Prime and $x = 0$ and $t = 0$ according to Mr. Unprime. (See Assumption #3.) Putting the space and time coordinates of such an event into Equations 1.6 and 1.7, we see that $C = 0$ and $F = 0$. Thus

$$x = Ax' + Bt' \tag{1.8}$$

and

$$t = Dt' + Ex' \tag{1.9}$$

We have still made no use of Assumption #2 or #4. The equation for a light pulse sent out from the origin to the right at the instant both observers pass is, in the primed coordinates, $x' = t'$. This equation must transform into $x = t$. Putting $x' = t'$ in Equations 1.8 and 1.9, we get

$$x = (A + B)t'$$

and

$$t = (D + E)t'$$

This gives $x = t$ only if

$$A + B = D + E \tag{1.10}$$

The equation of motion of a leftgoing light pulse sent out from the origin just as the Chief Observers pass is $x' = -t'$. This must transform into $x = -t$. This gives another equation relating the unknown constants

$$-A + B = -D + E \tag{1.11}$$

Equations 1.10 and 1.11 imply that $A = D$ and $B = E$. We have so far reduced our transformation to the form

$$x = Ax' + Bt'$$

$$\text{and} \quad t = At' + Bx'$$

Already we can see a rather strange symmetry between time and space variables. Also we can see that if Ms Prime determines that an event took place at $t' = 0$ at any place other than the origin, Mr. Unprime will not feel that that event took place at $t = 0$, even though they have synchronized their clocks on passing. Separated events that are simultaneous to one observer are not simultaneous to another. We now want to use Assumption #4. We set $x' = 0$. Our transformations should then give us $x = vt$. What we get is

$$x = Bt'$$

$$\text{and} \quad t = At'$$

or

$$\frac{x}{t} = \frac{B}{A}$$

This is the equation $x = vt$ only if $B = vA$.

We can now eliminate B and write the transformation as

$$x = A(x' + vt') \tag{1.12}$$

and

$$t = A(t' + vx')$$

Had we done just the same analysis to calculate the transformation from x and t to x′ and t′, the only difference would have been that the velocity of Mr. Unprime in Ms. Prime's system is −v rather than v. We would thus have obtained the result:

$$x' = A(x - vt)$$

$$t' = A(t - vx)$$

Substituting these relations for x′ and t′ into Equation 1.12 we get

$$x = A(A(x - vt) + vA(t - vx)) = A^2(1 - v^2)x$$

which determines the final constant A

$$A = \frac{1}{\sqrt{1 - v^2}}.$$

Thus the relationship between the space and time coordinates assigned to the same event by observers in the two frames is

$$x = \frac{1}{\sqrt{1 - v^2}}(x' + vt')$$

and

$$t = \frac{1}{\sqrt{1 - v^2}}(t' + vx')$$

We must now convert these formulas from the subsecond time unit to the usual time unit. We do this by noting that

$$t \text{ (in ss)} = c\,t \text{ (in seconds)}$$

and

$$v \text{ (in m/ss)} = (v/c) \text{ (in m/s)}.$$

Thus, simply replacing t, t′, and v by ct, ct′, and v/c does the trick. We then get

$$x = \frac{1}{\sqrt{1 - \beta^2}}(x' + \beta ct') \tag{1.13}$$

and

$$ct = \frac{1}{\sqrt{1 - \beta^2}}(ct' + \beta x') \tag{1.14}$$

where

$$\beta = v/c.$$

As we mentioned before, the transformation in the other direction is given by replacing v by −v in these formulas.

$$x' = \frac{1}{\sqrt{1 - \beta^2}}(x - \beta ct) \tag{1.15}$$

and

$$ct' = \frac{1}{\sqrt{1-\beta^2}}(ct - \beta x). \tag{1.16}$$

By methods similar to the ones we have used it can be shown that, if both primed and unprimed observers make measurements in all three space dimensions, the transformations involving their y and z coordinates are trivial. That is

$$y = y' \text{ and } z = z'$$

This transformation was first derived by H.A. Lorentz without a complete appreciation for its relationship to the principle of relativity. It is therefore called the *Lorentz transformation.*

QUESTION: Suppose, in the unprimed frame, all of Mr. Unprime's observers sneeze at $t = 0$. Where and when do the sneezes take place according to Ms Prime's observers? The velocity between the frames is 10^8 m/s.

ANSWER: According to the unprimed observers the space and time coordinates of the nth observer's sneeze are $x_n = n$ meters and $t_n=0$. The value of β is $10^8/3\times10^8 = 1/3$. Using Equations 1.15 and 1.16 we get the space and time coordinates that the primed observers ascribe to the sneeze of the nth unprimed observer.

$$x_n' = \frac{1}{\sqrt{1-(1/3)^2}}(n - 0) = 1.06n \text{ meters}$$

and

$$ct_n' = \frac{1}{\sqrt{1-(1/3)^2}}(0 - n/3)$$

or

$$t_n' = -1.18\times10^{-9} n \text{ seconds.}$$

The most startling aspect of this result is that the set of events are not simultaneous according to the primed observers. Of course, 1.18×10^{-9} is a very small number and 10^8 m/s is a very large velocity so that this effect would not be easy to observe.

1.6. TIME DILATION

We shall henceforth assume that all observers have ordinary clocks that tick once per second. The ticks of Ms Prime's clock are a sequence of events which happen at $x' = 0$ and $t' = 1, 2, 3$, etc. According to Mr. Unprime the nth tick of Ms Prime's clock takes place at the following place and time (Put $x' = 0$ and $t' = n$ into Equations 1.13 and 1.14).

$$x_n = \frac{vn}{\sqrt{1-\beta^2}} \text{ meters}$$

and

$$t_n = \frac{n}{\sqrt{1-\beta^2}} \text{ seconds}$$

Therefore, according to Mr. Unprime, the time interval between two successive ticks of Ms Prime's clock is

$$\Delta t = \frac{1}{\sqrt{1-\beta^2}} \text{ seconds}$$

This number is larger than one. In other words, as measured by Mr. Unprime, Ms Prime's clock is running slow. What is most surprising is that, if we do the same analysis using the inverse transformation (Equations 1.15 and 1.16) we find that, according to Ms Prime, Mr. Unprime's clock is running slow!

One's first reaction might be to say that "in reality" both clocks run at the same rate but due to some sort of illusion created by the velocities each observer's clock "appears" to be running slow to the other observer. The reason why the time dilation effect cannot be explained away as an illusion is that it can have real physical effects. For instance, suppose v, the relative speed between the two observers, is nine tenths the speed of light (that is, $\beta = .9$); then, according to Mr. Unprime, Ms Prime's clock is running slow by a factor of about 2.2. Thus, if Ms Prime lived to be 100 years old by her own clock, she would be more than 220 years old according to Mr. Unprime and she could therefore travel (at the speed she is going according to Mr. Unprime) a distance of about 200 light years. Since we shall find that nothing can travel at a speed larger than the speed of light, a distance of 200 light years would be out of the range of a human being if it were not for the time dilation effect. There is obviously no way of brushing aside as somehow unreal the fact of a person arriving someplace while still alive rather than a hundred years dead.

Although technology has not succeeded in moving human beings at .9c, there are many instances of elementary particles moving at velocities very close to c. Of particular interest in terms of the time dilation effect are the muons produced at the top of the atmosphere by high energy cosmic ray particles from space. When it is at rest a muon lasts only about two millionths of a second before spontaneously disintegrating into an electron and a neutrino. Thus, during this time, traveling at the speed of light, it could move a distance:

$$(3\times 10^8 \text{m/s})(2\times 10^{-6}\text{s}) = 600 \text{ meters}$$

Since the atmosphere is much thicker than 600 meters, almost none of these particles should reach the earth's surface. But in fact many of them do. The decay rate of the muons can be explained perfectly if one takes into account the fact that the lifetime of a moving particle is increased by a factor of $1/\sqrt{1-(v/c)^2}$ where v is its velocity.

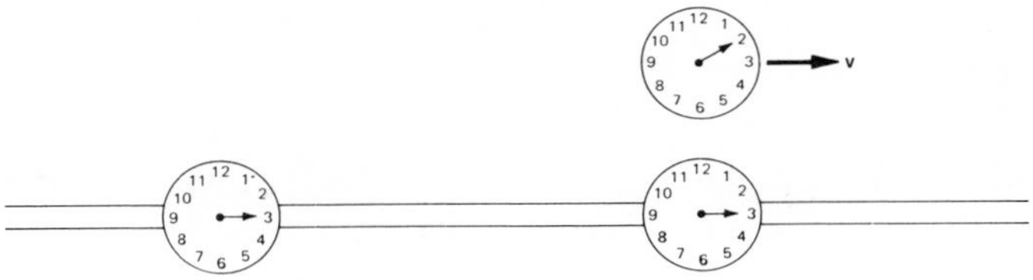

FIGURE 1.18 The time dilation effect. A clock moving through an inertial frame runs slow in comparison to the clocks in that frame.

In brief, if $\Delta t'$ is the time interval between two events which happen *at the same place in the primed frame*, then the time interval between the same two events (which will occur at different places) in the unprimed frame is

$$\Delta t = \frac{\Delta t'}{\sqrt{1-\beta^2}} \tag{1.17}$$

1.7. LENGTH CONTRACTION

Ms Prime sits at the origin of her meter stick. As we have seen the motion of Ms Prime through the unprimed coordinate frame is described by the equation

$$x = vt$$

Ms Prime's Junior Observer at the one meter mark has a trajectory in the unprimed frame which is given by setting x' equal to one in Equation 1.15. That is

$$1 = \frac{1}{\sqrt{1-\beta^2}}(x - vt)$$

or

$$x = vt + \sqrt{1-\beta^2}$$

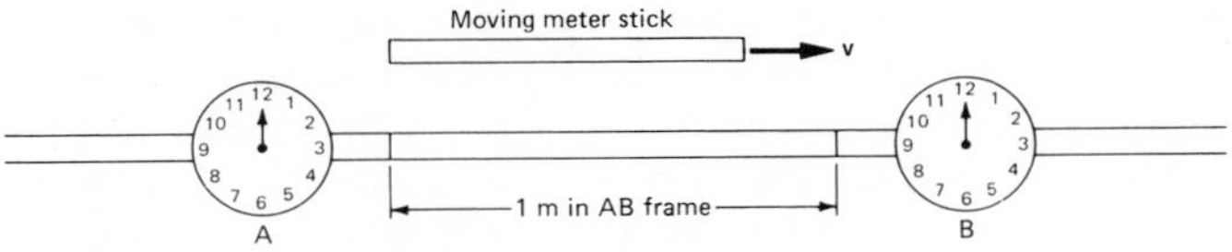

FIGURE 1.19 The length contraction effect. An object moving through an inertial frame is shortened (in the direction of its motion) in comparison to similarly constructed objects at rest in that frame.

Therefore, in the unprimed frame the distance between Ms Prime and her first observer is only $\sqrt{1-\beta^2}$ meters. Thus the length of meter rod between them, which is one meter in the frame in which it is at rest, appears to be shortened by a factor of $\sqrt{1-\beta^2}$ in the frame in which it moves at velocity v. In general, an object that has a length L_0 in its rest frame will appear shortened by a factor of $\sqrt{1-\beta^2}$ (in the direction of its motion) in a frame in which it moves at velocity $v = \beta c$. Thus it will have a length L in the new frame given by

$$L = \sqrt{1-\beta^2}\, L_0 \tag{1.18}$$

1.8. THE GEOMETRY OF SPACETIME

One can easily imagine the reader at this point muttering in frustration "I can follow the equations. If I'm careful I can even follow the logic, but I find it just impossible to form any clear picture of what this is all about". What the reader desires is a visualization of these space-time measurement processes which will agree with his or her fundamental intuitive notions of the geometry of space and the flow of time and which will reveal "in pictures" the geometrical meaning of the equations we have derived. That desire will never be completely satisfied because those fundamental intuitive ideas of space and time are wrong. Our powers of geometric visualization have been determined by a combination of a lifetime of experiences and a genetically determined intellectual apparatus formed in the crucible of evolutionary history. Since our usual concepts of space and time are fully adequate for describing ordinary experiences, and brain cells were not wasted in producing conceptual powers of no survival value, we have no ability to form pictures of geometries other than three dimensional Euclidean space. It is possible to mathematically define a four dimensional Euclidean space and to prove theorems about spheres and cubes and the like in that four dimensional space but nobody (not even the people who prove the theorems) can form a picture in his or her mind of any four dimensional object. All of our daily experiences can most efficiently be organized in terms of a picture of space and time equivalent to that described by Newton in his book, *Mathematical Principles of Natural Philosphy.*

> Absolute space, in its own nature, without relation to anything external, remains always similar and immovable.
>
> Absolute, true, and mathematical time, of itself, and from its own nature, flows uniformly without relation to anything external.

It is a world of objects located in a three dimensional Euclidean space. Their locations and their shapes vary as a function of a universal, absolute time. The basic geometrical element in this world is a *point in space.* The point in space may be occupied by a material body or it may be unoccupied. It exists throughout all time. This Euclidean-Newtonian picture of the world is different in certain fundamental ways from the geometrical picture of the world that physicists have been forced to adopt in order to construct a theoretical framework that agrees with experimental observations. The basic element in the new geometry is the *event.* Geometrically an event is *a certain place at a certain time*; that is a point in *spacetime.* Physically it is anything that happens at that particular place and time; a match lighting, two electrons bumping, or any other imaginable thing that can happen. An object is just the persistent sequence of events that happen at the location of that object. The reflection of light off the surface of the object; the whirling around of the electrons in the object, etc. There is always something happening at the location of an object. The collection of all events in spacetime forms a four dimensional set. Since we have no power to picture four dimensional sets we can best get a picture of the new geometry by the following device. We discard one space dimension. That is, we imagine that the spatial world is a plane rather than a three dimensional space. For convenience we assume that all the objects in our world are little particles whose positions can be represented by single dots. The Newtonian view of such a world could then be represented by having this collection of little dots swarm around on the two dimensional spatial plane as time progresses. We construct a model of the relativistic view of the world as follows. On a clear plastic sheet print dots representing the positions of all the particles at one instant of time. Do the same thing on a sequence of plastic sheets using equally and closely space instants of time. If we now stack the sheets together in sequence, we could easily see what appear to be the continuous trajectories of each particle. This is still a Newtonian model since we could easily pull out one sheet to determine exactly where all the particles were at a particular instant of universal time. To produce a relativistic model of the world, we stack all the sheets together carefully and heat them just enough so that they fuse and the sequences of dots become continuous curves. We now have our model of the relativistic spacetime. A solid piece of plastic with some curves imbedded in it to represent the trajectories of material particles (See Figure 1.20). Since spacetime is infinite in all dimensions, it is a somewhat unwieldy model. Let us take out just a big sphere of it and throw away the rest. The important difference between our spacetime and the Newtonian model we started with can be appreciated by considering what happens if we attempt to reverse the process of fusion. We bring our plastic sphere to a man with a bandsaw and we ask him to cut it back up into space slices. "Certainly", he says, "which way do

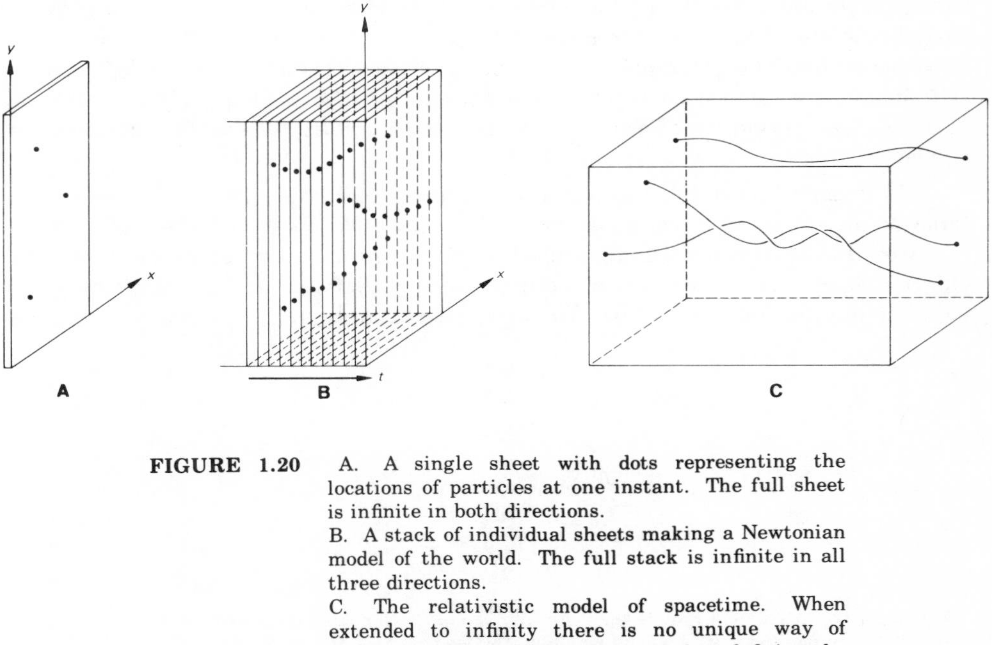

FIGURE 1.20 A. A single sheet with dots representing the locations of particles at one instant. The full sheet is infinite in both directions.
B. A stack of individual sheets making a Newtonian model of the world. The full stack is infinite in all three directions.
C. The relativistic model of spacetime. When extended to infinity there is no unique way of cutting it up into space slices and thus defining the collection of simultaneous events.

you want me to slice it?" "Perpendicular to the time direction", we tell him. "What time direction?" he asks, "All I see is a big ball of plastic with some wiggly lines in it." The realization suddenly comes to us that we have lost all track of which way the space sheets ran. It seems that we must be very careful in making the decision as to which way to cut because once we have chosen a particular slice we are making a statement that all events that happen in that slice are happening at the same time. We do not want to make such a decision without good reasons. We finally get a good idea. We shall investigate a particular method of slicing our spacetime into uniform sheets and see if, when we use that slicing, the spatial positions of the particles as a function of time satisfy the laws of physics. We choose a direction and try it. (Without actually feeding it into the bandsaw). To our delight we find we have luckily hit upon just the right direction because the laws of physics are satisfied exactly. Our delight fades when we discover that by choosing a different orientation for the time slices, we get a different description of the motion of the particles, but one that still satisfies all the laws of physics. In fact, we discover that the laws of physics are satisfied by any slicing within certain limits. The laws of physics are of no use in trying to discover the absolute time direction in spacetime. Since the motions of objects and the laws of physics are the only things we have to work with, we are forced to accept the conclusion that there simply is no absolute time direction. There are an infinite number of spacetime coordinate systems which can be used to describe the motion of things and every one is as good as every other one.

1.9. THE RELATIVISTIC ADDITION OF VELOCITY LAW

At the beginning of this chapter we saw that the law for addition of velocities, combined with the law of constant light speed, led to a contradiction of the principle of relativity. We have since then assumed that the law of constant light speed and the principle of relativity are both correct. The law of velocity addition that we had used before, must therefore be false. We shall now use the Lorentz transformation formulas to derive the correct law for addition of velocities.

Ms Prime is traveling through the unprimed coordinate system at velocity v. Suppose an object is traveling at velocity u′ down Ms Prime's meter stick. How fast would it be traveling as measured by Mr. Unprime? Let us assume that the object just crosses the origins of both coordinate systems as the Chief Observers pass one another (at t = t′ = 0). Then its initial coordinate in both frames is zero.

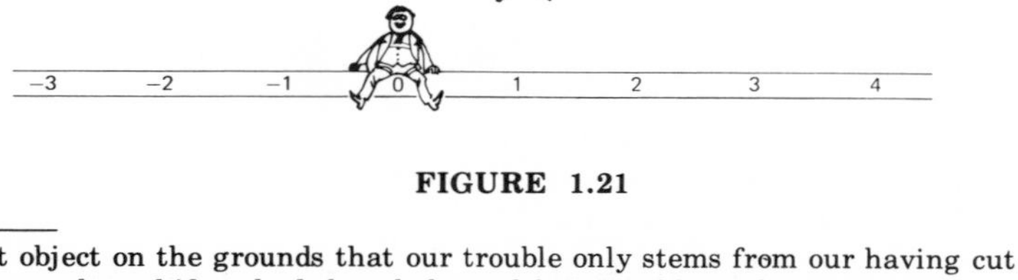

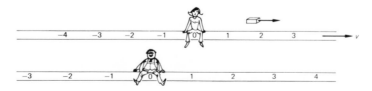

FIGURE 1.21

*One might object on the grounds that our trouble only stems from our having cut a spherical sample and if we had the whole model, we could see the direction of the surface. But remember, the original model was infinite in all directions as is real spacetime.

Consider any event that happens at the location of the object. According to Ms Prime the event will have spacetime coordinates (x′, t′) that satisfy the relation

$$x' = u't' \tag{1.19}$$

The same event will have unprimed coordinates (x,t) given by

$$x = \frac{x' + vt'}{\sqrt{1-\beta^2}} \tag{1.20}$$

and

$$ct = \frac{ct' + \beta x'}{\sqrt{1-\beta^2}} \tag{1.21}$$

If we use Equation 1.19 to write x′ in terms of t′ in Equations 1.20 and 1.21, we see that

$$\frac{x}{t} = \frac{u' + v}{1 + u'v/c^2}$$

or

$$x = ut$$

with

$$u = \frac{u' + v}{1 + u'v/c^2} \tag{1.22}$$

QUESTION: Suppose Ms Prime is moving at a velocity of .9c with respect to Mr. Unprime. A meteor is moving at a velocity of .9 c down the x′ axis as measured by Ms Prime. What is the velocity of the meteor with respect to Mr. Unprime? Is it larger than c?

ANSWER: We use the formula for velocity addition with v = .9c and u′ = .9c. Thus, the velocity of the meteor with respect to Mr. Unprime is

$$u = \frac{u' + v}{1 + u'v/c^2} = \frac{(1.8)\,c}{1 + (.9)^2} = .994\,c$$

This velocity is not larger than c. The addition of any two velocities less than c gives a velocity less than c. In fact, if object B is moving at .9 c with respect to object A and object C is moving at .9 c with respect to object B and object D is moving at .9 c with respect to object C, etc. etc., then, with respect to object A, the objects B, C, D, etc. move at velocities which get closer and closer to c as we go further out in the sequence but nothing moves at a velocity larger than c with respect to anything else.

QUESTION: Using the fact that Maxwell's equations for electric and magnetic fields are valid in any inertial frame it is easy to understand why no effect was detected in the Trouton-Noble and the Michelson-Morley experiments. What about the Fizeau experiment? Can we now understand the peculiar result found there?

ANSWER: Yes. We can understand the Fizeau result using the velocity addition law.

In the frame of reference of the water the light travels at speed $u' = c/n$. The water travels through the laboratory frame at velocity v. (We are considering the tube in which the water moves in the same direction as the light). By the velocity addition formula the speed of the light in the laboratory frame is

$$V = \frac{c/n + v}{1 + (c/n)\, v/c^2} = (c/n + v)(1 + v/nc)^{-1}$$

Since $v << c$ we can use the approximations

$$(1 + v/cn)^{-1} \approx 1 - v/cn$$

and

$$(c/n + v)(1 - v/cn) = c/n + (1 - 1/n^2)\, v - v^2/cn \approx c/n + (1 - 1/n^2)\, v$$

or

$$V \approx c/n + (1 - 1/n^2)\, v$$

1.10. THREE-DIMENSIONAL VELOCITY ADDITION

In the case considered above the moving object has only an x component of velocity. That is, as measured in the unprimed frame it is moving in the same direction as Ms Prime, although at a different speed. If, in the primed frame, the object is moving in some arbitrary direction with velocity components u'_x, u'_y, and u'_z , then the velocity transformation laws for the y' and z' components are different than that for the x' component. It is left as an exercise (Problem 1.14) to show that, according to Mr. Unprime the object will have velocity components u_x, u_y, and u_z given by

$$u_x = \frac{u'_x + v}{1 + u'_x v/c^2} \tag{1.23}$$

$$u_y = \frac{\sqrt{1 - \beta^2}\, u'_y}{1 + u'_x v/c^2} \tag{1.24}$$

and

$$u_z = \frac{\sqrt{1 - \beta^2}\, u'_z}{1 + u'_x v/c^2} \tag{1.25}$$

QUESTION: Is there any simple device that would help one to visualize these Lorentz transformations?

ANSWER: Yes, with a piece of white paper and a piece of clear paper one can make the following *Lorentz Transformer*. Mark axes on both papers and label them as shown in Figure 1.22. It is best to draw the primed axes in one color and the unprimed in another. Now pin them together through their origins. Each point on the plane now represents an event. To find the x and t coordinates of any particular event (that is, any particular point) one draws perpendicular lines to the x and ct axes

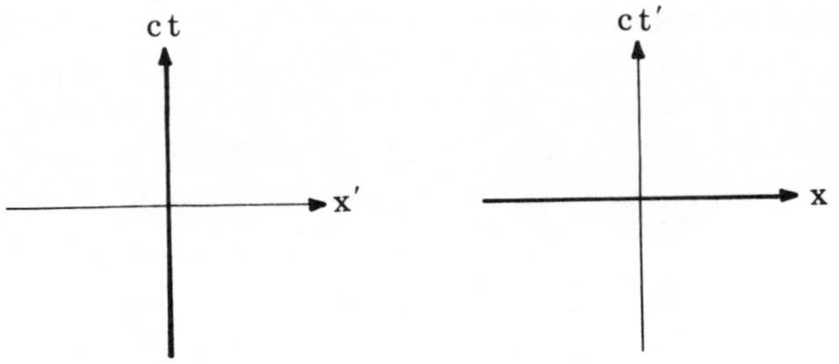

FIGURE 1.22 Making a Lorentz transformer.

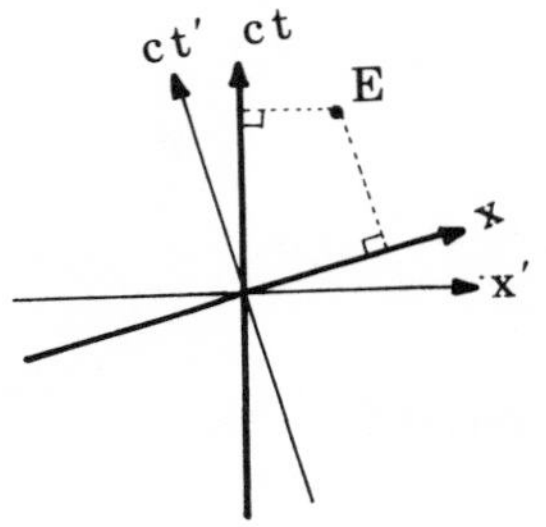

FIGURE 1.23
Measuring the spacetime coordinates of event E in the unprimed frame.

and reads off the values of x and ct (See Figure 1.23). Doing the same thing with the x′ and ct′ axes gives the primed coordinates of the same event. The two sets of coordinates are related by a Lorentz transformation. The value of $\beta = v/c$ for the Lorentz transformation is related to the angle between the two sheets by the formula

$$\beta = \sin\theta$$

In Figure 1.24 two events are shown which occur at the same place in the unprimed frame. That is, they have the same x coordinate. It is clear that the time interval between the two events assigned in the unprimed frame is smaller than the time interval in the primed frame. This illustrates the time dilation effect. The Lorentz contraction effect can be illustrated in a similar way (See Problem 1.16).

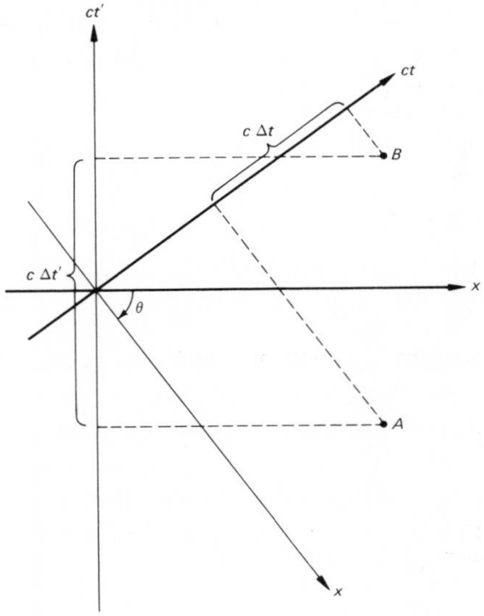

FIGURE 1.24
Events A and B happen at the same place in the (x,t) frame. The time interval between them is larger in the (x,t) frame than it is in the (x′,t′) frame.

1.11. THE DOPPLER EFFECT

It was known before the development of relativity theory that the motion of a light emitting source has an effect on the measured frequency of the light coming from the source. This effect is called the *Doppler shift* or the *Doppler effect.* We can understand the effect from a relativistic point of view as follows.

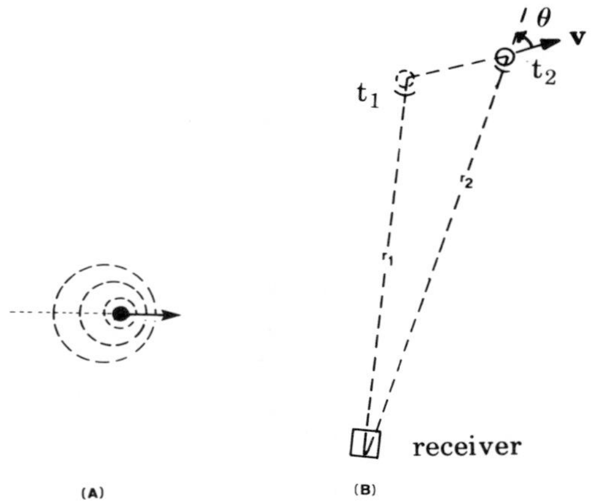

FIGURE 1.25 A. A moving source emits light waves. Because of the motion of the source the expanding spherical wave fronts are not concentric.
B. The distance between source and receiver changes between times t_1 and t_2 at which the source emits wave fronts.

Suppose an object is moving through the unprimed coordinate system at a speed v as shown in Figure 1.25. In its own frame of reference the object produces spherical light waves with frequency ν_s and period $P_s = 1/\nu_s$. (The subscript s means that these quantities are measured with respect to the *source.*) Thus the object can be pictured as emitting a sequence of wave fronts that move away from it in all directions. Let us calculate the arrival times of two of the wave fronts at the location of a *receiver* who is situated at the origin. As determined in the unprimed frame the wave fronts are emitted at times t_1 and t_2. They therefore arrive at the receiver at times T_1 and T_2 where

$$T_1 = t_1 + \frac{r_1}{c}$$

and

$$T_2 = t_2 + \frac{r_2}{c}$$

The apparent period of the light waves according to the receiver is therefore

$$P_r = T_2 - T_1 = t_2 - t_1 + \frac{r_2 - r_1}{c}$$

But, by the time dilation effect

$$t_2 - t_1 = \frac{P_s}{\sqrt{1-\beta^2}}$$

It is easy to see from Figure 1.25 that

$$r_2 - r_1 = \cos\theta\, v(t_2 - t_1) = \frac{\cos\theta\, v\, P_s}{\sqrt{1-\beta^2}}$$

Thus, the relationship between the period, as measured by the source and the period, as measured by the receiver is

$$P_r = \frac{(1+\beta\cos\theta)\, P_s}{\sqrt{1-\beta^2}} \tag{1.26}$$

Written in terms of the corresponding frequencies this is

$$\nu_r = \frac{\nu_s\sqrt{1-\beta^2}}{1+\beta\cos\theta} \tag{1.27}$$

Let us consider three important special cases.

Receding Doppler Shift ($\theta = 0°$)

$$\frac{\nu_r}{\nu_s} = \frac{\sqrt{1-\beta^2}}{1+\beta} = \left(\frac{1-\beta}{1+\beta}\right)^{1/2}$$

Approaching Doppler Shift ($\theta = 180°$)

$$\frac{\nu_r}{\nu_s} = \frac{\sqrt{1-\beta^2}}{1-\beta} = \left(\frac{1+\beta}{1-\beta}\right)^{1/2}$$

Transverse Doppler Shift ($\theta = 90°$)

$$\frac{\nu_r}{\nu_s} = \sqrt{1-\beta^2}$$

QUESTION: When heated the various chemical elements emit light with characteristic wavelengths. These particular wavelengths are commonly detected in the light reaching us from distant stars, an effect that allows one to determine the chemical composition of those stars. However, in light coming from extremely distant objects the characteristic wavelength patterns are still observed but the numerical values of the wavelengths are larger by some factor, α, than the corresponding wavelengths measured here on earth. That is, if λ_{earth} is the wavelength of some particular spectral line measured on earth and λ_{astro} is the wavelength of the same line detected in light from a distant astronomical source, then

$$\lambda_{astro} = \alpha\, \lambda_{earth}$$

where the *redshift parameter*, α, is different for different sources. As of 1982 the largest redshift parameter that has been found is $\alpha = 5.5$. If we interpret this effect as due to a receding Doppler shift, what would be the recession velocity of an object with such a redshift parameter?

ANSWER: Since $\lambda = c/\nu$ we see that $\nu_r = c/\lambda_{astro}$, $\nu_s = c/\lambda_{earth}$, and

$$\frac{\nu_r}{\nu_s} = \frac{\lambda_{earth}}{\lambda_{astro}} = \frac{1}{\alpha} = \left(\frac{1-v/c}{1+v/c}\right)^{1/2}$$

If we square this equation, and solve for v/c, we obtain

$$\frac{v}{c} = \frac{\alpha^2 - 1}{\alpha^2 + 1}$$

For $\alpha = 5.5$ this gives

$$v = .94c$$

SUMMARY

At the end of the Nineteenth century it was believed that the laws of mechanics were valid in all inertial frames (satisfied a relativity principle) but that Maxwell's equations were valid only in the ether frame. All experiments to detect the motion of the earth through the ether gave null results.

The theory of relativity assumes that all laws of physics have the same form in every inertial frame. With that assumption one can derive the *Lorentz Transformation* between the spacetime coordinates of an event in one inertial frame and the spacetime coordinates of the same event in another inertial frame. If the origins of the two frames coincide at $t = 0$ and $t' = 0$ and if the primed frame moves through the unprimed frame with velocity v in the x direction, the transformations are:

$$x = \frac{x' + vt'}{\sqrt{1-\beta^2}}$$

$$ct = \frac{ct' + \beta x'}{\sqrt{1-\beta^2}}$$

$$y = y'$$

$$z = z'$$

Time dilation - two events which happen at the same place in the primed frame and are separated by a time interval $\Delta t'$ will, in the unprimed frame, be separated by a longer time interval

$$\Delta t = \frac{\Delta t'}{\sqrt{1-\beta^2}}$$

Thus the primed clock will appear, to the unprimed observers, to run slow. The effect is symmetric; the unprimed clock appears to run slow according to the primed observers.

Lorentz contraction - an object that is of length L_0 in a frame of reference in which it is at rest will have a length

$$L = L_0\sqrt{1 - v^2/c^2}$$

in a frame in which it moves at velocity v in a direction parallel to its length.

Velocity addition - If an object moves through the primed frame with velocity components $u_x' = dx'/dt'$, $u_y' = dy'/dt'$, and $u_z' = dz'/dt'$ it will be moving through the unprimed frame with velocity components $u_x = dx/dt$, $u_y = dy/dt$, and $u_z = dz/dt$ where

$$u_x = \frac{u_x' + v}{1 + u_x' v/c^2}$$

$$u_y = \frac{\sqrt{1-\beta^2}\, u_y'}{1 + u_x' v/c^2}$$

$$u_z = \frac{\sqrt{1-\beta^2}\, u_z'}{1 + u_x' v/c^2}$$

Doppler effect - If the frequency of electromagnetic waves, as measured in the frame of reference of the source, is ν_s, then their apparent frequency as measured by a receiver at rest in the coordinate system is

$$\nu_r = \frac{\nu_s\sqrt{1-\beta^2}}{1+\beta\cos\theta}$$

where $\beta = v/c$, **v** is the velocity of the source, and θ is the angle between **v** and a line drawn from receiver to source.

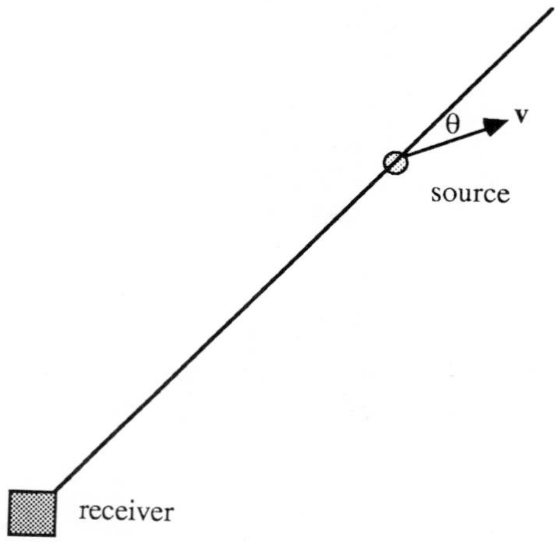

PROBLEMS

1.1 As in Figure 1.1 an observer with coordinate axes (x, y) is moving at a low velocity, $v = 2\,m/s$ in the x-direction with respect to another observer with coordinate axes (x′, y′). The velocities are low enough to use purely Newtonian physics. The frames coincide at time $t = 0$. In the unprimed frame a pebble is tossed in the air and moves with a trajectory $y(t) = (8 - g t^2/2)$ and $x(t) = 5t$, where t is measured in seconds and distances in meters. What is the trajectory in the primed frame?

1.2 With regard to the Michelson-Morley experiment show that, for $\beta = v/c << 1$, $T_{para} - T_{perp} = \beta^2 L/c$. Hint: $(1+x)^n \approx 1 + nx$ for $|x| << 1$.

1.3 Show that a contraction of the parallel arm of the Michelson-Morley apparatus by a factor of $(1 - v^2/c^2)^{1/2}$ would cause a null result for the experiment.

1.4 The motion of the earth around the sun has a speed of about 30 km/s. If the arms of the Michelson-Morley apparatus were one meter long and the light being used had a wavelength of $.4\,\mu m$ by what fraction of a wavelength would Michelson and Morley have expected the waves from the two arms to be shifted with respect to one another? (Hint: see Problem 1.2).

1.5* Suppose you were a Patent Examiner and someone submitted a patent application for a device that would measure the speed of a train (on a windless day) by timing the back-and-forth travel of a sound wave (See Figure P.1.5). Should he be granted a patent?

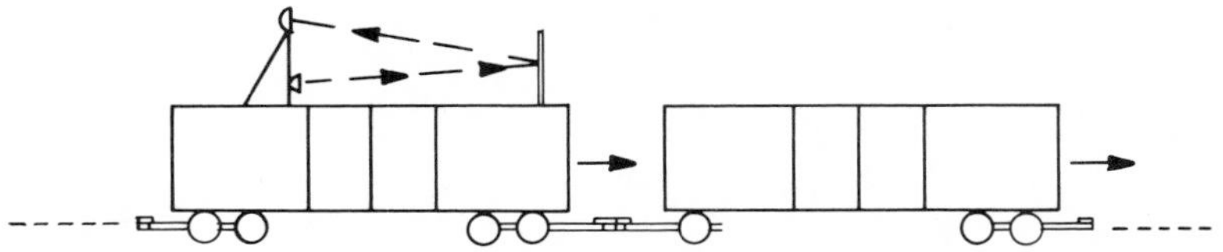

FIGURE P.1.5

1.6 The primed frame is moving at speed $v = .4c$ with respect to the unprimed frame. An explosion occurs at the time and place $x' = 6 \times 10^5\,m$ and $t' = 2\,s$. Where and when does the same event occur according to the unprimed observers?

1.7 Ms Prime is moving at speed c/3 with respect to Mr. Unprime. An event occurs in the primed frame at time $t' = 10\,s$ and position $x' = 4 \times 10^8\,m$. What spacetime coordinates does Mr. Unprime assign to the event?

1.8* Show that, according to the Lorentz transformation equations, for any values of x, y, z, and t, the following identity holds.

$$(ct)^2 - (x^2 + y^2 + z^2) = (ct')^2 - (x'^2 + y'^2 + z'^2)$$

1.9* A rod of length L = 200 m is at rest in the primed frame. Two explosions occur simultaneously at the ends of the rod. What is the distance between the explosions in the unprimed frame? $\beta = .6$. Explain why the space intervals between the two events in the two frames should or should not (take your choice) be related by a Lorentz contraction.

1.10 In the frame of reference in which they are at rest π^+ mesons have an average lifetime of 2.6×10^{-8} s. Neglecting relativistic effects, calculate the distance a newly created π^+ would travel before it disintegrated if it were moving at .99 c. Do the same, taking the time dilation effect into account.

1.11** In Ms Prime's frame a straight rod makes an angle θ' with respect to the x' axis. According to Mr. Unprime the rod is moving in the x direction at velocity v and is oriented at an angle θ with respect to the x axis. Determine the angle θ as a function of θ' and v.

1.12** A spaceship travels at velocity v m/s toward a distant star (as measured from the earth). The occupants of the spaceship send off a pulse of light in a direction which they determine to be perpendicular to the line joining the spaceship and the star. What would be the components of velocity of the light pulse as determined by an observer on the earth? (See Figure P.1.12). Show that the speed of the light pulse is c according to the earth observer.

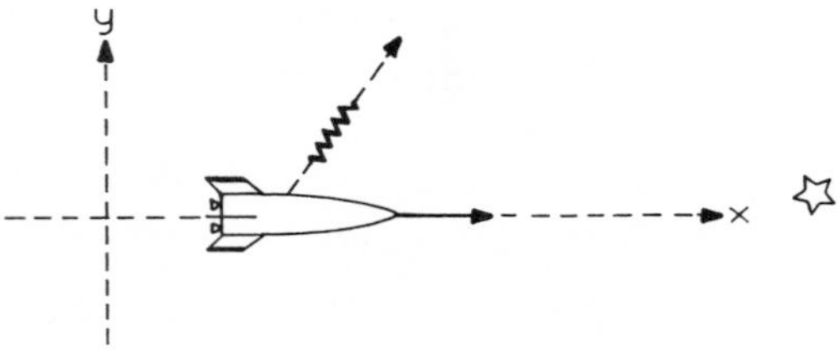

FIGURE P.1.12

1.13***A round disk is punched out of a very thin metal sheet leaving a hole in the sheet. The sheet is held stationary while the disk is moved through the hole as shown in Figure P.1.13. The disk is always parallel to the sheet and the angle of approach is small. The disk, being Lorentz contracted and therefore shorter than the hole, fits through easily. What does the motion look like when viewed by an observer moving with the disk? The hole will then be Lorentz contracted. Will the disk fit through?

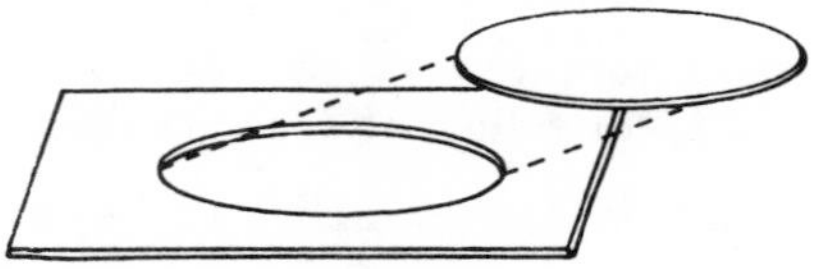

FIGURE P.1.13

1.14** Using the Lorentz transformation laws derive Equations 1.23, 1.24, and 1.25.

1.15***Near the surface of the earth, in Ms Prime's frame, a long straight iron bar is held up parallel to the x′ axis until time $t' = 0$ when all parts of it are simultaneously released. The bar remains parallel to the x′ axis as it falls. Any point on the bar has a trajectory

$$y'(t') = y'_0 \quad \text{for } t' < 0$$

and

$$y'(t') = y'_0 - g t'^2/2 \quad \text{for } t' > 0$$

By looking at the motion of the two end points and the center point of the rod in Mr. Unprime's frame, show that the rod bends in that frame.

1.16** On a diagram similar to Figure 1.24 use dashed lines to plot the trajectories of the two ends of a rigid rod of length L at rest in the unprimed frame. Take two events that happen simultaneously in the primed frame at the ends of the rod. By projection of those two events onto the x′ axis find the length, L′, of the rod in the primed frame. Indicate L and L′ on your diagram.

1.17* A satellite circles the earth at a distance of ten thousand kilometers from the earth's center. If it emits a constant radio signal of frequency 140 MHz, what will be the apparent frequency of its radio signal as measured by ground observers directly under the satellite? (Hint: The shift is small, so calculate $\nu_s - \nu_r$ rather than calculate ν_r directly.)

1.18** A continuous radio wave of frequency 100 MHz is sent by a stationary highway patrolman and reflected back by a directly approaching car. The car is travelling at 65 mi/hr. The patrolman has an electronic device that combines the transmitted and received signals at equal strengths to produce beats, a phenomenon obtained when two signals of slightly different frequencies are mixed. What is the frequency of the beats?

CHAPTER TWO

RELATIVISTIC DYNAMICS AND GENERAL RELATIVITY

2.1. MOMENTUM CONSERVATION

We now want to investigate whether the relativistic transformation laws are consistent with Newton's laws of dynamics. One of the simplest and most fundamental principles of Newtonian dynamics is the law of momentum conservation. Let us analyze a simple two-particle scattering experiment using the momentum conservation law and the relativistic velocity addition law (See Equations 1.23 through 1.25).

Suppose we are traveling with Ms Prime. Two identical particles are moving toward one another along the y' axis with velocities plus and minus u'. They have an elastic (energy conserving) collision and move off along the positive and negative x' directions, again with velocities plus and minus u' (See Figure 2.1 (A)). Using the Newtonian formula for momentum we see that momentum is obviously conserved in this collision.

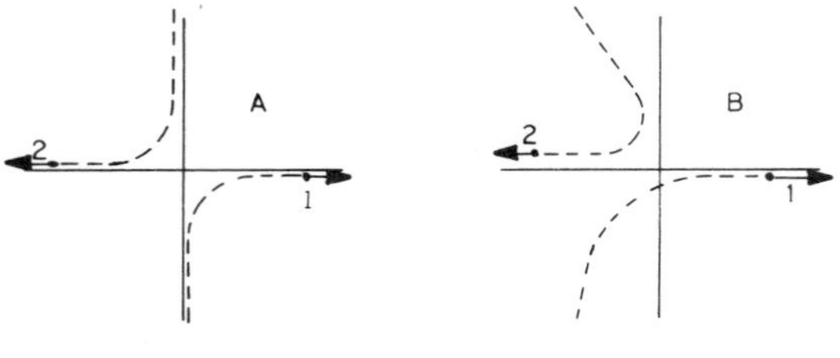

FIGURE 2.1 (A) A simple two-particle scattering process, viewed in the primed frame. (B) The same scattering process, as viewed by observers in the unprimed frame.

Now let us look at the same collision as seen by Mr. Unprime (See Figure 2.1(B)). In particular we want to see if the x-component of momentum is conserved. Before the collision the x-components of velocity of the two particles are obtained using the velocity addition formulas with $u_x' = 0$ and $u_y' = \pm u'$. The initial values of u_x for the two particles are

$$\#1 \quad (u_x)_{initial} = v$$

and

$$\#2 \quad (u_x)_{initial} = v$$

The final values of u_x are given by the velocity addition formulas with $u'_x = \pm u'$ and $u'_y = 0$. They are

$$\#1 \quad (u_x)_{final} = \frac{u'+v}{1+u'v/c^2}$$

and

$$\#2 \quad (u_x)_{final} = \frac{-u'+v}{1-u'v/c^2}$$

The x-component of the total momentum before the collision was

$$(P_x)_{initial} = m\,v + m\,v = 2m\,v$$

The x-component of the total momentum after the collision is

$$(P_x)_{final} = m\frac{u'+v}{1+u'v/c^2} + m\frac{-u'+v}{1-u'v/c^2}$$

Putting the fractions over a common denominator we obtain

$$(P_x)_{final} = 2m\,v\,\frac{1-(u'/c)^2}{1-(\beta u'/c)^2} = \frac{1-(u'/c)^2}{1-(\beta u'/c)^2}\,(P_x)_{initial}$$

where $\beta = v/c$. Since $1-(\beta u'/c)^2 > 1-(u'/c)^2$, the expression multiplying $(P_x)_{initial}$ in the above equation must be less than one. Therefore, according to Mr. Unprime

$$(P_x)_{final} < (P_x)_{initial}$$

We see that if momentum (using the Newtonian formula) is conserved in Ms Prime's coordinate frame, it is not conserved in Mr. Unprime's coordinate frame. Thus, *either the laws of Newtonian dynamics are wrong in all coordinate frames or the laws of dynamics are different in different coordinate frames.* The second possibility would violate the principle of relativity which states that the laws of physics are the same for all unaccelerated observers. Repeated and varied experiments have demonstrated that the relativity principle is true and that Newtonian mechanics must be modified. The reason why Newtonian dynamics cannot be made consistent with relativity is that it treats the time coordinate, t, completely differently from the space cordinates, x, y, and z. But we have already seen that there is a great deal of symmetry between space and time variables; one person's time interval is another person's space and time intervals. The cause of the lack of symmetry in Newtonian dynamics is that the space coordinates are expressed as functions of the time coordinate x(t), y(t), z(t). When one transforms the variables, the time and space coordinates get jumbled and the equations look completely different.

2.2. RELATIVISTIC VELOCITY

What we must do is write all four coordinates in terms of a parameter τ that would not change from one observer to another. We can see how to physically define such a parameter by considering the following experiment. Mr. Unprime takes some object and attaches to it a small standard clock (one that ticks, when stationary with respect to his clock, once per second). He then applies forces to the object so that it moves and accelerates. As the object moves through his coordinate system all his Junior Observers mark down, when it passes them, its space coordinates x, y, z, the time t and the reading on the clock attached to the object, τ. Because of the time dilation effect, the reading on the moving clock, τ, will not be the same as the time in the unprimed frame, t. After all this information is sent to the Chief, he compiles plots of the functions $x(\tau)$, $y(\tau)$, $z(\tau)$, and $t(\tau)$. That is, the position and time coordinates of the object when its own clock reads τ. The parameter τ we call the *proper time* of the object. We now define the *relativistic velocity* of the object by taking the rate of change of the coordinate with respect to the proper time. We shall write relativistic velocities with capital letters.

$$U_x = \frac{dx(\tau)}{d\tau}$$
$$U_y = \frac{dy(\tau)}{d\tau} \qquad (2.1.a)$$
$$U_z = \frac{dz(\tau)}{d\tau}$$

We can also define a *time component* of relativistic velocity.

$$U_t = c\,\frac{dt(\tau)}{d\tau} \qquad (2.1.b)$$

where the factor c has been added so that U_t will have the units of a velocity. The four components of relativistic velocity and the three components of "ordinary velocity" of an object are not unrelated. If we know the ordinary velocity components u_x, u_y, and u_z we can calculate the relativistic velocity easily enough. Consider a time interval dt. During that time interval a clock moving with the object (at speed u) would, since it runs slow, register a time interval

$$d\tau = dt\sqrt{1-u^2/c^2} \qquad (2.2)$$

Thus we see that

$$U_t = c\,\frac{dt}{d\tau} = \frac{c}{\sqrt{1-u^2/c^2}}$$

If, for any velocity u, we introduce the definition

$$\gamma_u \equiv \frac{1}{\sqrt{1-u^2/c^2}}$$

then U_t may be written in the form

$$U_t = \gamma_u c$$

We can also relate U_x, U_y, and U_z to u_x, u_y, and u_z. For instance

$$U_x = \frac{dx}{d\tau} = \frac{dt}{d\tau}\,\frac{dx}{dt} = \gamma_u u_x$$

Similarly $U_y = \gamma_u u_y$ and $U_z = \gamma_u u_z$. We can express all this by the two equations

$$U_t = \gamma_u c \quad \text{and} \quad \mathbf{U} = \gamma_u \mathbf{u} \qquad (2.3)$$

QUESTION: How does this relativistic velocity differ from the ordinary velocity of an object?

ANSWER: For objects that move with "reasonable" speeds (this does not include electrons in the Stanford Linear Accelerator or quasars receding from us at almost the speed of light) there is really no difference between **U** and **u**. For example, if $u = 300\,\mathrm{km/hr}$, which is a fairly high velocity using ordinary standards, then

$$U = 300.000000000023\,\mathrm{km/hr}$$

which makes it obviously indistinguishable from u. However, as the speed of an object approaches c, the factor γ_u increases rapidly, making the magnitude of **U** go to infinity. You can get some intuitive feeling for the meaning of U_x by imagining yourself in a rocketship, traveling along the x-axis. You want to measure your velocity with respect to the people in that frame. When the clock on your instrument panel says two o'clock exactly you stick your head out the window and ask the nearest Junior Observer what his coordinate is. He says 10 meters. You do the same thing again when your panel clock reads 15 seconds after two o'clock. Naturally you speak to a different Junior Observer this time who tells you his coordinate is 1530 meters. With this information you calculate U_x.

$$U_x = \frac{1530\,\mathrm{m} - 10\,\mathrm{m}}{15\,\mathrm{s}} = 101.33\,\mathrm{m/s}$$

Notice that with this completely natural procedure you get U_x and not u_x. It would seem completely absurd to use the frame time, t, in computing your velocity because, as far as you can tell the set of clocks in that frame are not synchronized. Therefore you certainly would not want to subtract a reading on one of them from a reading on another. Now let us see why U_x is not limited to having a numerical value less than c. Suppose you run your rocket engine until your speed, as measured by the x-frame observers, approaches the speed of light. (Do not ask how much fuel it would take to do such a thing. That would puncture the daydream very quickly by showing that relativistic rocket travel is utterly impractical.) As your speed increases, the x-axis, along with all the observers on it, becomes more and more Lorentz contracted. Eventually the once robust observers become tissue paper thin and separated by very small intervals. Since they are moving, with respect to you, at almost the speed of light, you find that, during one second of your instrument panel time huge numbers of these emaciated shooting targets pass your rocketship window. But the number of Junior Observers passing in one second is equal to U_x, which therefore increases without limit as they become more and more compressed due to Lorentz contraction.

2.3. RELATIVISTIC MOMENTUM

We are now going to define four components of *relativistic momentum* in terms of U_t *and* **U**.

$$P_t = m\,U_t \quad \text{and} \quad \mathbf{P} = m\,\mathbf{U} \tag{2.4}$$

We shall assume that, for any isolated system of particles (one without any external forces), each component of the total relativistic momentum remains constant. We can then show that if such a momentumm conservation law is true for one observer it must be true for every observer. It will turn out that the fourth component of momentum is related to the energy and so the energy conservation theorem becomes combined with the momentum conservation theorem in relativity. First we have to see how relativistic velocity is transformed from one coordinate system to another.

We picture an object, carrying a proper time clock, moving through space in some complicated way. That is, not necessarily at constant velocity. The ticks of the proper time clock are events and thus the spacetime coordinates assigned to those events by observers in two different inertial frames are related by a Lorentz transformation.

$$x(\tau) = \gamma_v(x'(\tau) + v\,t'(\tau))$$

where v is the constant relative velocity between the frames, not the velocity of the object. Differentiating this equation with respect to τ and recalling the definitions of U_x and U_t given in Equations 2.1(a) and 2.1(b), we get

$$U_x = \gamma_v(U'_x + \beta\, U'_t) \qquad (\beta = v/c)$$

If we multiply both sides of this equation by m we get one of the transformation equations for the components of the relativistic momentum of an object as measured in two different inertial frames.

$$P_x = \gamma_v(P'_x + \beta\, P'_t) \tag{2.5}$$

In a similar way we can get the other transformation equations.

$$P_t = \gamma_v(P'_t + \beta\, P'_x) \tag{2.6}$$

$$P_y = P'_y \quad \text{and} \quad P_z = P'_z \tag{2.7}$$

Comparing these transformation equations with those derived in the last chapter for the quantities x, y, z, and ct (Equations 1.13 and 1.14) we see that the transformation rule for the components of the relativistic momentum is exactly a Lorentz transformation. Any four physical observables, A_1, A_2, A_3, and A_4 which have the property that their values, as measured in two inertial frames, are related by a Lorentz transformation are collectively called a *four-vector*. Thus the quantities x, y, z, and ct are the components of the position four-vector of an event in spacetime. The relativistic velocity and relativistic momentum are also four-vectors.

2.4. CONSERVATION OF RELATIVISTIC MOMENTUM

When we considered the Newtonian momentum conservation law we found that assuming that Newtonian momentum was conserved in one inertial frame led to the conclusion that it was *not* conserved in other inertial frames. We now want to show that a relativistic momentum conservation law would not suffer from that problem. This will not prove that relativistic momentum is really conserved. Only experiment can do that. However, it will show that a law of conservation of relativistic momentum would be consistent with the principle of relativity that says that the physical laws must be valid in all inertial frames. We consider a two-particle scattering experiment as viewed in the primed frame. The particles are labelled A and B. The x component of the total relativistic momentum of the pair of particles is

$$P'_x = P'_{xA} + P'_{xB}$$

with similar equations for the y, z, and t components. The relativistic momentum components of each particle in the primed frame are related to the components in the unprimed frame by the Lorentz transformation. (Equations 2.5, 2.6, and 2.7) Since this is true for each particle it is obviously true for the components of the total relativistic momentum. Thus

$$P_x = \gamma_v(P'_x + \beta P'_t)$$

$$P_t = \gamma_v(P'_t + \beta P'_x),$$

$$P_y = P'_y, \quad \text{and} \quad P_z = P'_z$$

If the total relativistic momentum is conserved in the primed frame, then the value of any component before the collision is equal to its value after the collision.

$$(P'_x)_{\text{before}} = (P'_x)_{\text{after}},$$

$$(P'_t)_{\text{before}} = (P'_t)_{\text{after}}, \text{ etc.}$$

When this is combined with the Lorentz transformation one sees that the relativistic momentum must be conserved in the unprimed frame also. For instance

$$(P_x)_{\text{before}} = \gamma_v(P'_x)_{\text{before}} + \beta\gamma_v(P'_t)_{\text{before}}$$

$$= \gamma_v(P'_x)_{\text{after}} + \beta\gamma_v(P'_t)_{\text{after}}$$

$$= (P_x)_{\text{after}}$$

Now it is clear why we must assume that *all* components (including the time component) of relativistic momentum are conserved. If the time component were not conserved in the primed coordinate system then the x component would not be conserved in the unprimed coordinate system. The question we must now ask is: What is the physical meaning of the conservation of the time component of the relativistic momentum? Is it a new conservation law that did not exist in nonrelativistic mechanics? In terms of the ordinary velocity P_t is given by

$$P_t = \frac{mc}{\sqrt{1-u^2/c^2}} \tag{2.5}$$

For the typical velocities of material particles, which are much less than the velocity of light, u^2/c^2 is a very small number. We can then use the approximation

$$\frac{1}{\sqrt{1-u^2/c^2}} = (1-u^2/c^2)^{-1/2} \approx 1 + \tfrac{1}{2}u^2/c^2$$

We then see that

$$cP_t \approx mc^2 + \tfrac{1}{2}mu^2 \tag{2.6}$$

Except at velocities near the velocity of light, the time component of momentum of a particle is proportional to a constant plus the usual Newtonian kinetic energy of the particle. For the types of collision processes we have considered, the constant term, mc^2, can be ignored since it simply adds a fixed number to the initial value of P_t and the same number to the final value, thus having no effect on the conservation law. For velocities much less than c the law of conservation of P_t is completely equivalent to the law of conservation of Newtonian kinetic energy. Experiment has shown that, when u is not much less than c, the

Newtonian kinetic energy is not conserved but the time component of the total momentum is conserved. In fact, all components of the relativistic momentum are conserved.

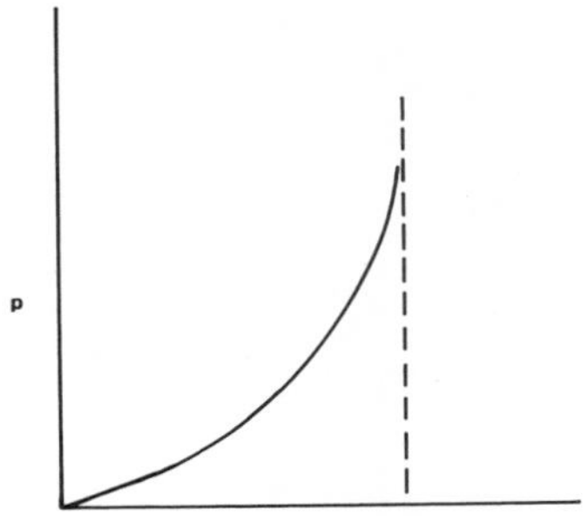

FIGURE 2.2
The momentum of a one kilogram mass as a function of its ordinary velocity.

QUESTION: Suppose a constant force f is applied to a particle of mass m. If the particle is initially stationary what will be its ordinary velocity u at time t?

ANSWER: We choose the x-direction as the direction of the force and call P_x simply P. (P_y and P_z are zero and we shall never have to consider P_t). Then, according to Equations 2.3 and 2.4,

$$P = m\,U_x = \frac{m\,u}{\sqrt{1-u^2/c^2}}$$

Squaring both sides of this equation we get

$$P^2 = \frac{m^2u^2}{1-u^2/c^2}$$

which gives, when solved for u, the formula

$$u = \frac{P/m}{\sqrt{1+P^2/m^2c^2}} \qquad (2.7)$$

(Remember that in this equation $P = |\mathbf{P}|$.) The force on a particle in relativistic mechanics has the same meaning that it does in Newtonian mechanics, namely the rate at which momentum is transferred to the particle. Since the momentum begins at zero and increases with the constant rate f, the momentum at time t must be

$$P(t) = f\,t$$

Therefore u(t) is given by

$$u(t) = \frac{f\,t/m}{\sqrt{1+(f\,t/m\,c)^2}} \qquad (2.8)$$

Let us consider what this formula means. The quantity ft is the total momentum that has been delivered to the particle by time t. If ft is much smaller than mc the particle undergoes constant acceleration as predicted by the nonrelativistic formula

$$u(t) = f\,t/m$$

As ft continues to increase the acceleration diminishes and the particle velocity approaches closer and closer to the velocity c. Thus, no matter how long we push on a particle, we can never get it to the velocity u = c although its momentum increases without bound (See Figure 2.3).

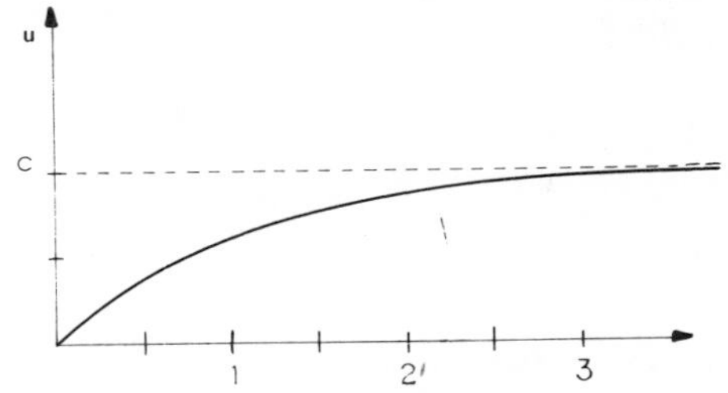

FIGURE 2.3 The ordinary velocity as a function of time for a particle subjected to a constant force. The x-axis shows the values of (F/mc)t.

QUESTION: Could the time dilation effect possibly be used to keep a person young?

ANSWER: It depends very much on what you mean by keeping a person young. As we shall see, it is possible for you, by keeping yourself moving at high velocity, to be still young and vigorous when everyone else is old and gray. (This assumes that you are presently young and vigorous.) Then everybody else would congratulate you on how wonderfully young you have remained. However, in terms of your personal experience you would not have lived an exceptionally long time at all. Traveling around in your spaceship you would only have had time to have the normal number of heartbeats, to think the normal number of thoughts, and to feel the normal number of emotions for a person of your apparent physical age. From your own point of view, all you would have accomplished by your incessant flitting about is to make everyone else old. Let us now consider the question in a more detailed mathematical fashion. It has been well confirmed by experiments involving unstable particles moving in accelerators that the proper time τ measures the rate at which the internal disintegration mechanism of the particle proceeds. This is true even when the particle is undergoing accleration due to external forces as long as the forces are not so violent as to disrupt the internal structure of the particle. We know by our everyday experience that a force of mg newtons applied to a person of mass m has no detrimental physical effects. Let us see what the effects would be of applying that gentle force over long periods of time. We picture a brother and sister, each ten years old, and at rest relative to one another. They are in separate airtight compartments in space. The compartments each contain a clock. We assume that the compartments have negligible mass. A steady external force of mg newtons in the positive x direction is applied to the sister's compartment, causing it to accelerate in a way we have analyzed in a previous question. The force is maintained for twenty years as measured by the brother's clock. The force is then reversed in direction and applied for another twenty years, causing the sister to gradually slow down and come to a stop on the brother's fiftieth birthday. She is, however, at a very great distance from him. The whole process is then reversed, bringing the sister back to the side of her aged brother as he celebrates his ninetieth birthday. We now calculate the time elapsed on her clock.

During the first twenty years (his time) her velocity as a function of his time is given by Equation 2.8 with f = m g.

$$u(t) = \frac{g\,t}{\sqrt{1+g^2t^2/c^2}} \tag{2.9}$$

During his time interval dt her clock changes by an amount $d\tau$ where

$$d\tau = dt\sqrt{1-u^2/c^2}$$

If we substitute the expression for u(t) from the first equation into the second, we obtain

$$d\tau = \frac{dt}{\sqrt{1+g^2t^2/c^2}} \tag{2.10}$$

Letting t range from 0 to t = T = 20 years = 6.3×10^8 s, we integrate to get the range of τ

$$\int_0^\tau d\tau = \int_0^T \frac{dt}{\sqrt{1+g^2t^2/c^2}}$$

which gives

$$\tau = \frac{c}{g}\log\left[g\,T/c+\sqrt{1+g^2T^2/c^2}\right] \tag{2.11}$$

Evaluating the right hand side for $T = 6.3\times 10^8$ s, $g = 9.8\,m/s^2$, and $c = 3\times 10^8\,m/s$, we get

$$\tau = 1.14\times 10^8 s = 3.6\,years$$

This is just one quarter of the sister's journey. Each of the other three quarters make equal contributions to her age, bringing her, upon her return to her brother's side, to the age of

$$(10+4\times 3.6)\,years = 24.4\ years$$

About the age of one of his granddaughters.

2.5. "DOES THE INERTIA OF A BODY DEPEND UPON ITS ENERGY CONTENT?"*

What about the constant term mc^2? Does its inclusion in the time component of momentum have any significance? We shall see that the constant term is exactly equal to the energy content of the object when it is at rest. To indicate what is meant by this peculiar statement, let us consider the following simple experiment, viewed in Mr. Unprime's frame.

We have two identical blocks, each of mass m. We also have a spring of mass

*This is a translation of the title of a two-page paper published by a junior patent clerk named Albert Einstein in 1905 [*Annalen der Physik* **17**(1905)].

m_s. We compress the spring between the blocks and connect them with a thin thread (See Figure 2.4). We measure the mass of the whole system. It is M.

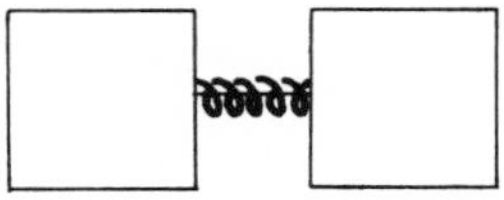

FIGURE 2.4
Two masses with a spring compressed between them are held together by a thin thread.

The system is stationary and therefore has time component of momentum

$$P_t = Mc$$

Now we carefully cut the thread in the center and the two blocks move in opposite directions with velocities +u and −u. The spring is left stationary (See Figure 2.5).

FIGURE 2.5
The thread has been cut and the two masses move away with speed u leaving the spring stationary and relaxed.

The time component of the total momentum is now

$$P_t = m_s c + 2\frac{mc}{\sqrt{1-u^2/c^2}}$$

Since P_t is conserved

$$M = m_s + \frac{2m}{\sqrt{1-u^2/c^2}} > m_s + 2m$$

Thus, before we broke the thread the mass of the system must have been more than the sum of the masses of its constituent parts! If the velocity u is much smaller than c, we can use the approximation

$$\frac{2m}{\sqrt{1-u^2/c^2}} \approx 2m(1+\tfrac{1}{2}u^2/c^2) = 2m + \frac{E_K}{c^2}$$

where $E_K = mu^2$ is the combined kinetic energy of the two blocks. But the kinetic energy of the blocks after the spring is released is equal to the potential energy that was stored in the spring while it was compressed. Thus

$$M = m_s + 2m + \frac{E_p}{c^2}$$

The mass of the system before the thread was broken was larger than the sum of the masses of the two blocks plus the extended spring by an amount equal to the energy stored in the compressed spring divided by c^2. In other words, if the internal energy of an object is increased by an amount ΔE its inertial mass also increases by an amount Δm where

$$\Delta E = \Delta m c^2 \qquad (2.12)$$

Thus if we take a block of metal and increase its internal energy by heating it, its mass (and weight) will increase in proportion to the heat energy added. Before the reader tries to confirm the effect with a pot of hot water on a bathroom scale, it must be pointed out that the change in mass is very small. If a million calories are added to a body, the extra mass comes to

$$\begin{aligned} \Delta m &= 4.18 \times 10^6/(3 \times 10^8)^2 \\ &= 464 \times 10^{-11}\mathrm{kg} \end{aligned}$$

This is much too small a change to be detected by weighing on a scale.

Although the mass changes associated with ordinary energy transfer processes, such as heating a body or stretching a spring, are too small to be detected, the mass changes involved in energy release by nuclear transformation processes are significant and easily measured. As an example, in the deuterium fusion process two atoms of deuterium combine to form one atom of helium with the release of a large amount of radiant energy. Deuterium is an atom made up of one electron, one proton, and one neutron. It has the same chemical characteristics as ordinary hydrogen. A helium atom contains two electrons, two protons, and two neutrons. Since the helium atom has exactly the constituents of two deuterium atoms, one might expect it to have twice the mass of a deuterium atom. However, the constituents in a helium atom are more tightly bound together. This is why energy is released in the transformation of deuterium to helium. The measured mass of a deuterium atom is $3.3436\times10^{-27}\,\mathrm{kg}$. The mass of a helium atom is $6.6447\times10^{-27}\,\mathrm{kg}$. Thus, according to Equation 2.12, the energy released when two deuterium atoms become one helium atom is

$$\begin{aligned} E &= (2m_D - m_{He})c^2 \\ &= 3.825 \times 10^{-12}\mathrm{J} \end{aligned}$$

To get a better idea of what this energy value means, let us calculate the energy released when Avogadro's number of deuterium atoms (about two grams of deuterium) are converted into half that number of helium atoms.

$$\begin{aligned} E &= \left(\frac{6.022 \times 10^{23}}{2}\right)(3.825 \times 10^{-12}\,\mathrm{J}) = 1.152 \times 10^{12}\,\mathrm{J} \\ &= 320000 \text{ kilowatt hours} \end{aligned}$$

This relation between energy release and mass change has been accurately verified by experiment.

QUESTION: The quantity $E = c\,P_t$ is called the relativistic energy of a body. At low velocities it is equal to the Newtonian kinetic energy plus the constant mc^2. Using the Newtonian formulas for momentum ($\mathbf{p} = m\mathbf{v}$) and kinetic energy ($E_K = \frac{1}{2}mv^2$), it is easy to derive the following relation between the momentum and kinetic energy of a particle moving at low speed.

$$E_K = p^2/2m \quad (\text{For } v << c) \tag{2.13}$$

What is the corresponding relation between the relativistic energy E and the relativistic momentum **P** of a moving particle?

ANSWER: Using Equations 2.3 and 2.4 we can express the relativistic energy and momentum in terms of the ordinary velocity $\mathbf{u}$ in the following way:

$$E = c\,P_t = \frac{m\,c^2}{\sqrt{1-u^2/c^2}} \tag{2.14}$$

and

$$\mathbf{P} = \frac{m\,\mathbf{u}}{\sqrt{1-u^2/c^2}} \tag{2.15}$$

We want to combine these two equations so as to eliminate the velocity $\mathbf{u}$. In order to do that we first square both equations in order to get rid of the radical sign.

$$E^2 = \frac{m^2c^4}{1-u^2/c^2}$$

$$\mathbf{P}\cdot\mathbf{P} = \frac{m^2u^2}{1-u^2/c^2}$$

We now write E^2 in terms of $\mathbf{P}\cdot\mathbf{P}$ by using the relation

$$\frac{1}{1-u^2/c^2} = \frac{1-u^2/c^2}{1-u^2/c^2} + \frac{u^2/c^2}{1-u^2/c^2} = 1 + \frac{u^2/c^2}{1-u^2/c^2}$$

This gives

$$E^2 = m^2c^4 + c^2\mathbf{P}\cdot\mathbf{P}$$

or

$$E = c\sqrt{m^2c^2 + \mathbf{P}\cdot\mathbf{P}} \tag{2.16}$$

which is the relativistic equivalent of Equation 2.13.

QUESTION: Is the change in the relativistic energy of a body equal to the work done on the body by external forces?

ANSWER: Yes. Consider a body of mass m being accelerated by a force $\mathbf{f}$. The rate at which work is being done on the body is equal to $\mathbf{u}\cdot\mathbf{f}$, where we define $\mathbf{f}$ by

$$\mathbf{f} = \frac{d\mathbf{P}}{dt}$$

We must therefore show that

$$\mathbf{u}\cdot\frac{d\mathbf{P}}{dt} = \frac{dE}{dt}$$

The relativistic energy of a body is related to its momentum by Equation 2.16. If we differentiate that equation with respect to t (using the rule for differentiation of the square of a vector, namely $d(\mathbf{P}\cdot\mathbf{P})/dt = 2\mathbf{P}\cdot d\mathbf{P}/dt$) we obtain

$$\frac{dE}{dt} = \frac{c\,\mathbf{P}\cdot d\mathbf{P}/dt}{\sqrt{m^2c^2+\mathbf{P}\cdot\mathbf{P}}} = c^2\,\frac{\mathbf{P}\cdot d\mathbf{P}/dt}{E}$$

By dividing Equation 2.15 by Equation 2.14 we can see that

$$c^2 \frac{\mathbf{P}}{E} = \mathbf{u} \tag{2.17}$$

Therefore

$$\frac{dE}{dt} = \mathbf{u} \cdot d\mathbf{P}/dt = \mathbf{u} \cdot \mathbf{f} \tag{2.18}$$

Thus, if a particle moves along a known trajectory under the influence of a known force, we may calculate its change in energy in the same way as we would in Newtonian dynamics.

$$(E)_{final} - (E)_{init} = \int \mathbf{f} \cdot d\mathbf{l} = \text{Work done by force along path} \tag{2.19}$$

2.6. PARTICLES OF ZERO MASS

The relationship between the energy and momentum of a moving particle (Equation 2.16) gives nonzero values of E even for the case of zero mass. That is, if we set m equal to zero in that equation, we get the following relation between E and **P**

$$E = c\,|\mathbf{P}| \tag{2.20}$$

It is an experimental fact that there exist physical particles which satisfy this relationship between E and P. The most important class of such zero mass particles is the set of particles called *photons*. In a manner which we shall describe in the chapters on quantum theory, photons are the carriers of the energy and momentum associated with electromagnetic waves.

Taking the magnitude of the vectors in Equation 2.17 we see that the velocity of any particle is related to its energy and momentum by

$$u = c^2 P/E \tag{2.21}$$

where $P = |\mathbf{P}|$. For a particle of zero mass (one which satisfies Equation 2.20) this relation gives

$$u = c^2 P/(c\,P) = c$$

Zero mass particles always travel at the velocity of light! If sufficient energy and momentum are available, a zero mass particle can be created but when it comes into existence, it is already moving with velocity c. They can be absorbed and have their energy and momentum transferred to the absorbing body. However, they can never be slowed down and collected. Zero mass particles can never be the permanent constituents of physical objects in the way that electrons, protons, and other particles of finite mass are. They are certainly rather bizarre but physically important solutions of the equations of relativistic dynamics. Some of the more common types of elementary particles of both zero and finite mass are listed in Table 2.1.

2.7. THE EQUIVALENCE PRINCIPLE

Imagine yourself inside a windowless space capsule orbiting the earth. You gently toss a pencil. Once it leaves your hand it slowly rotates about its center of mass while that center of mass moves through the air within the capsule at constant velocity. This is the phenomenon of weightlessness. A glass, placed in

TABLE 2.1

Masses and charges of some of the elementary particles

Particle Name	Symbol	Mass (u=at. mass unit)	Electric Charge (−e=electron charge)
Photon	γ	0	0
Electron	e^-	5.4860×10^{-4} u	−e
Positron	e^+	5.4860×10^{-4} u	+e
Proton	p	1.00728 u	+e
Neutron	n	1.00867 u	0
Pi plus	π^+	.149848 u	+e
Pi minus	π^-	.149848 u	−e
Pi zero	π^0	.144910 u	0
Mu minus	μ^-	.113432 u	−e
Mu plus	μ^+	.113432 u	+e
e neutrino	ν_e	0	0
μ neutrino	ν_μ	0	0

the air at rest, just sits there until you reach out and move it. Without a window to look out there is no way to tell that you are not moving at constant velocity in a distant empty region of space, far from any gravitating object. That is, there is no obvious way, by making local measurements, to distinguish the space capsule frame from a true inertial frame. It is important that we understand what physical law is at the root of this remarkable effect. The following line of reasoning will help us do so. We toss two objects, A and B, into the air. Let $\mathbf{r}_A$ and $\mathbf{r}_B$ be their position vectors as measured in a true inertial frame. The position of object B with respect to object A is described by the vector

$$\mathbf{r} = \mathbf{r}_B - \mathbf{r}_A \tag{2.22}$$

For simplicity let us assume that the velocities of both particles are much less than c so that we can use nonrelativistic dynamics to calculate their motion. Taking the second derivative of Equation 2.22 and using Newton's second law, we can easily see that

$$\frac{d^2\mathbf{r}}{dt^2} = \frac{\mathbf{F}_B}{m_B} - \frac{\mathbf{F}_A}{m_A}$$

If the force on each particle is precisely proportional to the particle's mass, as it is for gravitational forces, then

$$\frac{d^2\mathbf{r}}{dt^2} = 0$$

That is, the acceleration of any object with respect to any other object is zero which means that all objects move at constant velocity with respect to one another. This is the characteristic property of an inertial frame. Our imaginary space capsule, moving without rotation through the earth's gravitational field, is said to be freely falling. The underlying cause of the phenomenon of weightlessness within our freely falling capsule is that the gravitational force on any object within the capsule is proportional to that object's inertial mass. That is $\mathbf{F} = m\mathbf{g}$, where $\mathbf{g}$ is the gravitational field at the location of the capsule. In this analysis we have to assume that the capsule is much smaller than the earth (a reasonable assumption) so that $\mathbf{g}$ may be taken as a constant vector within the capsule.

In a number of experiments the exact proportionality between gravitational force and inertial mass has been confirmed to better than one part in 10^{12}. This may not seem very surprising. However we must remember that the inertial mass is affected by anything that changes the internal energy of an object. For example a parallel plate capacitor has a greater inertial mass when it is charged than when it is uncharged in spite of the fact that in charging it we simply move electrons from one plate to the other without changing the total number of particles in the system. The energy stored in the electric field between the plates must be hauled around with the capacitor and therefore affects its inertial mass. That the gravitational force remains proportional to the inertial mass means that a gravitational force must be exerted on the energy stored in the electrostatic field between the plates. But, if a gravitational force is exerted on electromagnetic forms of energy one should expect a light wave, which carries electromagnetic energy, to be deflected from a straight line path in passing near a massive object. Just such an effect has been unambiguously detected. (See Figure 2.6.) In fact a light ray in a gravitational field is deflected just enough so that, with respect to a freely falling object, the light ray appears to be going in a straight line. Nature seems to be trying to tell us something, and Einstein who was always a good listener when Nature spoke was the first one to get the message.

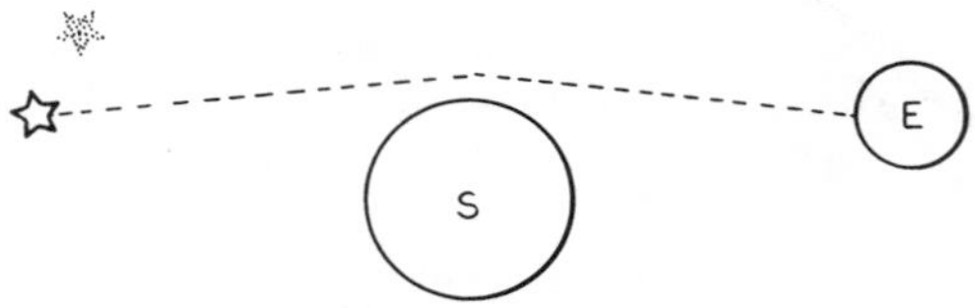

FIGURE 2.6 During a solar eclipse it is possible to observe stars whose light has passed very close to the sun. (At other times their faint light is overpowered by the brilliant sun.) When this is done it is found that the apparent position of the star is shifted due to bending of the light rays as they pass the sun.

Einstein realized that he could generalize the principle of relativity and the concept of an inertial frame in the following important way. A *local inertial frame* is either a constant velocity frame in a region without a gravitational field or a freely falling frame in a gravitational field. Actually one can simplify the definition of a local inertial frame by just saying that it is a freely falling frame and using the fact that a freely falling frame moves at constant velocity in a zero gravitational field. His generalized principle of relativity, known as the *principle of equivalence,* is that all the laws of physics, as applied to local experiments, are the same in any local inertial frame. Thus, the principle of equivalence says that it is absolutely impossible, by carrying out experiments within the space capsule, to distinguish between inertial motion in empty space and free fall motion in a gravitational field.

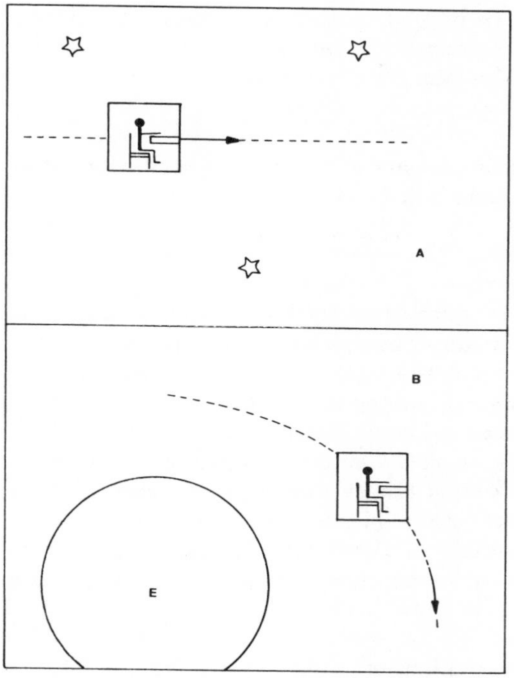

FIGURE 2.7 The scientists in the constant velocity inertial laboratory (A) and the free-fall laboratory (B) find the same laws of physics.

2.8. GENERAL RELATIVITY

Using the principle of equivalence as a guide, Einstein was able to construct a new theory of gravitation. Because it involves a generalization of the principle of relativity the new gravitational theory is called the theory of *general relativity.* It was already known that Newton's law of gravitation describes gravitational effects extremely well. Therefore any new gravitational theory would have to agree with the Newtonian theory over the large range of phenomena for which that theory is known to work. General relativity agrees with Newton's law of gravitation if two conditions are satisfied.

1. *The gravitational field is weak.* We say that the gravitational field at a point is weak if the work that would have to be done in moving a particle of mass m from that point to infinity against the force of gravitational attraction is much less than $m c^2$, the energy equivalent of the particle mass.

2. *The velocities of all particles in the problem under consideration are much less than the velocity of light, c.*

QUESTION: How weak is the sun's gravitational field at the position of Mercury, the closest planet? That is, what is the ratio of the work needed to remove a mass m near Mercury's orbit to $m c^2$?

ANSWER: The work needed to move a mass m to infinity is equal to the difference in the potential energy at infinity (namely, zero) and the potential energy at the location of the mass ($V = -m M G/r$). That is

$$\text{Work} = m M G/r$$

where M is the mass of the sun and G is the gravitational constant. The radius of Mercury's orbit is 5.8×10^{10} m. Thus

$$\text{Work}/m c^2 = M G/r c^2 = \frac{(2 \times 10^{30})(6.67 \times 10^{-11})}{(5.8 \times 10^{10})(9 \times 10^{16})} = 2.6 \times 10^{-8}$$

Since this ratio is so much smaller than one, the Newtonian theory of gravitation accounts for Mercury's orbital motion quite accurately. However, at the time that the theory of general relativity was introduced, astronomers had been puzzling over a small discrepancy between the calculated orbit of Mercury and the orbit that was really observed. Newtonian theory predicts that the orbit of a planet around the sun should be a closed ellipse that remains fixed in space. Now the gravitational pull of other planets will actually perturb the orbit somewhat, but the effects of the other planets on Mercury's orbit could be accurately predicted using Newton's laws of motion and gravitation. After those corrections had been made it was still found that the direction of the perihelion of Mercury's orbit (see Figure 2.8) carried out an unaccounted for motion of about one hundreth of a degree per century. (More exactly, 43 seconds of arc per century.) That such a

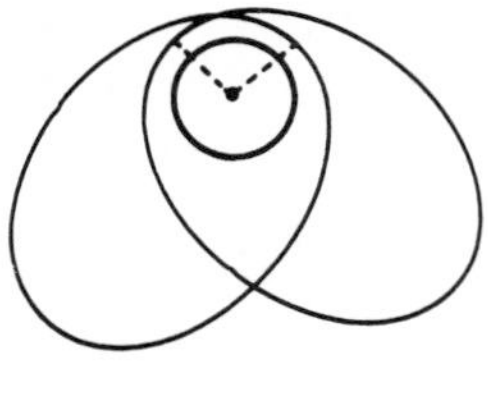

FIGURE 2.8
The position on Mercury's orbit at which the planet is closest to the sun is called the perihelion. During each century the position of the perihelion along with all the other points on the orbit undergo an unexplained (by Newtonian theory) rotation of 43 seconds of arc.

ridiculously small discrepancy between theory and observation was noticed is attributable to the near fanaticism that astronomers have always exhibited on questions relating to the precision and correctness of their observations. One of the first triumphs of general relativity was that it perfectly accounted for the observed discrepancy.

QUESTION: If no significant experimental violation of Newton's law of gravitation was known to Einstein, what motivated him to spend years of difficult labor in constructing a new gravitational theory?

ANSWER: According to Newton's law the gravitational field at location **r** and time t depends on the distribution of mass in the universe at the same instant. Moving a mass at one location instantaneously changes the field everywhere else. Gravitational effects propagate with infinite speed according to the Newtonian theory. It is easy to show that such a law is incompatible with the principle of relativity. It was natural to assume that Newton's law was related to the, as yet undiscovered, true law of gravitation in much the same way as Coulomb's law of electric force is related to the more complex laws of electromagnetism devised by Maxwell. The simple Coulomb law also gives instantaneous propagation of electric effects. But notice that the Maxwellian modification of the law, which yields a finite velocity of propagation of electromagnetic fields, also includes a much richer collection of physical phenomena, such as magnetic effects, induction, and electromagnetic waves. Thus it could be guessed that the complete gravitational theory would not be a trivial modification of the Newtonian theory but would be a much richer theory, predicting complex and novel gravitational effects. That guess has been definitely confirmed. The theory of general relativity describes physical effects never conceived of in the Newtonian theory. These include gravitational waves, black holes, the expansion of the universe, and many others. The resulting theory, which has been dazzlingly successful, was well worth the labor of its gestation.

2.9. THE MOVING CLOCK PROBLEM

Before we go further into general relativity we shall look at a little problem that does not seem to have anything to do with gravitational theory. However, once we obtain the solution of the problem, we will combine it with the equivalence principle to draw a very important conclusion.

We picture Mr. Unprime at the origin of his meter stick. He is in completely empty space. That is, his coordinate frame is an inertial frame in the old ungeneralized sense. Besides his regular clock, which remains always at rest at $x = 0$, he has another clock. We will call it the proper time clock. The proper time clock and the regular clock both read zero. Mr. Unprime wants to deliver the proper time clock to the Junior Observer at $x = X$ by frame time $t = T$. How should he move it so that it arrives there with the smallest possible discrepancy between it and the Junior Observer's frame clock?

First we have to show why there should be any discrepancy at all. The proper time clock, whose reading we shall call τ, is initially synchronized with the frame clocks and it is identical to them in structure. However, the frame clocks will all remain at rest in the frame but the proper time clock, if it is going to arrive at position X by time T will have to be moved sometime. As soon as it is moved, because of the time dilation effect it will start to run slowly. Therefore, when it gets to position X, the proper time clock will be somewhat behind the frame clocks. Since the discrepancy is always of the form, τ is less than t, we get the smallest discrepancy if we make the final reading on the proper time clock as large as possible. Thus we can restate the problem by asking how the proper time clock should be moved from $(x,t) = (0,0)$ to $(x,t) = (X,T)$ in order to maximize its final reading τ.

If the clock is moved very slowly, its retardation due to the time dilation effect is small, which suggests that Mr. Unprime should move it to $x = X$ as gradually as possible. But there is a limit to how slowly he can move it and still get it there by time T. The smallest possible velocity he can use is $v = X/T$. If he uses that velocity, then the clock will be running slowly during the full time interval $0<t<T$. This suggests that he might try another strategy. That is, to keep the proper time clock stationary at the origin, running at full speed, until the last possible moment, and then move it to position X at the maximum possible speed, almost the speed of light. Of course, in that case the clock would hardly run at all while it was moving but it would only be moving for a short time. In Problem 2.20 it is shown that the first strategy is the best one. In fact, it can be proved that *of all the possible ways of moving the clock from* $(x,t) = (0,0)$ *to* $(x,t) = (X,T)$ *the one which gives the maximum proper time interval is the path of constant velocity.*

But we also know that the constant velocity path is the path of any freely-moving object that passes through the spacetime points (0,0) and (X,T). We can combine these two facts to form what we shall call the *spacetime geometrical principle*, namely: *the paths of freely moving particles are paths of maximum proper time.* The reason why this is called a geometrical principle is because of its similarity to the Euclidean characterization of straight lines as paths of minimum length.

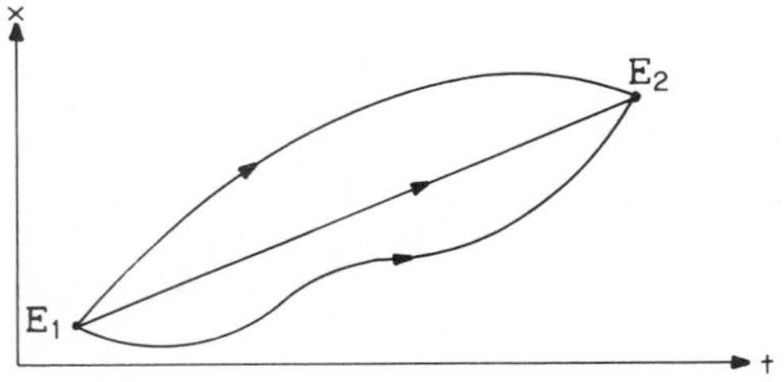

FIGURE 2.9 Any way of moving a clock from event E_1 to event E_2 can be plotted on a spacetime graph. Of all the possible ways, the one that will get the clock there with the maximum proper time is constant velocity motion. This is also the motion that a free particle would exhibit in going from one event to the other.

Now we shall see that surprising things result when we combine the spacetime geometrical principle with the equivalence principle. We picture a team of scientists in a closed laboratory that is falling toward the earth on a parabolic trajectory. They observe the motion of a small clock that is freely falling along with the laboratory. According to the equivalence principle their freely falling laboratory acts like an isolated inertial frame and therefore, with respect to the laboratory walls, the clock remains at rest. But besides observing the motion of that clock they also take a few identical clocks and move them along other paths (not free-fall trajectories) that begin and end on points of the freely falling clock's trajectory. (See Figure 2.10). If their freely-falling laboratory is really indistinguishable from a true inertial frame, then they must find that the freely falling clock registers a larger proper time interval than any

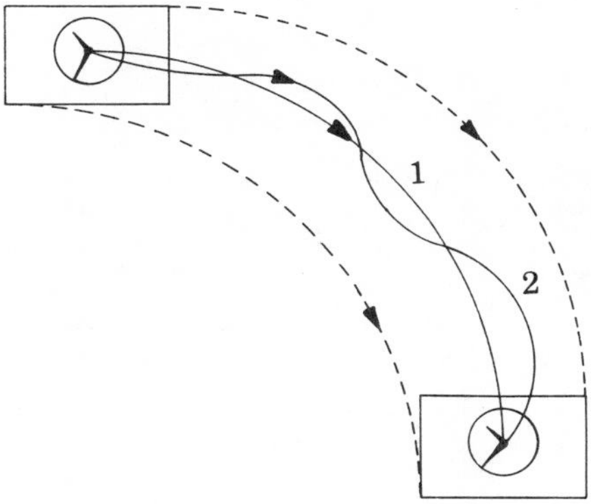

FIGURE 2.10
The actual free fall trajectory of the clock (1) maximizes its proper time interval in comparison with any other trajectory (2) that begins and ends at the same points.

of the other clocks. Thus we find that the geometrical principle, as stated above is true, *even in the presence of gravitating bodies.* The free fall trajectories, which in the earth's inertial frame are no longer constant-velocity paths, must still be paths of maximum proper time. In general relativity the geometrical principle plays the part of Newton's equation of motion in determining the trajectory of any object moving in a gravitational field.

2.10. THE SPACETIME METRIC

In order to obtain a more detailed idea of the mathematical structure of general relativity we must consider how one would actually calculate the proper time interval along an arbitrary trajectory. We will continue to consider only motions that are parallel to the x axis so that the only spacetime coordinates involved are x and t. We will also again introduce a time unit, the subsecond, that is chosen to make the speed of light equal to one. Let the trajectory be described by the function x(t). Using Equation 2.2, with c set equal to one, we obtain the following formula for total proper time interval along the trajectory.

$$\tau = \int_0^T \sqrt{1-(dx/dt)^2}\,dt \tag{2.23}$$

We can begin to see why people talk about the "geometry" of spacetime if we compare this formula with the formula for the length of a curve, y(**x**), that connects the points (0,0) and (X,Y) in a Euclidean x-y plane.

$$l = \int_0^X \sqrt{1+(dy/dx)^2}\,dx \tag{2.24}$$

The two formulas differ only by a sign. It is common to use the term *geodesic* to describe both a trajectory of maximum proper time in spacetime and a curve of minimum length in the Euclidean plane. For both cases the geodesics are straight lines. If we consider a surface that is not flat, then we can still define a geodesic as that curve on the surface that connects two given points and has the minimum total length. However, Equation 2.24 for the length of an arbitrary curve can no longer be used. That formula is valid only if x and y are Cartesian coordinates on a flat surface. For a curved surface there is some metric function $F(x,y,y')$ (where $y' = dy/dx$) which depends on the detailed shape of the surface, such that

$$l = \int_0^X F(x,y,y')\,dx \tag{2.25}$$

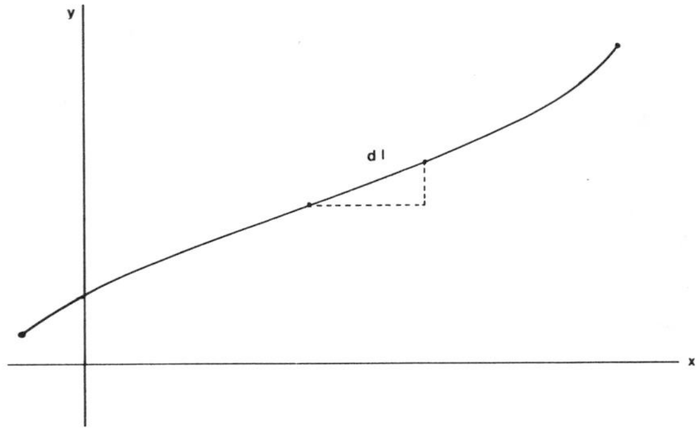

FIGURE 2.11 Calculating the total length of a curve by using the Pythagorean Theorem on each element of the curve.

The problem of calculating the metric functions of various surfaces is a problem in differential geometry with which we will not concern ourselves. However, the basic idea of a metric function as being the integrand in the integral formula for curve length plays an important part in general relativity.

At this point we are faced with a paradox. If we assume that the trajectory of a particle falling in a gravitational field is a path of maximum proper time (i.e., a geodesic) and we use Equation 2.23 to calculate the proper time interval along any path, then we are led to the conclusion that the particle moves in a straight line at constant velocity, because that is the trajectory that maximizes the proper time. The conclusion is obviously false. Particles in gravitational fields do not move at constant velocity. Where have we gone astray? The equivalence principle was based on the experimental observation that the gravitational force on anything is precisely proportional to the thing's inertial mass. The equivalence principle leads directly to the geometrical principle that free fall trajectories are trajectories of maximum proper time. But if the geometrical principle is combined with Equation 2.23 we immediately get a false result. After carefully thinking about this problem Einstein convinced himself that the equivalence principle was correct and that the false link in the logical chain was Equation 2.23. In general relativity it is assumed that freely falling particles follow geodesic curves in spacetime. However, Equation 2.23 is used for calculating those geodesics only for trajectories far from any massive bodies. The presence of massive bodies causes the metric function to differ from its empty space form. If, in the neighborhood of a massive body a clock is moved along a curve $\mathbf{r}(t)$ with velocity $\mathbf{v}(t) = d\mathbf{r}/dt$ during the time interval $0 < t < T$, then the reading on the clock will change an amount τ, which can be calculated using a *spacetime metric function* $M(\mathbf{r}, \mathbf{v})$ in the equation

$$\tau = \int_0^T M(\mathbf{r},\mathbf{v})\,dt \tag{2.26}$$

The specific form of the metric function M can be calculated from a knowledge of the distribution of masses by using a set of differential equations called the Einstein equations. Although, for weak gravitational fields, the trajectories predicted by general relativity agree with those predicted by Newtonian gravitational theory, the structure of the theory is very different. The Newtonian gravitational field **g** is replaced by the spacetime metric function M while the Newtonian equation of motion $d\mathbf{v}/dt = \mathbf{g}$ is replaced by the geometrical principle that falling particles follow spacetime geodesics.

When the gravitational field is strong, as it is in the neighborhood of dense, compact stars, the predictions of general relativity disagree radically with those of Newtonian gravitational theory. We shall discuss some of the strong-field effects in the next section. Even in weak fields general relativity predicts small effects that are different in kind from anything predicted by Newtonian theory. For instance, it is predicted that a clock that is at rest a distance r from a mass m will run slow, in comparison with a clock placed at a very great distance from the mass, by a factor $(1-2mG/c^2r)^{1/2}$. That is, if the clocks are initially synchronized and we call the first clock reading τ and the distant clock reading t, then

$$\tau = \left(1 - 2\frac{mG}{c^2r}\right)^{1/2} t \qquad (2.27)$$

In other words, the people on the ground floor of a tall apartment building live slower but longer than the people on the top floor. The gravitational time dilation effect has been confirmed in both astronomical and terrestrial experiments. Certainly the simplest and best standard clock is an atom that emits radiation of some known fixed frequency that depends only upon its internal structure. The oscillations of the electromagnetic waves being emitted can be considered as the "ticks" of the clock. If the clock runs slower near a very massive body, then the observed frequency of such emitted radiation coming from the surface atoms of a star of mass M and radius R should be lower (i.e., shifted toward the red end of the spectrum) by a factor of $(1-2MG/c^2r)^{1/2}$. This effect is known as the *gravitational redshift*. For the surface of the sun the redshift factor is .99999894 which makes the shift difficult but not impossible to detect. However, there are many compact stars whose masses are close to the sun's mass but whose sizes are very much smaller. For a number of those stars the gravitational redshift has been clearly detected. (For other ones, bright nearby stars or other factors prevent astronomers from clearly resolving the atomic radiation spectrum.)

2.11. THE POUND-REBKA EXPERIMENT

The obstacle to making accurate measurements of the gravitational redshift is the very small magnitude of the effect. General relativity theory predicts that an atomic clock in the basement of a building should run slow in comparison with an identical clock on the roof. The predicted frequency difference, $\Delta\nu$, is given by the simple formula (which can be derived from Equation 2.27)

$$\frac{\Delta\nu}{\nu} = \frac{gh}{c^2}$$

where h is the height of the building, g is the gravitational acceleration, and ν is the frequency of the atomic clock. Since $g = 9.8\,m/s^2$ and a typical building is about 20 m, the fractional shift in the frequency should be on the order of 10^{-15}. In order to make a measurement of a frequency shift of one part in 10^{15} one should have both a source and a detector with the same order of frequency resolution. If the frequency of your source drifts unpredictably by more than a part in 10^{15} or if the radiation given off by the source is composed of a mixture of frequencies with a fractional spread of more than 10^{-15}, then one would expect the small gravitational redshift to get lost in these other uncertainties. Actually by using statistical analysis of the data and a number of experimental tricks, one can use sources and detectors of much poorer frequency resolution, namely about one part in 10^{13}. However, even that frequency resolution is much better than the best atomic clock or the sharpest atomic spectral line.

In the late 1950's due to the work of R.L. Mossbauer, sources and detectors with the required frequency resolution became available. Using the new Mossbauer sources and detectors, R.V. Pound and G.A. Rebka, Jr., were able to accurately confirm the gravitational frequency shift caused by a twenty meter difference in elevation between source and detector. The predictions of the general theory of relativity were confirmed within one percent.

2.12. THE CURVATURE OF SPACE

What is probably the most remarkable prediction of general relativity is that the ordinary space geometry near massive bodies is not Euclidean. In particular, the Pythagorean theorem is not exactly true near a massive body. For instance, if the object shown in Figure 2.12 is constructed from twelve identical rods, the angle opposite the hypotenuse will not be exactly 90 degrees. Near the surface of the earth the expected deviation is about 10^{-10} radians which is too small to measure by any presently known means. One might object to the idea that the Pythagorean theorem can possibly be wrong on the grounds that it can be logically derived from the axioms of Euclidean geometry. But this objection is not really valid. All that the logical derivation does is to show that if the axioms of Euclid are true, *then* the Pythagorean theorem is also true. This presents no difficulty to the theory of general relativity which states that the Pythagorean theorem and the axioms of Euclid are both false.

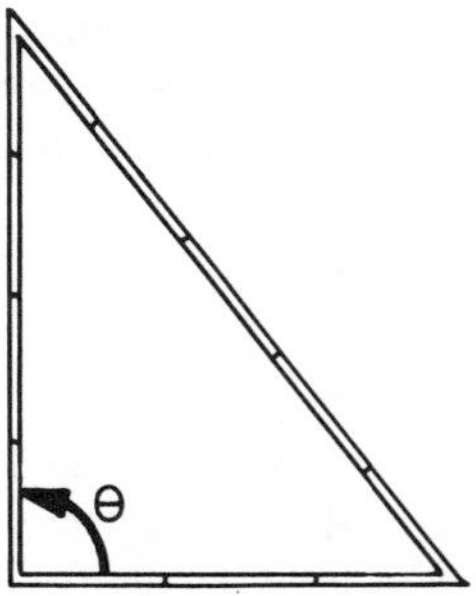

FIGURE 2.12
θ is not equal to 90° if this object is near a gravitating body even though $5^2 = 3^2 + 4^2$.

In discussing the experimental verification of the Euclidean axioms one has to be very careful because they contain terms that are not experimentally defined. For example, suppose we want to test the axiom that says that through two given points there is one and only one straight line. If we are going to check whether there is actually more than one straight line through some particular two points we must first decide what things qualify as straight lines. Certainly the definition found in the *Elements*: "A straight line is a line that lies evenly with the points on itself" will be of little use. The property of a Euclidean straight line that can most easily be adopted as an operational definition of one is the property of minimum length. We can experimentally determine the length of any curve by laying out small identical rulers end-to-end along the curve. Thus we can, in a reasonably unambiguous way, restate the Euclidean axiom in the following experimentally testable form.

Through any two points there is a unique curve of minimum length.

All the experimental evidence seems to indicate that the axiom in this form is false. For example, the deflection of light near massive objects, combined with the fact that light always travels along a path of minimum length shows that there must be more than one such path through some pairs of points (See Figure 2.13). If the axioms and postulates of Euclid are taken as a set of statements that

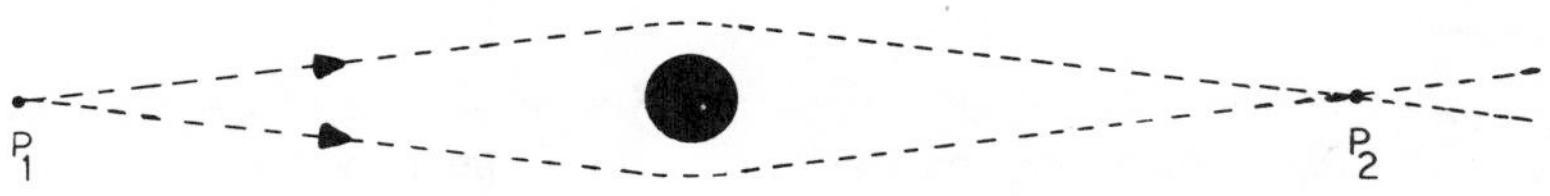

FIGURE 2.13 Because light is deflected by a massive body there may be more than one light path between P_1 and P_2. But, since light travels on a path of minimum length, there may also be more than one such path.

define some abstract mathematical objects called lines, points, planes, etc., then one cannot really discuss the question of whether they are really true. However, as soon as they are restated as physical laws about measurable quantities, they may be found to be either right or wrong. Countless "experiments" over thousands of years have shown that they are at least extremely good approximations to the true geometrical relationships in the real world. General relativity predicts that near any ordinary massive body there should be small deviations from those laws and near certain extremely compact massive objects such as black holes and neutron stars Euclidean geometry is not even a good approximation.

2.13. BLACK HOLES

Stars are formed by the condensation of large volumes of gas. The young star, like the interstellar gas that gave birth to it, is composed primarily of the element hydrogen. It is the gravitational force that pulls the gas molecules together and leads to the creation of one or more compact stars from what was initially a very diffuse gas cloud. During the lifetime of the star, that same gravitational force constantly acts to draw the star together and make it ever more compact. After a relatively short period of initial condensation, the gravitational force is exactly balanced by the tremendous outward pressure of the hot stellar material and the star maintains a fixed equilibrium size. But that equilibrium might be compared with the equilibrium of a person holding up a heavy weight. In the end the weight is likely to win out, because no expenditure of energy is required for the weight to maintain the downward force, but the person must constantly use energy to maintain the lifting force. In a similar way the star can only remain hot as long as energy is being released by thermonuclear processes going on within it. It must eventually cool down, since the total energy available to it, although huge, is finite. If the star's mass is less than some critical mass, estimated to be about 1.4 times the sun's mass, it will, when it eventually cools, reach one of two stable, extremely dense, configurations. One possible stable end point of a burned out star is a cold white dwarf. White dwarfs are very compact stars almost totally depleted of hydrogen. A typical white dwarf has about one solar mass but a diameter only one one-hundredth of the sun's. This gives it an average density of 1.4×10^9 kg/m^3. At this tremendous density, 10^5 times the density of lead, even at zero temperature the material of the star can withstand the enormous gravitational forces trying to compress it. A stable white dwarf undergoes no further evolution. When it finally cools to the point at which its radiation can no longer be detected, it becomes a chunk of dead invisible mass, noticeable only by its gravitational effects on the other stars in its neighborhood.

Although the mass densities common in white dwarfs are so great that they are beyond any meaningful comparison with those densities encountered on earth, the matter in a white dwarf is absolutely diffuse in comparison to that in a neutron star. The mass density in a white dwarf is typically 10^5 times the density of lead, but the density in a neutron star is 10^8 times the density in a white dwarf. The structure of neutron stars will be described in more detail in a later chapter covering nuclear physics. Although it is difficult to believe that real objects with such incredible densities could exist, the accumulated evidence that the astronomical objects called pulsars are simply rapidly rotating neutron stars is so strong that it is practically irrefutable.

If a star has a mass greater than the critical mass there is no known process by which it could permanently withstand the crushing gravitational forces that would develop in its final compact stages. Its radius would eventually approach the *black hole radius*, R_o, related to the star's mass by

$$R_o = 2GM/c^2$$

What happens to the star after its radius is reduced below R_o is a question that cannot be answered with any confidence. However, it is also a question with no physical consequences. Once a star reaches the black hole radius all possible communication with it is cut off. No object, particle, or radiation can escape through the gravitational fields at the black hole radius. Therefore no further light from the collapsing star reaches any outside observer. The star, in effect, vanishes from the universe. But its gravitational effect remains. Any object that falls into the black hole radius after the star has collapsed suffers the same disappearance as the original star. There are no particle trajectories or light rays that intersect the spherical surface $r = R_o$ twice.

QUESTION: For a star of 1.5 solar masses, what is the black hole radius?

ANSWER: A solar mass is 2×10^{30} kg. Therefore

$$R_o = \frac{2GM}{c^2} = \frac{(2)(6.67 \times 10^{-11})(3 \times 10^{30})}{(3 \times 10^8)^2} = 4.45 \times 10^3 \text{m} = 4.45 \text{km}$$

This should be compared with the radius of an ordinary star of 1.5 solar masses, which is about 10^6 km. The radius of a neutron star of the same mass is about 10 km. The black hole radius is not much smaller.

SUMMARY

Newtonian dynamics must be modified in order to be made consistent with the principle of relativity. This modification can be accomplished by writing everything in terms of four-vectors.

A *four-vector* is a set of four physical observables whose measured values in one inertial frame are related to their measured values in another inertial frame by a Lorentz transformation.

The *relativistic velocity* of an object is a four-vector whose components are

$$U_x = dx/d\tau,\ U_y = dy/d\tau,\ U_z = dz/d\tau, \text{ and } U_t = c\,dt/d\tau$$

τ is the time that would be registered by a clock moving with the object. It is called the *proper time.*

The relativistic velocity is related to the ordinary velocity, $\mathbf{u} = d\mathbf{r}/dt$, by

$$\mathbf{U} = \gamma_u \mathbf{u} \quad \text{and} \quad U_t = \gamma_u c$$

where

$$\gamma_u = \frac{1}{\sqrt{1-u^2/c^2}}$$

The *relativistic momentum* of an object is a four-vector that is equal to the mass of the object times its relativistic velocity.

$$\mathbf{P} = m\mathbf{U} = m\gamma_u \mathbf{u}$$

and

$$P_t = m\,U_t = m\,c\,\gamma_u$$

All four components of the total relativistic momentum are conserved for an isolated system.

The time component of momentum is related to the relativistic energy of the particle.

$$E = c\,P_t = \frac{m\,c^2}{\sqrt{1-u^2/c^2}}$$

where $\mathbf{u}$ is the ordinary velocity of the particle.

At velocities much less than c, the relativistic energy is equal to the mass energy, $m\,c^2$, plus the Newtonian kinetic energy

$$E \approx m\,c^2 + \tfrac{1}{2} m\,u^2$$

If the primed frame moves with velocity v in the x direction with respect to the unprimed frame, then the measurements of the relativistic momentum of any object in the two frames are related by

$$P_t = \gamma_v (P'_t + \beta P'_x)$$

$$P_x = \gamma_v (P'_x + \beta P'_t)$$

$$P_y = P'_y$$

and

$$P_z = P'_z$$

If the internal energy of an object changes by an amount ΔE, then its inertial mass will change by an amount Δm, where

$$\Delta E = \Delta m c^2$$

Particles of zero mass always travel at the velocity of light. They have energy and momentum which are related by

$$E = cP$$

where

$$P = (P_x^2 + P_y^2 + P_z^2)^{1/2}$$

The relationship between momentum **P** and energy E for a particle of mass m is

$$E = c\sqrt{m^2c^2 + \mathbf{P}\cdot\mathbf{P}}$$

General relativity is based on two fundamental principles:

1. The equivalence principle, which states that within any freely falling laboratory all the laws of physics have the same form that they do in an inertial system in empty space.
2. The geometrical principle, which states that the trajectory of any freely falling particle is that trajectory which maximizes the proper time interval.

The proper time that would elapse on a clock moved along the trajectory $\mathbf{r}(t)$ during the time interval $O \leqslant t \leqslant T$ is always given by an equation of the form

$$\tau = \int_O^T M(\mathbf{r},\mathbf{v})dt$$

The metric function M has the simple form

$$M = \sqrt{1 - v^2/c^2}$$

only if the motion takes place very far from any massive bodies.

Near gravitating bodies the metric function has a more complicated form that can be calculated from a set of differential equations called the Einstein equations.

General relativity predicts that the rate at which a clock runs depends upon where it is located within a gravitational field. Clocks placed more deeply in the field (at lower gravitational potential) run more slowly than those placed further from the gravitational mass. This effect causes a gravitational redshift in the spectral lines of radiating atoms near a massive body.

General relativity also predicts that the laws of Euclidean geometry are violated close to massive bodies.

The theory also correctly predicts the bending of light rays that pass close to massive bodies.

PROBLEMS

2.1 Consider the collision shown in Figure 2.1. If $u = v = c/2$ and $m = 10^{-27}$ kg, what are the values of the total momentum before and after the collision, using the Newtonian formula?

2.2* Describe the range of possible values that the following quantities can take: (A) u_x, (B) U_x, (C) $u = |\mathbf{u}|$, (D) $U = |\mathbf{U}|$, (E) U_t. Assume $m > 0$.

2.3* For a certain particle $x(0) = 0$, $U_x(\tau) = \alpha$, and $U_y = U_z = 0$. Find $x(t)$.

2.4* Show that, for any value of $\mathbf{u}$

$$U_t^2 - \mathbf{U} \cdot \mathbf{U} = c^2$$

2.5* In Ms Prime's frame an object is moving with relativistic velocity components $U'_x = .5c$, $U'_y = U'_z = 0$. (A) Calculate U'_t. (B) Calculate U_x, U_y, and U_t in Mr. Unprime's frame. The speed of Ms Prime through the unprimed frame is $v = .7c$.

2.6** A particle of mass m is initially stationary. It is struck by another particle of mass m, traveling at speed v. Both particles combine to form a particle of mass M moving at speed V. No radiation is given off. Calculate M and V in terms of m and v.

2.7* At the distance of the earth (1.5×10^{11} m) the density of radiant energy from the sun is about 1.5 kw/m^2. Calculate the rate at which mass is being converted to energy by the sun. Write your answer in metric tons per year. (1 metric ton = 10^3 kg).

2.8* The Stanford Linear Accelerator produces a beam of electrons which have an energy of 20 GeV. (One Giga-electron-volt = 10^9 eV = 1.6×10^{-10} J.) What is their speed as a fraction of c?

2.9* A particle of mass m is initially stationary at $x = 0$. Starting at $t = 0$ it is subjected to a constant force $f = dP/dt$. Find x as a function of t.

2.10* From the point of view of the elctrons traveling through it, how long is the Stanford "two-mile" accelerator? A typical electron energy is 20 GeV.

2.11*** An instantaneous rest frame for a particle undergoing accelerated motion is any inertial frame in which the particle is instantaneously at rest. Suppose a particle begins at rest in Mr. Unprime's frame at $t = 0$. It then moves in such a way that, in its instantaneous rest frame, its acceleration is always g (and in a constant direction down the x axis).

a) Calculate the object's ordinary acceleration, velocity, and position as determined by Mr. Unprime.

b) How far would the object get during the time interval, $t = 0$ to $t = 10^9$ s, which is about 32 years of Mr. Unprime's time?

c) How far would the object get during a proper time interval of 10^9 s?

d) What would be the value of t when $\tau = 10^9$ s?

2.12** Consider two planets, both in circular orbits of radius r about the sun. The orbits lie in the same plane. One planet moves according to Newtonian mechanics and one moves according to relativistic mechanics. The gravitational forces on the two are the same.

a) Which planet has the greater angular velocity?

b) If r is the mean radius of the earth's orbit, how many years (of the slower planet) would it take for the faster planet to gain one complete revolution?

2.13* If a constant force of one thousand newtons (about 200 pounds) is applied to a person of 100 kg mass, how long would it take for the velocity of the person to reach .9 c? The time should be given in the frame in which the person is initially at rest.

2.14* How much work would be done by a force of mg newtons applied for twenty years to a person of mass m = 75 kg? The person is initially at rest. Determine the average rate at which work is being done in kilowatts.

2.15** A particle of mass m and charge e moves within a region of uniform magnetic field. The magnetic field has magnitude B and points in the z direction. The particle moves in the x-y plane.

a) Show that the particle moves in a circle and that the relationship between the angular frequency of the circular motion ω and the radius of the circle R is

$$\omega = \lambda/\sqrt{1+(\lambda R/c)^2}$$

where $\lambda = eB/m$.

b) If the magnetic field is 12 tesla and a proton moves in a circle with radius R = 80 cm, what is the energy of the proton in units of m_pc^2?

2.16 A typical spring weighs 50 grams, is ten centimeters long, and has a force constant of one newton per centimeter. All these values are for the spring at equilibrium. The spring is stretched by 5 cm. How much does its mass increase? The mass increase corresponds to the mass of how many electrons?

2.17* An atom of ^{8}Be is composed of 4 protons, 4 neutrons, and 4 electrons. It has a mass of 8.005308 u (1 u = 1 Nuclidic Mass Unit = 1.66043×10^{-27} kg). An atom of ^{4}He is composed of 2 protons, 2 neutrons, and 2 electrons. It has a mass of 4.002605 u. Notice that two ^{4}He atoms have less mass than one ^{8}Be atom even though they are composed of the same particles. Left to itself an atom of ^{8}Be will spontaneously disintegrate into two ^{4}He atoms. The excess mass is converted into the kinetic energy of the helium atoms. Assuming the ^{8}Be atom is initially at rest, calculate the final velocity of the ^{4}He atoms.

2.18** A K^+ meson at rest decays into a π^+ and a π^0. What is the velocity of the π^0? What is the energy (including the rest energy, mc^2) of the π^+? (The mass of a K^+ is .53033 u.)

2.19** As viewed in a certain frame, which we call the center of mass frame, two protons approach one another with equal but opposite velocities. The speed of each is .999c. Calculate the relativistic energy of the protons in the center of mass frame. What we call the laboratory frame is a frame of reference in which one of the protons is at rest. Calculate the relativistic energy of the protons in the laboratory frame. What is the ratio of E_{lab} to

E_{cm}? This analysis is applicable to high-energy proton-proton scattering experiments in which one proton, that is part of a stationary target, is struck by another proton from a proton accelerator. The range of elementary particle reactions that can be obtained depends upon the value of E_{cm}. One of the problems faced by high-energy physicists is that E_{cm} is much smaller than E_{lab} for high-energy particles.

2.20* Mr. Unprime wants to move a clock from the spacetime point (0,0) to the spacetime point (X,T) where $X = 3 \times 10^8$ and $T = 3s$. Calculate the reading on the face of the clock when it arrives at (X,T) after being moved along the two possible trajectories shown in Figure P.2.20.

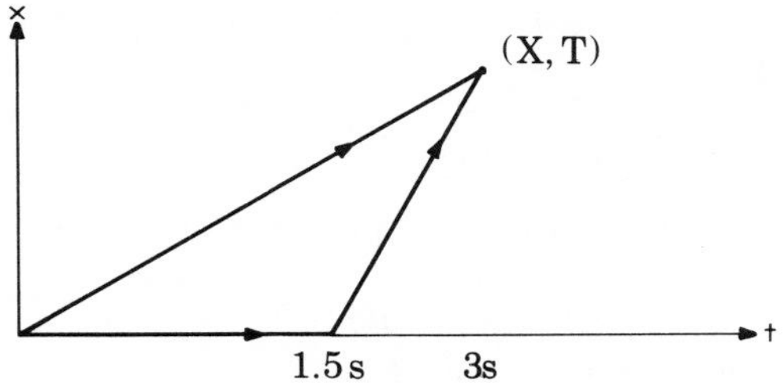

FIGURE P.2.20

CHAPTER THREE

THE ORIGINS OF QUANTUM PHYSICS

3.1. CLASSICAL PHYSICS

According to the ideas generally accepted during the late nineteenth century, physical reality consists of two types of entities: particles and fields. A particle is something that has a mass, a charge, and a location in space. A field has none of those properties. A field is everywhere at once. The momentum and energy carried by the field are distributed throughout the space in which the field exists. This picture of the world is usually referred to as classical physics. The classical theory is able to describe and accurately predict an incredibly wide range of phenomena. It can be used to calculate the trajectories of the planets with great precision. It can account for light, heat, sound, magnetism, and electricity. Of course the theory of relativity produced significant changes in the classical picture, but it did not alter the fundamental separation of the elements of reality into particles and fields. Nor did relativity theory change the general concept of a particle and a field. Starting in the late nineteenth century there was another major line of experimental and theoretical work which eventually led to a revision of our fundamental ideas of the structure of the world that was even more radical than the revision brought about by relativity theory. This line of work, which led to what is now called *quantum theory*, (or *quantum mechanics*) was for the most part independent of relativity theory.

The relation between classical physics and quantum theory is similar to that between Newtonian mechanics and relativistic mechanics. Newtonian mechanics is an approximation to relativistic mechanics that can be used only when all velocities in the problem are small in comparison with the velocity of light. In a similar way classical physics is an approximation to quantum theory that can be used only when the size of the system is much larger than a typical atomic size. Both rules regarding the conditions in which the approximate theory is valid are

FIGURE 3.1
A closed container with reflecting walls contains an object at temperature T. What is the radiant energy density when this system is in equilibrium?

only unreliable "rules of thumb". There are cases in which relativistic effects can be detected even though all the objects in the system are moving at low velocities. There are also cases in which large size objects show distinct quantum mechanical effects.

3.2. CAVITY RADIATION

In the late nineteenth and early twentieth century it was realized that in spite of its general success the classical picture of the physical world contained certain flaws. One of the most troublesome problems arose when people attempted to combine Maxwell's electromagnetic theory with thermodynamics. To see how the problem arises let us consider a completely evacuated container in which there is an object at some finite temperature (See Figure 3.1). The walls of the container are perfectly reflecting mirrors. The object, being composed of charged particles in thermal motion, will radiate energy into the container. It will also absorb radiant energy that has been reflected from the walls. Since no energy leaves the container the whole system will eventually come into equilibrium at some temperature T. The problem we shall consider is to calculate the density of radiant energy within the container as a function of the equilibrium temperature.

First let us show that the equilibrium radiant energy density cannot depend at all on the details of the structure of the body such as its color or chemical constitution. Suppose two different bodies, A and B, are both at temperature T and at equilibrium with the radiation within their separate enclosures. Suppose further that the radiant energy density in enclosure A is greater than that within enclosure B. If we now open a small window between the two containers (See Figure 3.2), a greater radiant energy flux will be incident upon the window from container A than from container B, due to the higher radiation density in A. There will be a net flow of radiation from A to B, causing the radiant energy density of B to increase while that of A decreases. Bodies A and B will no longer be in equilibrium with the radiation within their containers. Body A will begin to radiate energy into its container and its temperature will decrease. Body B will absorb some of the increased radiant energy and its temperatuure will increase. This process will continue until the radiation energy densities within the two containers are equal at which time the net flux of energy through the window will become zero. The two bodies will be left at unequal temperatures, T_A and T_B, with $T_B > T_A$. During the time interval after the window had been opened energy would have been flowing from the cooler to the hotter body in obvious violation of thermodynamic laws.

The only way to avoid this contradiction of the laws of thermodynamics is to assume that the density of radiant energy in a container at equilibrium depends only on the temperature. This conclusion is in agreement with experimental

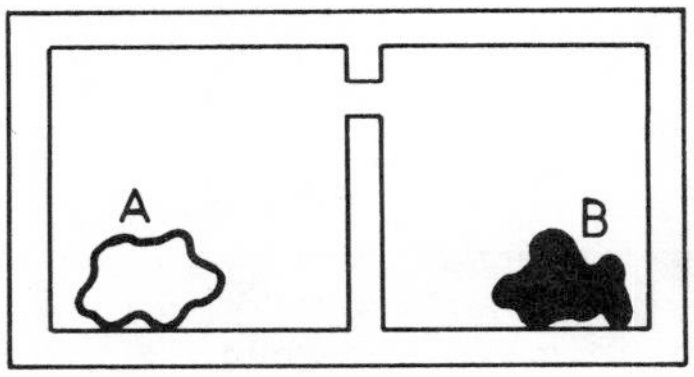

FIGURE 3.2
At equilibrium the radiant energy density in the two containers must be the same regardless of the nature of objects A and B.

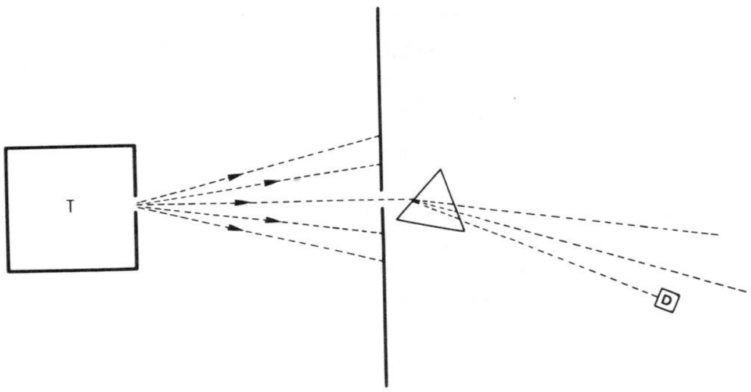

FIGURE 3.3 A setup for measuring the cavity radiation distribution function.

measurements of the radiant energy density within containers at various temperatures. By making a small window in the container and measuring the radiation passing through the window it is possible to determine the energy density inside. In fact, one can measure not only the radiant energy density but much more detailed characteristics of the radiation inside the cavity. In Figure 3.3 is shown a typical setup for making such measurements. An evacuated oven or cavity is maintained at some temperature T. The cavity contains a small window. The radiation from that window is collimated and then passed through a prism. The prism separates the radiation into its various frequency components and then a radiant energy meter (a bolometer) measures the radiant energy flux within the frequency range $d\nu$. With some analysis, that we will omit, one can then determine the energy density (energy per unit volume) within the cavity of those waves within the frequncy range $d\nu$. Call this $D(\nu)d\nu$. The units of the *frequency distribution function* $D(\nu)$ are Joules per cubic meter per Hertz, or $J{\cdot}s/m^3$. The experimentally determined function $D(\nu)$ is shown in Figure 3.4 for a

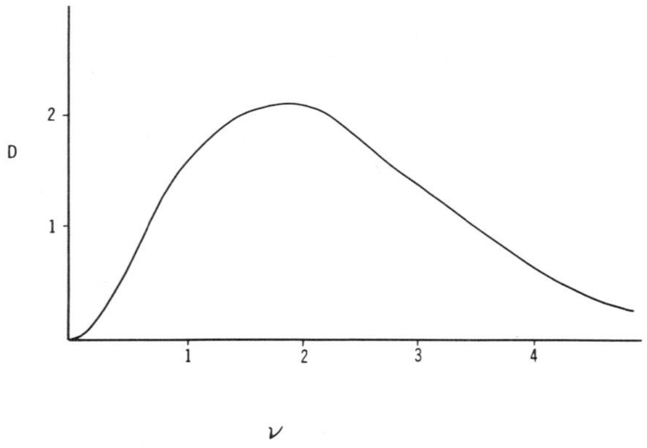

FIGURE 3.4 The experimental frequency distribution function for a cavity at 300 K. The y-axis gives $D(\nu)$ in units of $10^{-19}\,J{\cdot}s/m^3$. The x-axis is the frequency ν in units of 10^{13} Hz.

temperature of 300 K. This distribution of radiant energy within a cavity at equilibrium is called *cavity radiation.* It is also called *blackbody radiation* because the radiation emitted by a small hole in a cavity at equilibrium can be shown to be indistinguishable from the radiation emitted by a black body at the same temperature. A black body in this context is a hypothetical object that absorbs every bit of radiation that falls on it.

The reason why the study of cavity radiation generated such interest at the time it took place was that there is a simple argument that shows that the function $D(\nu)$ can never be derived by any analysis using classical physics. Naturally the phrase, classical physics, was unknown before the introduction of quantum theory, and therefore the preceding statement would have read that the function $D(\nu)$ cannot be derived using the laws of physics. The argument involves the dimensions (that is, the units) of all the quantities that we might possibly have to work with in our analysis. If we are going to carry out a dimensional analysis we should first write all our physical variables in one system of units. Naturally we will use the MKS system. The only thing that is obviously not in MKS units is the temperature T which is usually expressed in the somewhat illogical unit of Kelvin degrees. The conversion factor needed is Boltzmann's constant k_B. The MKS unit of temperature is the joule and k_B gives the number of joules per kelvin degree. Thus if θ is the temperature in joules then

$$\theta = k_B T$$

where T is the Kelvin temperature. That the correct unit for temperature is the joule is shown by the ideal gas law which states that

$$\theta = p\,V/N$$

The right hand side contains purely mechanical quantities with the units

$$(N/m^2)(m^3) = N{\cdot}m = J$$

If we introduce a notation [x] which means *the units of x*, then we can write that

$$[D(\nu)] = J{\cdot}s/m^3$$

We assume that there is some formula for $D(\nu)$. What physical quantities can appear in that formula? Since we have already said that $D(\nu)$ is completely independent of all of the properties of the container enclosing the cavity, including its size, and its composition, none of those things can appear in our formula or else we could change $D(\nu)$ by changing those things. Certainly θ and ν would appear in our formula since D depends on both of them. Also the velocity of light, c, might appear in the formula since D describes the energy density of electromagnetic waves. The only other fundamental constant in classical physics is G and if we make the reasonable assumption that gravitation is in no way involved in cavity radiation then G will not appear in our formula. The units of θ, ν, and c are

$$[\theta] = J$$

$$[\nu] = 1/s$$

$$[c] = m/s$$

First let us assume that the formula is of the form

$$D = \alpha\, \theta^K \nu^L c^M$$

where α is dimensionless. Taking the units of both sides we get

$$J{\cdot}s/m^3 = \frac{J^K m^M}{s^{L+M}}$$

The only solution of this is

$$K = 1,\ M = -3,\ L = 2$$

It is also easy to show that there is no combination of θ, ν, and c that is dimensionless. Therefore α cannot depend on some dimensionless combination of the variables but must be a purely mathematical constant. Thus we see that the only dimensionally correct formula is

$$D = \alpha \frac{\theta \nu^2}{c^3}$$

When detailed (and complicated) calculations are carried out, using Maxwell's equations and the laws of thermodynamics, one unavoidably gets the formula

$$D = 8\pi \frac{\theta \nu^2}{c^3}$$

A comparison of this formula with the experimental $D(\nu)$ is shown in Figure 3.5.

Clearly, the classical formula for $D(\nu)$ is absurd. The total electromagnetic energy density at temperature θ is obtained by integrating $D(\nu)$ from $\nu = 0$ to $\nu = \infty$. But the integral of ν^2 is infinite. However, one can see from Figure 3.5 that the classical prediction agrees with the experimental results for low frequencies. The obviously false prediction of infinite energy density at finite temperature was known as the *ultraviolet catastrophe.* It was naturally no real catastrophe, but just a catastrophic embarassment for theoretical physicists. There was clearly no way of getting the theoretical curve to bend over and give reasonable results unless one had another fundamental constant with different units to work with.

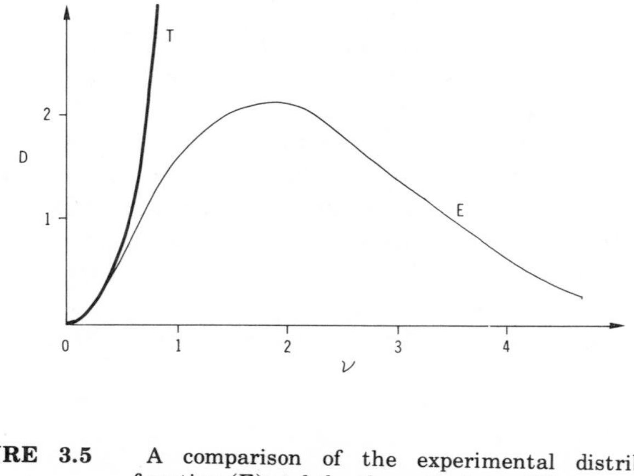

FIGURE 3.5 A comparison of the experimental distribution function (E) and the theoretical one (T).

This was where things stood when the German physicist Max Planck took up the problem around the year 1900. Planck first found, by trial and error, a completely empirical formula that fit the best experimental results accurately. It was what is now called the *Planck distribution.*

$$D(\nu) = \frac{8\pi h \nu^3}{c^3}(e^{h\nu/\theta} - 1)^{-1} \tag{3.1}$$

where $\theta = k_B T$ and h is a new fundamental constant, with the units of J-s, whose value was chosen so as to get an accurate fit to the experimental curve. This new constant, which will be found on almost every page for the rest of the book, is called *Planck's constant.* Its value is

$$h = 6.6256 \times 10^{-34}\,\mathrm{J{\cdot}s}$$

Having thus discovered the conclusion he needed, Planck was faced with the problem of finding a reasonable set of postulates and an argument that would lead to it. He had no more understanding of where this strange new constant came from than the reader is likely to have at this point. While the details of Planck's argument are too complicated to be presented here the essential ideas can be understood without too much mathematical analysis and are very important.

We picture an empty box in the form of a one meter cube. It is made of metal with a high electrical conductivity. We now ask, what are the possible electromagnetic fields that one can have inside such a box without worrying about how they are actually generated. (They will in fact be generated by the random thermal motion of the electrons in the metal.) It turns out that, according to Maxwell's equations, the possible electromagnetic fields are in the form of standing waves with various frequencies that can be calculated. The calculation is very similar to the calculation of the resonant frequencies of the sound waves in an air-filled cavity with rigid walls or the calculation of the frequencies of standing waves on a segment of string with fixed ends. Just as two electromagnetic waves can pass through one another without disturbing each other, so the various electromagnetic standing waves in a cavity can be simultaneously excited to any amounts without interacting with one another. The result of the electromagnetic standing wave calculation that is important for our present purposes is that, if we group the standing waves according to their frequencies, then

$$\begin{matrix}\text{the number of different} \\ \text{standing waves with} \\ \text{frequencies within the} \\ \text{range } \nu \text{ to } \nu + d\nu\end{matrix} = \frac{8\pi\nu^2}{c^3}d\nu$$

If we compare this with Equation 3.1 we see that the Planck distribution law tells us that, within the cavity at temperature T, the average energy of any *one* of the standing waves of frequency ν is

$$\overline{E}_\nu = h\nu(e^{h\nu/\theta} - 1)^{-1}$$

Planck showed, by a thermodynamic argument, that this is exactly the formula one would get if one made the following strange assumption:

> The electromagnetic field energy of a standing wave of freqency ν can only have the possible values 0, $h\nu$, $2h\nu$, $3h\nu$, $\cdots$ and not any value in between those ones.

We say that the energy associated with these waves is quantized with the value of the *energy quantum* being $h\nu$. Planck understood perfectly well that this was no "explanation" of the cavity radiation formula at all since the quantization assumption disagreed completely with the predictions of Maxwell's electromagnetic equations. One could not simply replace Maxwell's equations by the quantization hypothesis, because the quantization rule by itself was not at all a complete theory. In carrying out the theoretical analysis of cavity radiation, Planck had to use Maxwell's equations throughout the calculation and only impose the strange quantization rule at the last stage of the calculation.

QUESTION: What is the total radiant energy inside a cubic meter cavity at temperature 27°C (300 K)?

ANSWER: We obtain the total radiant energy density by integrating $D(\nu)$ over all frequencies.

$$\frac{E}{V} = \int_0^\infty D(\nu)d\nu$$

$$= \frac{8\pi h}{c^3}\int_0^\infty \frac{\nu^3 d\nu}{e^{h\nu/\theta} - 1}$$

Introducing a new variable of integration $x \equiv h\nu/\theta$ with $dx = \frac{h}{\theta}d\nu$ we get

$$\frac{E}{V} = (8\pi\theta^4/h^3c^3)\int_0^\infty \frac{x^3 dx}{e^x - 1}$$

The integral can be evaluated using the Integral Table at the back of the book. We get

$$\frac{E}{V} = \frac{8\pi^5\theta^4}{15h^3c^3}$$

The total energy density is thus proportional to the fourth power of the absolute temperature. If we write the temperature in conventional units the proportionality constant is given by

$$\frac{E}{V} = aT^4$$

where

$$a = \frac{8\pi^5 k_B^4}{15h^3c^3} = 7.54 \times 10^{-16}\, J/m^3K^4$$

In the case at hand $V = 1\,m^3$ and $T = 300\,K$, which gives a total radiant energy of

$$E = 6.11 \times 10^{-6} J$$

We see that it does not take much energy to heat up a vacuum to room temperature.

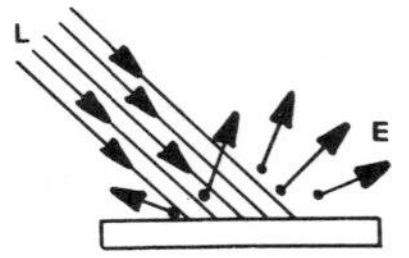

FIGURE 3.6
The photoelectric effect. Incoming light (L) causes the ejection of electrons (E) from a clean metal surface.

3.3. THE PHOTOELECTRIC EFFECT

The evidence for the quantization hypothesis was made stronger, but no less puzzling, by Einstein's theoretical analysis of something called the *photoelectric effect.* When light shines on a clean metal surface electrons are ejected from the surface with some distribution of velocities that depends upon the frequency and intensity of the light and the nature of the metal (See Figure 3.6.). Using the following assumptions, Einstein made some definite predictions regarding the dependence of the electron velocities on the frequency and intensity of the light.

Assumptions

1. The energy in the light is carried and delivered to the electrons in discrete quanta of amount $h\nu$, where ν is the frequency of the light, and h is Planck's constant.
2. An electron within the metal is bound to the metal by an energy W, which depends on the nature of the metal. W is called the *work function* of the metal. That is, it requires an amount of work, W to remove an electron from inside the metal.
3. The probability of one electron absorbing more than one quantum of energy is negligibly small at ordinary light intensities. (For very high intensity light one can see the effects of multiple quantum absorption.)

If an electron in the metal absorbs one quantum of energy, it will have a kinetic energy of $h\nu$. If its velocity is directed out of the metal, then it will fly out of the surface. If, on its way out, it does not lose any energy by collisions with other particles in the metal, then it will have an energy of

$$E = h\nu - W \tag{3.2}$$

when it gets far from the metal surface. The work W is the work that the electron had to do against the attractive forces tending to pull it back into the metal. Many of the electrons, after absorbing a quantum of energy, will lose some of the energy due to collisions before leaving the surface. Thus the energy given by Equation 3.2 is the maximum possible energy that ejected electrons can have.

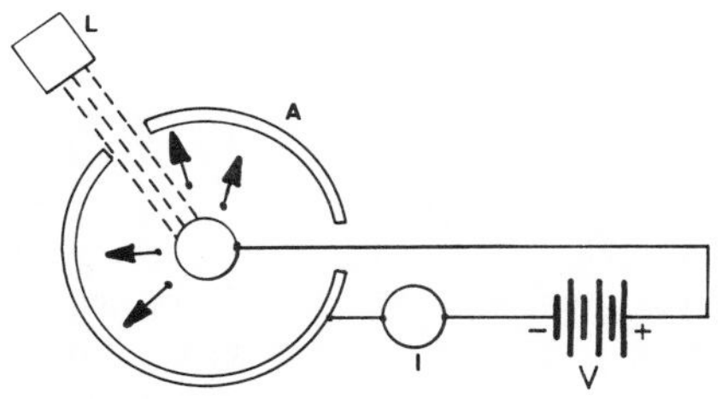

FIGURE 3.7
If eV is greater than the maximum energy of the electrons being ejected then none of them reach the spherical anode (A) and the current (I) is zero. (L) is a source of monochromatic light.

In Figure 3.7 is shown an experimental setup for measuring the maximum kinetic energy of electrons emitted by the photoelectric effect. A small sphere of the test metal is put at the center of a larger spherical cathode that contains a hole through which a beam of light can be focused on the test metal. If there is no electric field opposing the motion of the ejected electrons, they will all eventually strike the cathode. If the electrostatic potential difference, V, between the test metal and the cathode satisfies the inequality

$$eV < h\nu - W$$

then some of the electrons will reach the cathode. One can then adjust V to find the cutoff voltage at which there is no photoelectric current. According to the quantization hypothesis the cutoff voltage should be a linear function of the frequency of the light with a slope equal to h/e and it should go to zero when $h\nu$ is equal to the work function of the metal. The cutoff voltage should be independent of the intensity of the light. The intensity of the light, according to the quantization hypothesis, tells how many quanta of light energy strike the surface per second and therefore would be proportional to the magnitude of the photoelectric current whenever V is less than the cutoff voltage. All these simple predictions have been confirmed in many experiments. They disagree completely with anything that might be expected from a classical picture of the photelectric process. In the classical picture the intensity of the light is proportional to the square of the electric field strength associated with the light wave. Therefore increasing the intensity should increase the force on the electrons which would increase the energy of the ejected electrons.

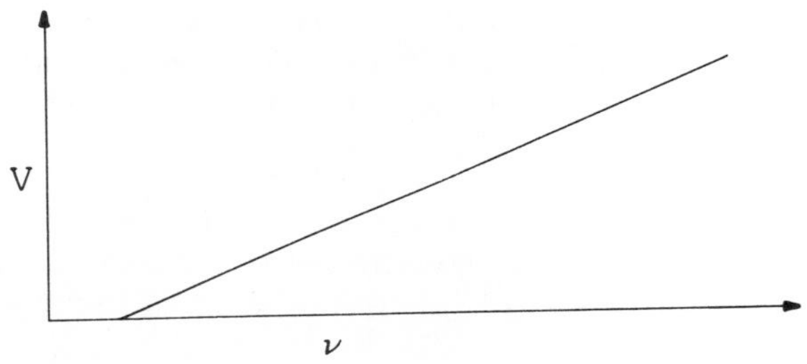

FIGURE 3.8 The cutoff voltage (V) as a function of the light frequency (ν). According to Einstein's theory $V = (h\nu - W)/e$ which means that the cutoff voltage should be proportional to the light frequency.

3.4. THE COMPTON EFFECT

That electromagnetic waves are capable of carrying energy and momentum was a prediction of the classical theory that had been experimentally confirmed. This fact was not disturbed by the quantization hypothesis. The quantization hypothesis only stated that the energy carried in the wave came in units or quanta equal to $h\nu$. But the momentum carried by electromagnetic waves was known to be related to the energy carried by the waves by the equation

$$p = E/c$$

If the energy is quantized in units of $h\nu$ then it seems unavoidable that the momentum carried by electromagnetic waves must also be quantized. The quantum units of momentum would be

$$p = h\nu/c = h/\lambda \qquad (3.3)$$

where λ is the wavelength. The quantity p gives the magnitude but not the direction of the single quantum of electromagnetic momentum. Similarly λ gives the wavelength but not the direction of propagation of the wave. We can rectify both of these deficiencies in the following way. We introduce a *wave vector* **k** whose manitude is related to λ by

$$k = 2\pi/\lambda$$

and whose direction is defined as the direction of propagation of the wave. Because the direction of the electromagnetic momentum, **p**, is the same as the direction of **k**, we can rewrite Equation 3.3 as the vector equation

$$\mathbf{p} = (h/2\pi)\mathbf{k} \equiv \hbar\mathbf{k} \qquad (3.4)$$

where the quantity $\hbar$ (pronounced "aitch-bar") is defined as

$$\hbar = h/2\pi = 1.0545 \times 10^{-34}\ \text{J·s}$$

$\hbar$, which contains the same information as h, is also referred to as Planck's constant.

Assuming that these energy and momentum quanta exist one might picture them as some sort of particles, commonly called *photons*, that are associated with the electromagnetic wave. The energy and momentum of a photon satisfies the relationship

$$E = cp$$

This is the relationship derived in the last chapter for particles of zero mass.

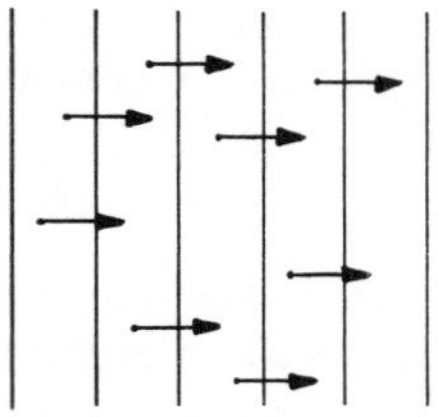

FIGURE 3.9
The energy and momentum of an electromagnetic wave are carried by some kind of zero mass particles, called photons.

In 1922 an experiment was performed by A.H. Compton that clearly illustrated the quantization of electromagnetic momentum and energy. The experiment can be described as the scattering of a zero mass photon by an electron. The electron is initially stationary (See Figure 3.10).

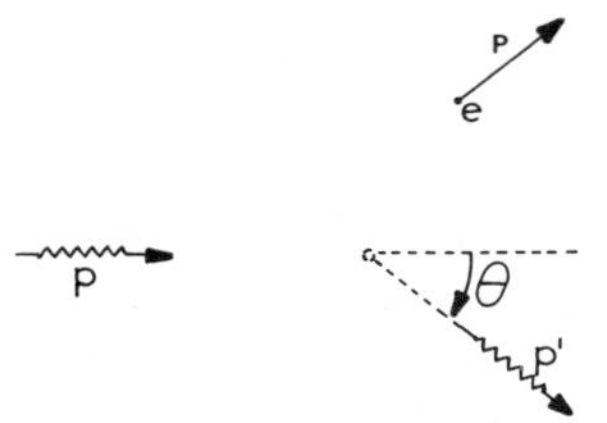

FIGURE 3.10
The scattering of a photon by an electron. Photons are traditionally drawn as wavy lines to indicate that they simutaneously exhibit both wave and particle properties.

Before the collision, the photon has momentum **p** and energy cp. The electron has momentum zero and relativistic energy mc^2. After the collision the photon has momentum $\mathbf{p}'$ and energy cp' while the electron is left with momentum **P** and a relavistic energy of $E = c\sqrt{P^2+m^2c^2}$. The momentum conservation law tells us that

$$\mathbf{P} = \mathbf{p} - \mathbf{p}' \tag{3.5}$$

The energy conservation law states that

$$c\sqrt{P^2+m^2c^2} + cp' = mc^2 + cp \tag{3.6}$$

Using Equation 3.5 to eliminate P in Equation 3.6 gives

$$\sqrt{(\mathbf{p}-\mathbf{p}')^2+m^2c^2} = p - p' + mc \tag{3.7}$$

From Figure 3.10 we can see that

$$(\mathbf{p}-\mathbf{p}')^2 = p^2 + p'^2 - 2pp'\cos\theta$$

If we use this in Equation 3.7, square both sides, cancel identical terms on the two sides of the resulting equation, and then divide it by $mcpp'$, we obtain

$$\frac{1}{p'} = \frac{1}{p} + \frac{1-\cos\theta}{mc} \tag{3.8}$$

This is an equation for the momentum of the photons associated with the scattered electromagnetic radiation. If the scattered photons have momentum p′ then the electromagnetic wave associated* with them must have a wavelength $\lambda' = h/p'$. Thus we can rewrite Equation 3.8 as an equation giving the wavelength of the radiation that is scattered by an angle θ.

$$\lambda' = \lambda + (1-\cos\theta)\lambda_c \tag{3.9}$$

where the quantity

$$\lambda_c = h/mc \tag{3.10}$$

is called the *Compton wavelength*. It has the dimensions of a length but is not actually the wavelength of anything.

Equation 3.9 predicts that the radiation scattered at an angle θ with respect to the incoming beam will have a wavelength larger than the incoming radiation

*The author realizes that the very unclear "association" between the photon particles and the electromagnetic waves must be very annoying to the reader. The exact relationship between particles and fields will be described when we present modern quantum theory. He is convinced that the policy of presenting the exact theory right from the start is not the best one. Quantum theory is sufficiently abstract and weird as to be largely incomprehensible without a certain amount of preliminary experimental background. That background can best be obtained by reviewing the major historical experiments.

by an amount $(1-\cos\theta)\lambda_c$. In the experiment performed by Compton a beam of X-rays of wavelength about .7 Å was scattered by a sample of graphite. The wavelength of the scattered radiation was measured and compared with the angle of scattering. It was found that the scattered radiation had two components (See Figure 3.11). One component had a wavelength which satisfied Equation 3.9 very well. This component was due to a scattering process in which an electron, weakly bound to one of the graphite carbon atoms, scattered a photon and was ejected from the carbon atom in the process. Since a photon of wavelength .7Å has an energy of about 20,000 eV, which is much larger than the binding energy of the electron in the carbon atom, the electron could be considered as essentially free and the analysis leading to Equation 3.9 could be applied. The other component of the scattered radiation had the same wavelength as the incoming radiation. It was due to another kind of scattering process in which the energy and momentum are transferred to the whole carbon atom rather than being taken up by a single electron. One could still apply Equations 3.9 and 3.10, but in this case the mass to be used in Equation 3.10 is the mass of a carbon atom rather than the mass of an electron. The carbon atom has such a large mass that the wavelength shift predicted by Equation 3.9 is undetectable.

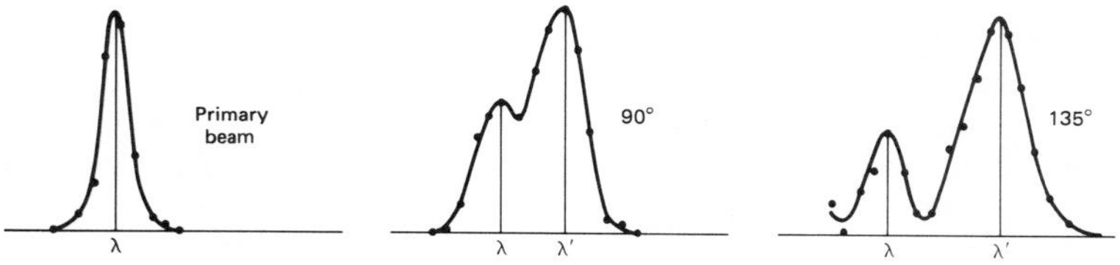

FIGURE 3.11 The intensity of scattered X-rays as a function of their wavelengths. (A) The wavelength distribution in the incoming X-ray beam. (B and C) The scattered radiation at 90° and 135° respectively. The values of λ′ predicted by Equation 3.9 are shown by heavy vertical lines while the average wavelength of the incoming beam λ is shown by light vertical lines.

The net effect of all the experiments which exhibited the particle-like or quantum properties of electromagnetic radiation was to produce an extremely bewildering situation in physics. On the one hand the classical theory of electromagnetism was being used with more and more success to describe the propagation, reflection, and diffraction of electromagnetic waves. On the other hand the photoelectric effect, the Compton effect, and other experiments we have not mentioned seemed clearly to show that the energy and momentum transported by an electromagnetic wave is somehow carried by zero mass particles. No one had any idea how to make these conflicting pictures of electromagnetic radiation consistent.

During the interval of time in which the above experiments involving light quanta were being carried out (about 1900 to 1925) another line of experiments and theoretical work was producing a similar paradox regarding the structure of the atom. We shall now give a brief desciption of some of the more important of those experiments and theoretical investigations.

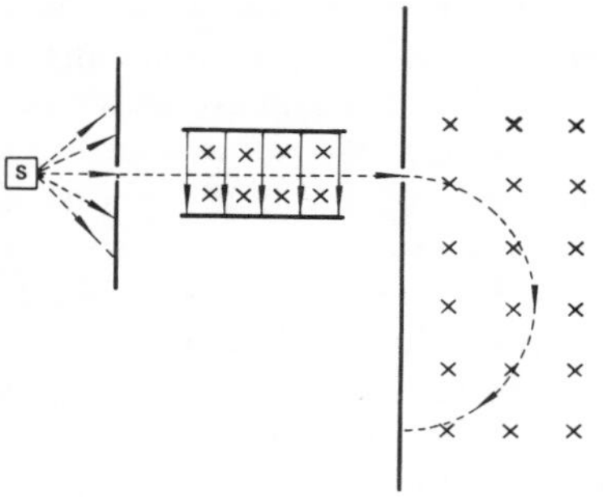

FIGURE 3.12
The identification of particles by means of their motion through electric and magnetic fields. In the velocity selector (VS) the electric field points downward and the magnetic field points into the paper. A particle will pass through without deflection only if $qE = qvB$ or $v = E/B$. In the uniform magnetic field to the right the particle (of known velocity) moves in a circle of radius $R = mv/qB$. This allows a determination of m/q, from which one can identify the particle.

3.5. ALPHA-PARTICLE SCATTERING

Certain radioactive elements were known to emit a type of nonelectromagnetic radiation called α-radiation. In a later chapter on nuclear physics we shall describe the physical process involved in α-radiation in detail. By means of experiments in which a thin pencil of "α-rays" were observed passing through electric and magnetic fields it could be determined that the radiation consisted of a shower of high energy positively charged particles. The particles, called α-particles, had a mass nearly equal to that of a helium atom and a positive charge equal to 2e, where the charge of an electron is −e. At the Manchester Laboratory of Ernest Rutherford two of his assistants, Geiger and Marsden, carried out an experiment in which an extremely thin gold foil was bombarded by a stream of α-particles of known energy. The positively charged α-particles were deflected due to the microscopic electric fields within the gold atoms and passed out of the gold foil moving in all possible directions. The flux of α-particles at a given angle with respect to the incident beam was measured by means of a scintillation counter (See Figure 3.13).

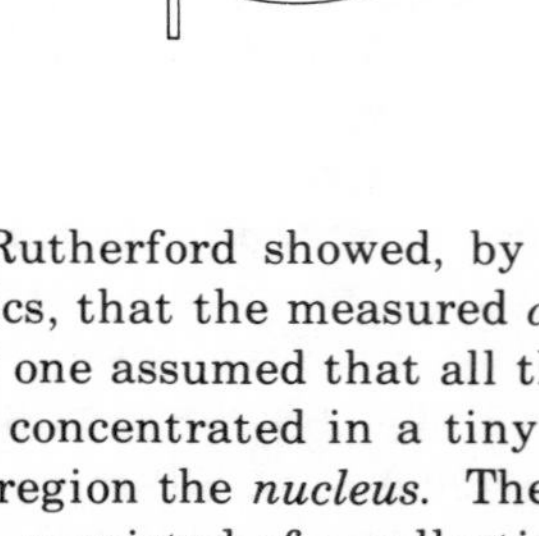

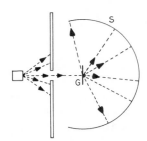

FIGURE 3.13
The scattering of α-particles. The scintillation screen (S) gives off a little flash of light when struck by an α-particle that has been scattered by the gold foil (G).

Rutherford showed, by means of an argument which used purely classical physics, that the measured α-particle flux at all angles could be simply accounted for if one assumed that all the positive charge and almost all the mass of an atom were concentrated in a tiny region in the center of the atom. Rutherford called that region the *nucleus*. The negative charge, which was needed to neutralize the atom, consisted of a collection of electrons that were somehow distributed within the rest of the space occupied by the atom. The size of the nucleus was extremely small, even in comparison with the tiny size of the whole atom. A typical atom has a diameter of about $1\,\mathring{A} = 10^{-10}$m. A typical nucleus has a diameter of 1 Fermi $= 10^{-15}$m. Thus the volume of the atom is 10^{15} times larger than the volume of the nucleus and, since most of the mass resides in the nucleus, the mass density within a nucleus is about 10^{15} times the mass density of the ordinary material we are used to.

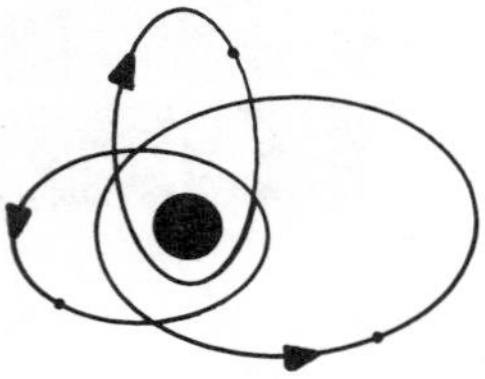

FIGURE 3.14
The planetary atom.

According to classical physics such a planetary atom as is shown in Figure 3.14 should be completely unstable. The moving electrons would be expected to produce electromagnetic waves that would carry away energy. As the electrons lost energy they would spiral ever closer to the attracting nucleus. The whole assemblage would contract to a point, with a burst of radiation, in a short time. Even if one ignored the problem of stability there was no way of undersanding, within the planetary atom model, the most fundamental property of atoms, namely the fact that every atom of an element is like every other atom of the same element. Using the planetary picture of, say, a helium atom, one would expect the size of the atom to depend upon the radii of the electron orbits around the nucleus. But there is nothing in classical physics that would prevent different helium atoms from having different electron orbits. It is not necssary to belabor the obvious point that an entity as changeable and mushy as a collection of electrons orbiting a nucleus bears little resemblance to the picture of rigid, identical, indestructible atoms which is necessary to explain the chemical and physical properties of matter. In spite of the difficulty physicists experienced in trying to harmonize Rutherford's planetary model of the atom with the known properties of atoms, later repetitons of his experiment and a number of different experiments clearly confirmed that the basic structure of an atom involves a heavy positive nucleus surrounded in some way by light negative electrons.

3.6. ENERGY QUANTIZATION IN ATOMS

At the time that the idea of electromagnetic energy quanta developed it had already been known for a long time that isolated atoms, when energized by thermal collisions, radiated their excess energy in the form of electromagnetic waves of definite discrete frequencies. The collection of possible radiation frequencies, called the *emission spectrum* of the atom, is different for different elements (See Figure 3.15). Before the introduction of the Rutherford planetary model of an atom the existence of discrete emission frequencies did not seem difficult to understand in principle. It was assumed that the atom was a fairly rigid but slightly deformable structure made up of electrically charged material. As such it would be expected to have certain normal modes of vibration with associated discrete normal mode frequencies. It seemed reasonable that these vibrational frequencies of the semi-rigid atomic structure were observed as the emission frequencies of electromagnetic waves.

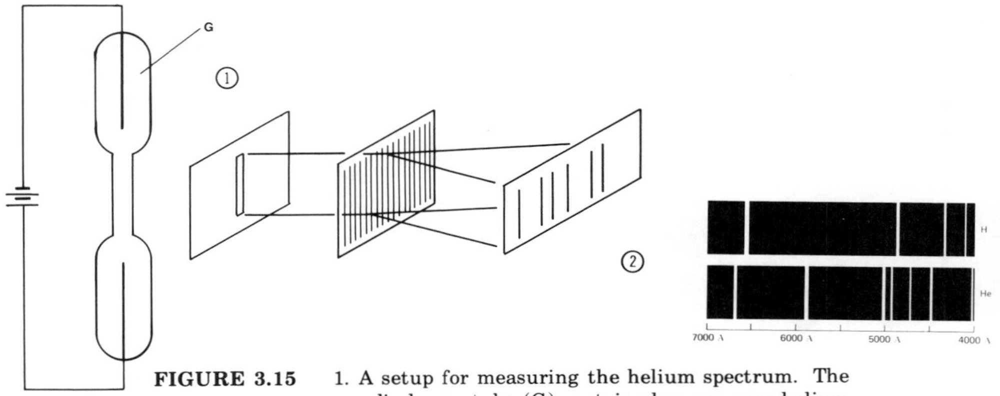

FIGURE 3.15 1. A setup for measuring the helium spectrum. The gas discharge tube (G) contains low pressure helium gas which glows when an electric current is passed through it. The screen (S) has a thin slit for collimating the light. The light is dispersed when it passes through the diffraction grating (D) and produces a series of sharp lines on the strip of photographic film (F).
2. The visible portion of the emission spectra of hydrogen and helium.

After Planck's postulate of electromagnetic energy quantization Einstein suggested a complete reinterpretation of the emission spectrum of an atom. He reasoned that if the electromagnetic field energy could have only quantized values, then when an atom emitted electromagnetic radiation it could only do so in discrete, whole quanta. He viewed the process of radiation by an atom as one in which the atom emitted a complete quantum or photon of radiation. The photon carried away an amount of energy $h\nu$, where ν is the frequency of the emitted radiation. Since the photon carries away an amount of energy $h\nu$, the energy state of the atom must change by a corresponding amount.

$$E_{before} - E_{after} = h\nu$$

where E_{before} and E_{after} are the energies of the atom before and after the photon is emitted. The fact that a particular kind of atom only radiates photons of those discrete frequencies that are in its spectrum means that the energy changes which such atoms undergo are always of certain amounts. An obvious way to explain such discrete energy changes by an atom is to assume that the energy of the atom itself can only have certain particular discrete values. Let us temporarily assume that this is so and see what consequences we can draw from the assumption. We assume that the energy of a particular atom can, for some unknown reason, have only the values E_1, E_2, E_3, $\cdots$ and no value in between these discrete values. The set of numbers $E_1, E_2, E_3, \cdots$ we call the *energy spectrum* of the atom, and any one of the numbers in it, E_n, we call an *energy level.* The lowest energy level, E_1 we call the *ground state energy.* If the atom is in a state with energy E_m and then emits a photon of energy $h\nu$, in the process changing to the energy state E_n, then by energy conservation the photon energy must be equal to the difference in the two atomic energy levels.

$$h\nu = E_m - E_n$$

If we call the frequency of the photon emitted in this process ν_{mn}, then

$$\nu_{mn} = (E_m - E_n)/h \tag{3.11}$$

To see that this assumption leads to a simple verifiable prediction, let us look at all the possible energy transitions among the three levels E_1, E_2, and E_3. (We always order the levels so that $E_1 \leqslant E_2 \leqslant E_3$, etc.) There are three possible transitions $E_2 \rightarrow E_1$, $E_3 \rightarrow E_2$, and $E_3 \rightarrow E_1$, leading to the following values for the radiation frequencies.

$$(E_2 \rightarrow E_1) \qquad \nu_{21} = (E_2 - E_1)/h$$

$$(E_3 \rightarrow E_2) \qquad \nu_{32} = (E_3 - E_2)/h$$

$$(E_3 \rightarrow E_1) \qquad \nu_{31} = (E_3 - E_1)/h$$

From these equations it is a trivial matter to see that

$$\nu_{31} = \nu_{32} + \nu_{21}$$

In general, for any three levels, $E_k \leqslant E_m \leqslant E_n$ we get

$$\nu_{nk} = \nu_{nm} + \nu_{mk} \tag{3.12}$$

We are thus led to expect, in the emission spectrum of every element, many instances of two emission frequencies which, when added, yield another frequency in the spectrum. Because these emission frequencies can usually be measured with great precision, this is a very strong prediction. That many combinations of spectral lines satisfy Equation 3.12 was already known at the time of Einstein's analysis. It was called the *Ritz combination principle.*

3.7. THE FRANCK-HERTZ EXPERIMENT

In 1914 J. Franck and G. Hertz reported on an experiment that strongly corroborated Einstein's idea that the discrete emission spectrum of atoms was caused by the existence of discrete atomic energy levels. A description of their apparatus is given in Figure 3.16. A metal, if it is heated sufficiently, will emit electrons through its surface. The effect is called *thermionic emission.* It was of great importance when electronic devices were made with vacuum tubes rather than transistors. The basic mechanism is quite understandable. Due to thermal agitation some of the electrons in the metal gain enough energy to overcome the work function at the surface and fly off. A typical thermionic electron has an energy of about $E \approx k_B T$ where T is the absolute temperature of the metal. For usual hot wire temperatures (about 10^3 K) the energy of thermionic electrons is about .1 eV.

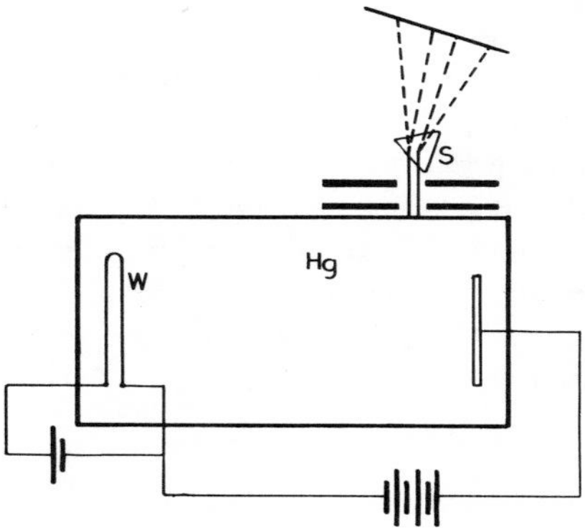

FIGURE 3.16
The Franck-Hertz experiment. The hot wire emits electrons that are accelerated toward the cathode. By collisions they excite mercury atoms in the dilute vapor (Hg) which then emit radiation. The wavelength of the radiation from one region in the vapor is then measured with the spectrometer (S).

In the Franck-Hertz apparatus electrons are emitted by a hot wire into a very dilute mercury vapor. An accelerating potential, V, draws the electrons toward the cathode. If an electron does not collide with a mercury atom on the way, it will have an energy of eV electron volts by the time it reaches the cathode and is absorbed. Thus, by colliding with the mercury atoms it is possible for the electrons to impart to the atoms any energy up to, but not exceeding, eV. If the mercury atoms are excited by this electron bombardment, they will give off their extra energy as radiation. Franck and Hertz observed the emitted spectrum as the accelerating potential, V, was varied. They found that:

a. the frequency of any line in the emission spectrum was independent of V, but,
b. any particular line would only appear if the accelerating potential V was larger than some critical value that was different for different lines.

They also showed that the "turn-on potential", V, for any line could be calculated in the following way.

1. Using the measured spectrum of Hg and the Einstein hypothesis regarding the source of the discrete spectrum, one constructs an energy spectrum for the mercury atom.
2. A particular line, ν_{mk}, associated with the transition $E_m \rightarrow E_k$, would appear in the observed spectrum if and only if the potential V satisfied the inequality

$$eV \geqslant E_m - E_1$$

The interpretation of this effect within the Einstein picture of radiative emission is very simple. Unless the electrons have sufficient energy to excite an atom from its ground state to the energy state E_m no photons associated with transitions from that energy state will be emitted.

If one tries to relate the emission spectrum to some sort of normal modes of vibration of the atom, there is no obvious way to understand the Franck-Hertz effect. Thus the Franck-Hertz experiment very much strengthened the belief in the existence of discrete energy levels in atoms. Of course, it did not at all explain why there should be discrete energy levels; it only convinced people that there really were such levels.

3.8. THE HYDROGEN ATOM

As one can easily understand, active research workers became very excited but also very puzzled about this strange quantization phenomenon. As physicists always do, they looked for the simplest system in nature which exhibited it. The obvious candidate for that position was the hydrogen atom. Experiments had already confirmed that a hydrogen atom was composed of a single proton and a single electron. If they could not understand hydrogen, there seemed to be little chance that they would understand more complex atoms involving many interacting electrons. Not only was the picture of a hydrogen atom simple, but it was known that, using the Einstein interpretation of the emission spectrum, the energy levels of hydrogen satisfied a tantalizingly simple formula. The nth energy level has the value

$$E_n = - A/n^2 \qquad (3.13)$$

where the *Rydberg constant* $A = 2.18 \times 10^{-18} J = 13.6 eV$.

The challenge to theorists could be presented in a very direct form. Construct a theory that will yield the constant, A, in terms of the basic parameters of the system, such as the electron and proton masses, their charges, the velocity of light, and Planck's constant. The Danish physicist, Niels Bohr was the first to meet that challenge.

The Bohr theory begins with a completely classical picture of the hydrogen atom. It is an electron orbiting, according to Newton's laws of dynamics, around a proton. Since the proton is so much heavier than the electron we shall neglect its "wobble" and consider it as fixed in space at the origin of our coordinate system. For simplicity Bohr considered only circular orbits of the electron. The theory was later extended by Arnold Sommerfeld to include elliptical orbits as well. An electron moving at speed v in a circle of radius r has a centripetal acceleration equal to v^2/r. According to Coulomb's law, the force attracting the electron to the proton is equal to $k\,e^2/r^2$, where k is the Coulomb force constant. By Newton's second law we get

$$\frac{m\,v^2}{r} = \frac{k\,e^2}{r^2} \tag{3.14}$$

We now add to this, the definitions of the angular momentum, L, and the energy, E.

$$L = m\,v\,r \tag{3.15}$$

$$E = \tfrac{1}{2}m\,v^2 - \frac{k\,e^2}{r} \tag{3.16}$$

Multiplying Equation 3.14 by $m\,r^3$ we get

$$r = \frac{L^2}{m\,k\,e^2} \tag{3.17}$$

Using this to eliminate r in Equation 3.15 yields

$$v = k\,e^2/L \tag{3.18}$$

These two equations allow us to write the energy in terms of the angular momentum in the form

$$E = -\frac{m\,k^2 e^4}{2\,L^2} \tag{3.19}$$

If we assume that this relation between E and L, which has been derived using Newtonian mechanics and Coulomb's law, is true in each of the quantized energy states of the hydrogen atom, then the angular momentum value in the nth energy state can be obtained by replacing E by $-A/n^2$ in Equation 3.19 and solving for L. What results is

$$\begin{aligned} L &= n\,k\,e^2(m/2\,A)^{1/2} \\ &= n\,(9\times 10^9\,\frac{N\cdot m^2}{C^2})(1.6\times 10^{-19}\,C)^2\left(\frac{9.11\times 10^{-31}\,kg}{2\times 2.18\times 10^{-18}\,J}\right)^{1/2} \\ &= n\times 1.05\times 10^{-34} kg\cdot m^2/s \end{aligned} \tag{3.20}$$

Since the unit kg-m^2/s is the same as the unit J-s, the multiplying factor in this equation can be recognized as Planck's constant, $\hbar$. Thus Bohr showed that a planetary picture of the hydrogen atom with discrete circular orbits whose energies have the observed values implied a very simple quantization rule for the angular momentum of the electron in any of the allowed orbits. Namely

$$L_n = n\hbar \qquad \text{with } n = 1, 2, 3, \cdots \tag{3.21}$$

Using this relation in Equation 3.20 it is possible to write the Rydberg constant, A, in terms of other fundamental constants

$$A = \frac{m k^2 e^4}{2\hbar^2} \tag{3.22}$$

Using Equations 3.17 and 3.18 one can calculate r_n, the radius of the nth allowed orbit, called the nth *Bohr orbit*, and the velocity of the electron in that orbit.

$$r_n = n^2 \frac{\hbar^2}{m k e^2} \equiv n^2 a_o \tag{3.23}$$

and

$$v_n = \frac{k e^2}{n\hbar} \tag{3.24}$$

The radius of the smallest Bohr orbit is referred to as the *Bohr radius.* Its value is

$$a_o = \frac{\hbar^2}{m k e^2} = 0.528\,\text{Å} \tag{3.25}$$

Other evidence, such as the distance between the nuclei in the H_2 molecule (.74 Å), shows that a_o is a reasonable estimate of the radius of a hydrogen atom.

According to Equation 3.24 the greatest orbital speed is attained in the smallest orbit. We can estimate the errors involved in using nonrelativistic mechanics by calculating the ratio of v_1 to the velocity of light. This ratio is called the *fine structure constant*, α.

$$\alpha = \frac{v_1}{c} = \frac{k e^2}{\hbar c} \approx \frac{1}{137}$$

Since the relativistic corrections to nonrelativistic formulas are generally proportional to $(v/c)^2$ it is clear that they will be quite small in hydrogen.

The essential elements of the Bohr theory of the hydrogen atom are the following:

1. There exist only certain discrete allowed orbits for the electron. These orbits are determined by the quantization condition that the angular momentum be equal to $n\hbar$ where n is any positive integer.
2. While the electron moves in one of the allowed orbits it does not radiate. (This, of course, contradicts Maxwell's equations.)
3. A hydrogen atom that was not in its ground state might, at any time, jump from the state it was in into some lower energy state. In doing so it would emit the excess energy as a single photon. The theory made no prediction when this would happen, nor did it explain how the electron moved in going from one allowed orbit to the other.

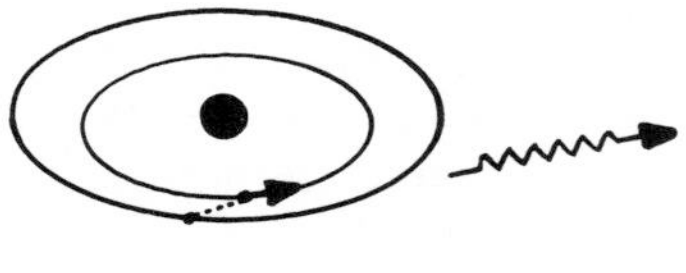

FIGURE 3.17
The emission of a photon by a hydrogen atom is accompanied by the simultaneous movement of the electron from one allowed orbit to another in the Bohr theory.

There is no question about the fact that Bohr never considered this assemblage of classical concepts and arbitrary quantization conditions as a real theory of the hydrogen atom. He knew that it was only a useful step toward a complete self-consistent physical theory of atomic structure. Shortly after Bohr presented his first work, Arnold Sommerfeld was able to modify the theory slightly so as to include elliptical orbits and relativistic effects. All attempts to extend the theory to atoms containing more than a single electron failed.

As we shall see later, there is no such thing as an orbit of an electron in an atom according to the correct quantum theory of the atom. In spite of the fact that an erroneous concept (the electron orbit) plays such a fundamental role in the theory, the Bohr theory does predict many properties of single-electron atoms correctly. The ions, He^+, Li^{++}, etc. are all one-electron systems to which the Bohr theory can be applied. The only modification which would have to be made in the analysis in order to apply it to an atom consisting of one electron moving around a nucleus of charge Ze would be to replace one of the factors of e in each equation by Ze. For example, the energy levels of such a system would be

$$E_n = -\frac{m k^2 Z^2 e^4}{2\hbar^2 n^2} \tag{3.26}$$

Thus the energy levels of He^+ and Li^{++} should be respectively four times and nine times those of hydrogen. This prediction is confirmed experimentally.

3.9. DE BROGLIE WAVES

In 1924 Louis de Broglie, a doctoral candidate at the Faculty of Sciences of the University of Paris, presented a thesis in which some very bold, but at the time unsubstantiated, conjectures were made. His studies concerned the relationship between waves and particles in the quantum theory. He described the initial stages of his work as follows.

> The necessity of assuming for light two contradictory theories - that of waves and that of corpuscles - and the inability to understand why, among the infinity of motions which an electron ought to be able to have in the atom according to classical concepts, only certain ones were possible: such were the enigmas confronting physicists at the time I resumed my studies of theoretical physics.
>
> When I started to ponder these difficulties two things struck me in the main. Firstly the quantum theory of light could not be regarded as satisfactory since it defines the energy of a light corpuscle by the relation $E = h\nu$ which contains a frequency ν. Now a purely corpuscular theory does not contain any element permitting the definition of a frequency. This reason alone renders it necessary in the case of light to introduce simultaneously the corpuscle concept and the concept of periodicity.
>
> On the other hand, the determination of the stable motions of electrons in an atom involves whole numbers, and at the time the only phenomena in which whole numbers had been involved in physics were those of interference and of normal mode vibrations. That suggested the idea to me that electrons themselves could not be represented as simple corpuscles either, but that a periodicity had to be assigned to them too.

> I thus arrived at the following overall concept which guided my studies: for both matter and radiations, light in particular, it is necessary to introduce the corpuscle concept and the wave concept at the same time. In other words, the existence of corpuscles accompanied by waves has to be assumed in all cases. However, since corpuscles and waves cannot be independent because, according to Bohr's expression, they constitute two complementary facets of reality, it must be possible to establish a certain parallelism between the motion of a corpuscle and the propagation of the associated wave. My first objective was, therefore, to establish this correspondence.

Thus de Broglie assumed that, just as there were particles associated with light waves, there were waves of some yet undetected variety associated with massive particles such as electrons. What he pictured was a particle moving within a wave packet. A *wave packet* is a small region of waves that move together through space. One can create a wave packet by taking a flashlight that has a very narrow beam (a laser would be much better) and turning it on and off within an extremely short time (See Figure 3.18). If a packet of light waves

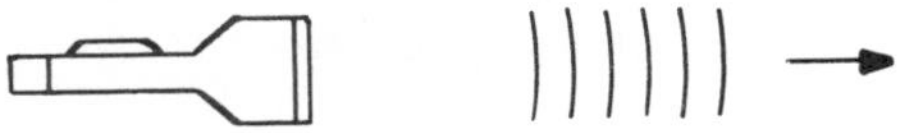

FIGURE 3.18 A well-focused flashlight turned on and off in a ridiculously short time interval would produce a small wave packet.

travels through a region in which the index of refraction varies smoothly from place to place, it will not propagate in a straight line. Nor will it move at constant speed. The packet of light waves would follow some curved trajectory at nonuniform speed. De Broglie realized that if a particle were somehow confined to, or guided by a packet of "matter waves", then the particle would execute all the motions of the wave packet. But it was already known that a particle accelerated along a curved path whenever it passed through a region of nonconstant potential, $V(x,y,z)$. He was thus faced with the problem of finding some relationship between the characteristics of the wave packet, such as its wavelength and frequency, and the characteristics of the particle, such as its momentum and the potential energy at its location. This relationship had to have the property that the motion of the wave packet, which one could calculate using the laws of geometrical optics, would agree exactly with the motion of the particle which could be calculated using Newton's laws of dynamics.

Although de Broglie considered the more general case of a smoothly varying potential and even used the correct relativistic equations of dynamics, we shall take only the simple case of a nonrelativistic particle moving from one two dimensional region of constant potential into another region in which the potential is also constant but different from that in the first region. One could make a model of such a system by considering a particle that was sliding on one of two smooth horizontal surfaces at different levels with a rounded off step between them. The particle would get a rapid change in its direction and speed whenever it went from one region to the other (See Figure 3.19A).

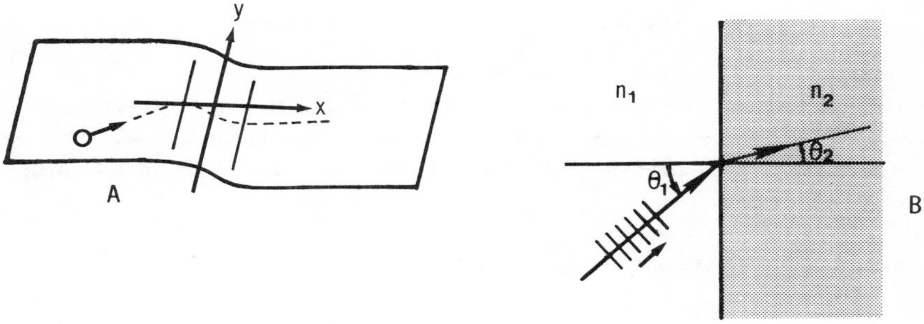

FIGURE 3.19 The wave packet (B) changes its speed and direction of motion in the same way as the particle (A)

If we draw x- and y-axes as shown in the figure, it is clear that the y-component of the force on the particle will always be zero. Since, according to Newton's equation of motion, $dp_y/dt = F_y$ it is clear that the y-component, but not the x-component, of momentum will remain constant. In moving from region 1 into region 2 the particle's momentum will change from $\mathbf{p}_1$ to $\mathbf{p}_2$. The change in the direction and magnitude of the particle's momentum can be determined from the conservation laws for the energy and the y-component of momentum.

$$\frac{p_1^2}{2m} + V_1 = \frac{p_2^2}{2m} + V_2 \tag{3.27}$$

$$p_{1y} = p_{2y} \tag{3.28}$$

In a similar way, when a wave packet moves from a region in which the phase velocity (the velocity of the wave fronts) has one value, v_1, into a region in which it has another value, v_2, the wave packet is refracted according to Snell's law (See Figure 3.19B). The change in wavelength and direction of propagation of the wave packet can be determined by two rules. One is that the frequency in the new medium is the same as the frequency in the old. This is an immediate consequence of the fact that the number of wave fronts that approach any point on the interface of the two regions in one second must equal the number of wave fronts that leave the same point. Wave fronts do not disappear at the interface. Thus

$$\nu_1 = \nu_2 \tag{3.29}$$

The other rule is Snell's law, which states that

$$\frac{\sin\theta_1}{v_1} = \frac{\sin\theta_2}{v_2} \tag{3.30}$$

If we multiply corresponding sides of Equations 3.29 and 3.30, and use the fact that the magnitude of the wave vector $k = 2\pi\nu/v$ we get the relation

$$k_1 \sin\theta_1 = k_2 \sin\theta_2$$

or

$$k_{1y} = k_{2y} \tag{3.31}$$

Comparing Equation 3.27 and 3.28 with Equations 3.29 and 3.31 shows that the particle and the wave packet will have the same trajectory if we assume the following relationships between the properties of the particle and the properties of the wave

(1) ν is proportional to $E = \dfrac{p^2}{2m} + V$

and

(2) $\mathbf{k}$ is proportional to $\mathbf{p}$

Comparison with the case of photons certainly suggests that the constant of proportionality is Planck's constant. In order to use only one version of Planck's constant we introduce the angular frequency $\omega = 2\pi\nu$ and assume that

$$\frac{p^2}{2m} + V = \hbar\omega \tag{3.32}$$

and

$$\mathbf{p} = \hbar\mathbf{k} \tag{3.33}$$

These relations, which were first suggested in de Broglie's thesis, are therefore called the de Broglie relations. It can be shown that the de Broglie relations also lead to agreement between the particle and wave packet motion for the case of a continuously varying potential.

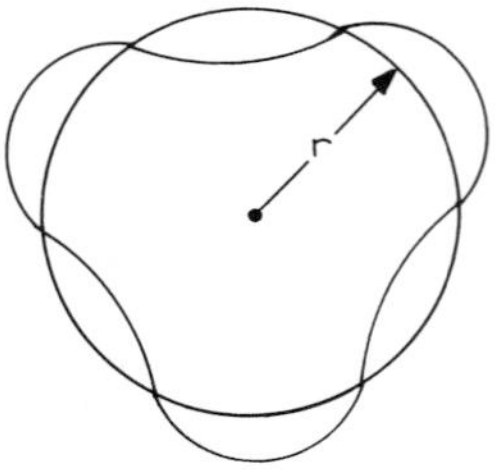

FIGURE 3.20
De Broglie could obtain the Bohr quantization condition by demanding that the electron's orbit contain an integral number of matter wavelengths.

De Broglie also pointed out that the concept of matter waves led to a simple interpretation of the Bohr quantization condition for hydrogenic orbits. The allowed Bohr orbits turn out to be exactly those orbits for which the de Broglie wave has an integral number of wavelengths along the orbital path (See Figure 3.20). To see this we multiply Equation 3.14 by mr and get

$$p^2 = \frac{mke^2}{r}$$

But Equation 3.33 implies that

$$p^2 = h^2/\lambda^2$$

This gives

$$\frac{m k e^2}{r} = \frac{h^2}{\lambda^2} \tag{3.34}$$

In order to fit an integral number of wavelengths exactly around the orbit we would need to have

$$2\pi r = n\lambda \tag{3.35}$$

Using Equation 3.35 to eliminate λ in Equation 3.34 we get the usual equation for the radius of a Bohr orbit

$$r_n = \frac{\hbar^2 n^2}{m k e^2}$$

The condition that the length of the orbit be equal to an integral number of wavelengths is exactly the condition that would be needed for some kind of standing waves to be set up along the orbit. This is how de Broglie interpreted the Bohr quantization condition.

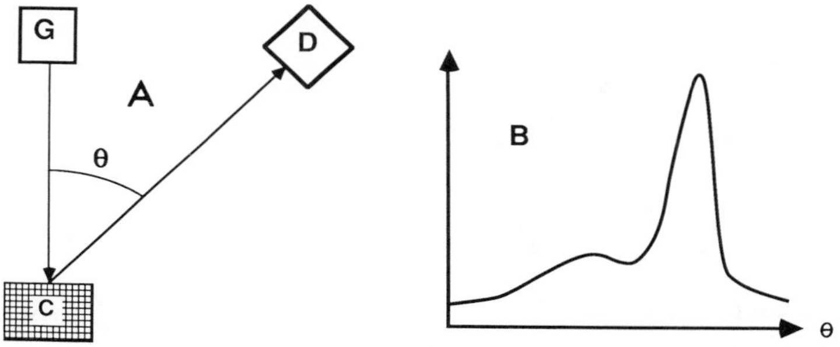

FIGURE 3.21 (A) The Davisson-Germer experiment. Electrons from the electron gun (G) are scattered by the crystal (C) and detected by the electron detector (D). (B) The scattered electron intensity as a function of angle shows strong peaks for certain angles.

3.10. THE DAVISSON-GERMER EXPERIMENT

At the time of publication of de Broglie's work two physicists at the Bell Laboratories in New York, C. J. Davisson and L. H. Germer, were puzzling over some strange results they had obtained from an electron scattering experiment. They had aimed a beam of electrons of known energy, E, at a single crystal of nickel and observed the intensity of those scattered electrons whose energies were close to the energy of the incoming electrons. Scattering without energy loss is called *elastic scattering.* The intensity of elastically scattered electrons was measured as a function of the angle of scattering for a number of different orientations of the electron detector with respect to the crystal planes of the nickel crystal. Pronounced maxima were found for certain scattering angles and energies (See Figure 3.21).

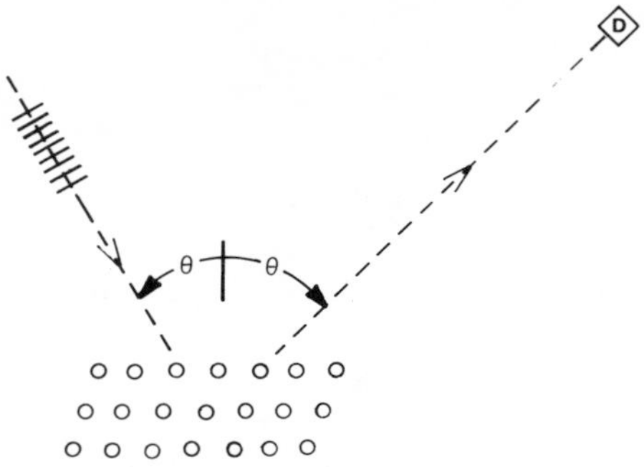

FIGURE 3.22 The microscopic view of electron wave scattering. The incoming electron waves are scattered by atoms in the crystal. The outgoing electrons are detected at D.

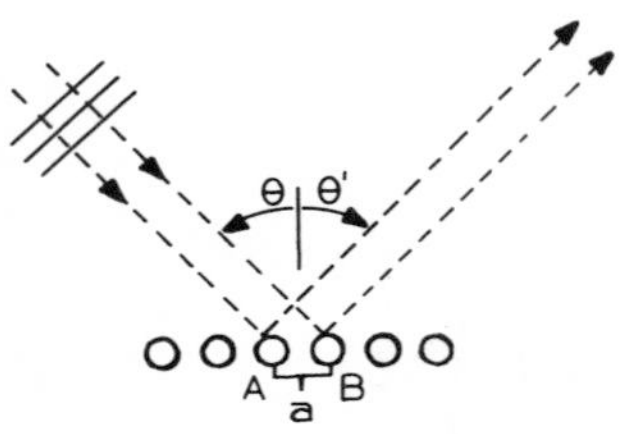

FIGURE 3.23
The wave fronts have to travel an extra distance $a \sin\theta$ to reach atom B in comparison with atom A. Thus the scattered waves leave B later than they leave A. But the waves from A must travel an extra distance $a \sin\theta'$ to get to the detector. If $\theta = \theta'$ the two effects cancel and the waves from A and B arrive in phase.

During a trip to England, Davisson discussed the peculiar results of his electron scattering experiments with a group of physicists who suggested that they might be explained in terms of de Broglie's electron wave hypothesis. Upon returning to New York Davisson discovered that his experiments did indeed supply a clear and simple confirmation of de Broglie's conjecture. The analysis needed is rather simple.

Let us assume that the hypothesized electron waves are in some way scattered by the nickel atoms in the crystal. We consider a perfect crystal upon which a uniform plane electron wave (whatever that is) is incident at an angle θ with respect to a certain set of crystal planes (See Figure 3.22). If we were to add the amplitudes of the waves scattered, in some particular direction, θ', by all the atoms in the crystal we would be very likely to obtain a small resultant wave amplitude because the waves coming from the different atoms would have different relative phases and would, for the most part, cancel one another. We now ask ourselves under what conditions will the waves from all atoms add in phase. In Figure 3.23 it is shown that the waves from all the atoms of a single plane add in phase if $\theta' = \theta$. Even if all the waves originating from a single plane do add in phase one will still get cancellation unless the waves from consecutive planes also add in phase. The condition that this be so is shown in Figure 3.24 to be

$$2 d \cos\theta = n \lambda \tag{3.36}$$

Since Davisson and Germer knew the spacing between adjacent crystal planes in nickel, they could use Equation 3.36 to determine the wavelength of the scattered electron waves. The results of their experiments confirmed the de Broglie relations very well (See Figure 3.25).

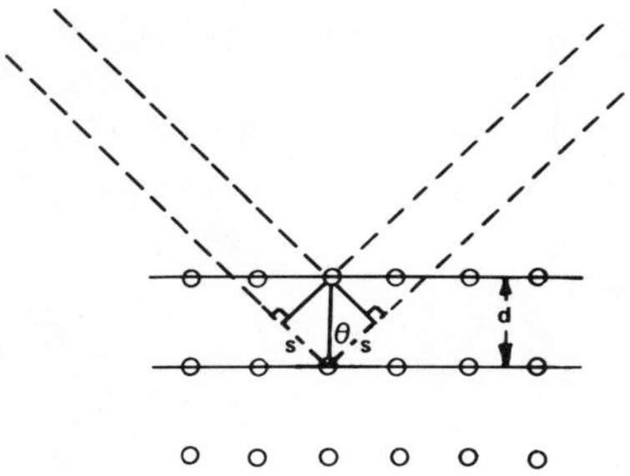

FIGURE 3.24

The waves scattered from the second plane have to travel an extra distance $2s = 2\,d\cos\theta$ in comparison with those scattered from the first plane. All waves will arrive in phase if $2\,d\cos\theta = n\,\lambda$ ($n = 1, 2, \cdots$).

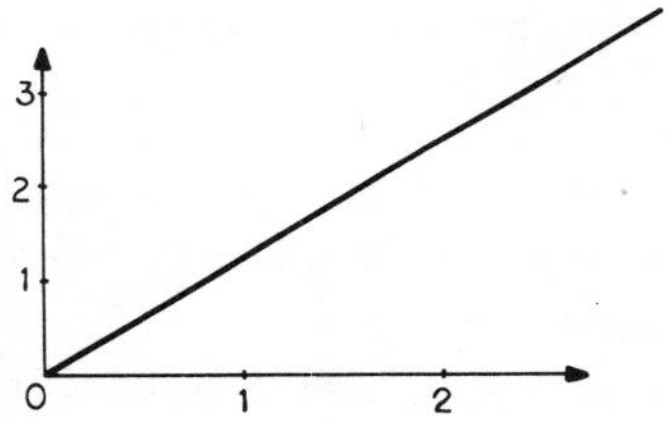

FIGURE 3.25

The relationship between the measured de Broglie wavelength and the potential through which the electrons have been accelerated. The y-axis gives λ in units of $\mu\text{-}m$ while $V^{-1/2}$ is shown on the x-axis. According to the de Broglie relation $p = h/\lambda$. Combining this with $p^2/2\,m = e\,V$ gives $\lambda = h/(2meV)^{1/2} = 1.22 \times 10^{-9}\,V^{-1/2}$.

SUMMARY

Within an empty cavity at temperature T, the radiant energy density of those waves within the frequency range ν to $\nu + d\nu$ is $D(\nu)d\nu$, where

$$D(\nu) = \frac{8\pi h \nu^3}{c^3(e^{h\nu/kT} - 1)}$$

and $h = 6.6 \times 10^{-34}$J·s.

The energy and momentum carried by an electromagnetic wave can be pictured as being carried by zero mass particles, called *photons*. The energy and momentum of any photon associated with a wave of frequency ν and wavelength λ are given by

$$E = h\nu$$

and

$$p = h/\lambda$$

If we introduce another form of Planck's constant,

$$\hbar = \frac{h}{2\pi} = 1.05 \times 10^{-34} \text{J·s}$$

we can write the momentum vector, **p**, of the photon in terms of the wave vector, **k**, of the wave by

$$\mathbf{p} = \hbar \mathbf{k}$$

The photoelectric effect is the name given to the property of metals of emitting electrons when irradiated by light of sufficiently high frequency.

The maximum energy of the photoelectrons that are emitted is related to the frequency of the incident light ν and the work function of the metal W by

$$E_{max} = h\nu - W$$

The number of photoelectrons is proportional to the light intensity, but E_{max} is not related to the light intensity.

One can understand the photoelectric effect by assuming that each photoelectron absorbs one photon, and thereby receives an amount of energy $h\nu$. The photoelectron must do an amount of work W against attractive forces at the metal surface in order to escape from the metal.

The *Compton effect* is the wavelength shift in light scattered by free electrons. The light scattered by an angle θ with respect to the incident beam has a wavelength λ' given by

$$\lambda' = \lambda + (1 - \cos\theta)\lambda_c$$

where λ is the wavelength of the incident light and λ_c is the Compton wavelength,

$$\lambda_c = h/mc$$

The Compton effect can be derived by picturing the light scattering process as the scattering of photons by electrons.

Rutherford showed that one could understand the scattering of α particles by atoms by assuming that the atoms were composed of an extremely small nucleus surrounded by light electrons.

An isolated atom can radiate electromagnetic waves only at certain discrete frequencies. The set of possible frequencies is called the emission spectrum of the atom. The frequencies in the emission spectrum are given by

$$h\,\nu_{mn} = E_m - E_n$$

where E_m and E_n are the discrete energies of the atomic states before and after the transition.

The set of possible energies of an atom is called the atom's energy spectrum. The lowest energy level is the ground-state energy.

Franck and Hertz excited the atoms in a mercury vapor by electron collision and found that any given frequency ν_{mn} in the mercury spectrum was not emitted until the energy of the bombarding electrons exceeded $E_m - E_1$, the energy needed to excite the atom to the higher energy discrete state from which it emitted the photon according to the Einstein picture of atomic radiation.

Niels Bohr showed that one could derive the hydrogen energy levels by using Newtonian mechanics plus the assumption that the angular momentum of the atom could have only the values

$$L_n = n\hbar \qquad (n = 1, 2, 3, \cdots)$$

From this he obtained the experimentally observed energy levels

$$E_n = -A/n^2$$

where $A = m k^2 e^4 / 2\hbar^2 = 2.18 \times 10^{-18}\text{J} = 13.6\ \text{eV}$.

The radius of the nth Bohr orbit in a hydrogen atom is

$$r_n = n^2 a_o$$

where the *Bohr radius*, a_o, is given by

$$a_o = \hbar^2/m k e^2 = .53 \times 10^{-10}\text{m} = .53\ \text{Å}$$

De Broglie showed that if one assumed that some kind of matter wave was associated with a particle of mass m and that the frequency and wave vector of the matter waves were related to the energy and momentum of the particle by the de Broglie relations

$$\frac{p^2}{2m} + V = h\,\nu$$

and

$$\mathbf{p} = \hbar\mathbf{k}$$

then, according to geometrical optics, the motion of a matter wave packet would satisfy Newton's equations of motion for a particle of mass m. De Broglie also showed that one could derive the Bohr angular momentum quantization conditions by demanding that the electron orbit contain an integral number of matter wavelengths.

The existence of electron matter waves was confirmed by Davisson and Germer in experiments on electron scattering by crystals.

PROBLEMS

3.1** Write an integral expression for the density of photons in a cavity at temperature T.

3.2 The temperature at the center of the sun is estimated to be about 3 million degrees Kelvin. What would be the energy density in a cavity at that temperature?

3.3* The translational kinetic energy of an ideal gas at temperature T is given by $E = \frac{3}{2} N k_B T$. Consider one mole of gas in a cubic meter enclosure.

(a) As a function of T calculate the ratio of the kinetic energy of the gas particles to the energy of the equilibrium radiation that also permeates the enclosure.

(b) At what temperature would the radiation contribute one percent of the total energy?

3.4* The heat capacity of an object is defined by $C = dU(T)/dT$ where U(T) is the internal energy of the object at temperature T. What is the heat capacity of one cubic meter of vacuum at one million degrees?

3.5 At what maximum velocity would electrons be ejected from a metal with work function 1.3 eV by light of wavelength .5μm? ($1\,\mu m = 10^{-6}m$)

3.6* In a photoelectric effect experiment it is found that $V_{cutoff} = 7.18\,V$ when $\lambda = 10^{-7}m$ and $V_{cutoff} = 1\,V$ when $\lambda = 2 \times 10^{-7}m$. From this data determine the value of Planck's constant and the work function of the metal.

3.7 At what rate are photons being emitted by a 60 watt light bulb? Approximate the nature of the light as monochromatic light of wavelength .7μm.

3.8* At what rate do photons hit a square centimeter of surface from a 100 watt monochromatic source of wavelength .6μm situated at one kilometer from the surface and radiating equally in all directions?

3.9** Positrons are particles that are identical to electrons in all their properties except their electric charge. They have a charge of +e rather than −e. When a positron and an electron come together it is possible for the pair of particles to disappear completely at the same instant that two photons are created. The process, called *pair annihilation*, conserves relativistic energy and momentum. Suppose a stationary electron, situated at the origin of the x-axis is struck by a positron moving at speed .5 c in the positive x-direction and the pair annihilate, giving rise to two photons which move off along the x-axis in opposite directions. Calculate the wavelengths of the two photons.

3.10** Consider the following imaginary process. An electron of momentum $\mathbf{p}$ absorbs a single photon and is left with a new momentum $\mathbf{p}'$. Using relativistic conservation laws, show that such a process is impossible.

3.11* A photon is *back-scattered* by an initially stationary electron. (The final direction of motion of the photon is opposite to its initial direction.) The final velocity of the electron is .5 c. What are the initial and final energies of the photon?

3.12*** An object of mass m is traveling in the x-direction at velocity v when it emits a photon in the opposite direction. The object's velocity and mass change to v + dv and m + dm. (dm must be negative since the object must lose inertial mass by emitting energy.)

(a) Assuming that both dm and dv are very small, show that

$$\frac{d\beta}{1-\beta^2} = -\frac{dm}{m}$$

where $\beta = v/c$.

(b) The above equation is the equation of motion for a *radiation rocket.* (A rocket whose exhaust consists of radiation rather than massive particles. It can be shown that this is the most efficient rocket possible.) By integrating the equation, calculate the ratio of the initial mass to the final mass for a rocket that, starting from rest, attains a final velocity of .5 c.

3.13 Light, of initial wavelength 10^{-7}m is scattered by an angle of 45° by an electron. What is the wavelength of the scattered light?

3.14 A hypothetical atom has only four discrete energy states. They have energies $E_1 = 0$, $E_2 = 1\,eV$, $E_3 = 2\,eV$, and $E_4 = 2.5\,eV$. What are the maximum and minimum wavelengths found in its emission spectrum?

3.15* A system has an infinite number of discrete energy states with the nth energy level being given by $E_n = \alpha n^2$. Describe its emission spectrum.

3.16* A hypothetical molecule has an emission spectrum consisting of the three frequencies 1.2×10^{13} Hz, 3.2×10^{13} Hz, and 4.4×10^{13} Hz. Assuming its ground state energy is zero, what are its other energy levels?

3.17* If a Franck-Hertz experiment were carried out with hydrogen rather than mercury, at what accelerating potential would the emission line of frequency $\nu_{32} = (E_3 - E_2)/h$ appear?

3.18 Consider a rotating sphere of radius 1 cm and mass 50 gm. Its initial angular velocity is 50 rad/s. If it loses angular momentum at a rate of $10^6\hbar$/s, how long will it take to stop? (The moment of inertia of a sphere is $\frac{2}{5}MR^2$.)

3.19 Muonium is a hydrogen-like system in which the electron has been replaced by a μ^-. (A μ^- is a particle with a charge $-e$ and a mass 1.88×10^{-28} kg.)

(a) Write a formula for the energy levels of muonium.

(b) What is the shortest wavelength in the emission spectrum? In what range does it lie? (Infrared, ultraviolet, etc.)

3.20 According to the quantum theory the angular momentum of the earth orbiting about the sun is also quantized in units of $\hbar$. For that system, estimate the value of n in the formula $L_n = n\hbar$. Does this explain why we do not notice the effects of discrete quantization in the motion of macroscopic objects?

3.21** At the interface between an optical medium with index of refraction n_1 and another medium with index of refraction $n_2 < n_1$ a light ray will be completely reflected if the angle it makes with the normal to the interface is greater than the *angle of internal reflection* $\theta_0 = \sin^{-1}(n_2/n_1)$.

(a) Show that $\theta_0 = \sin^{-1}(\lambda_1/\lambda_2)$ where λ_1 is the actual wavelength of the light and λ_2 is the wavelength of light of equal frequency in medium 2.

(b) Consider a particle of energy E approaching the interface between two regions of different potential, V_1 and V_2. Show that, according to Newtonian mechanics, the particle will be reflected at the interface if its trajectory makes an angle greater than θ_0 with the normal, where $\theta_0 = \sin^{-1}(p_2/p_1)$ and p_1 and p_2 are the momentum values of a particle of energy E in the two regions.

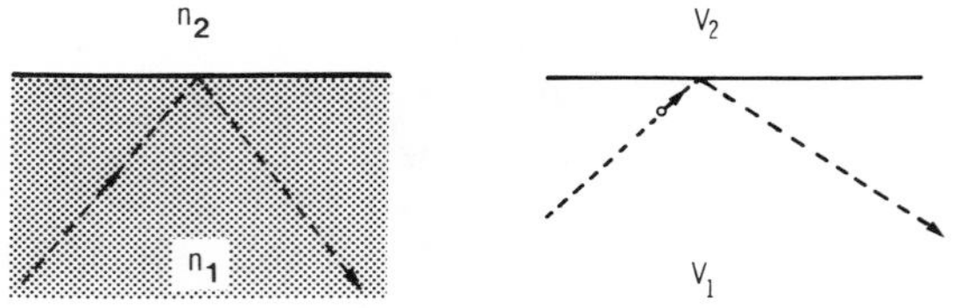

FIGURE P.3.21

CHAPTER FOUR

THE WAVE EQUATION

4.1. THE DIFFRACTION OF LIGHT AND ELECTRONS

In the last chapter the reader was subjected to a rather heavy dose of bewildering experimental effects and half-baked theoretical explanations of those effects. In this chapter we shall present the elements of the theory that finally brought order into the very puzzling situation we have described. At the risk of trying the reader's patience, we will describe still another experiment. It is a somewhat simplified, idealized experiment that is designed to allow a fairly straightforward transition to the theoretical ideas that follow. The experimental apparatus is pictured in Figure 4.1. It consists of a source, an opaque screen with a single slit, and a phosphorescent screen. The source can be made to emit either a beam of monochromatic ultraviolet light or monoenergetic electrons by the flick of a switch. The intensity of the beam (in either case) may be adjusted at will. The slit width can also be varied. The screen glows when irradiated by electrons or UV light.

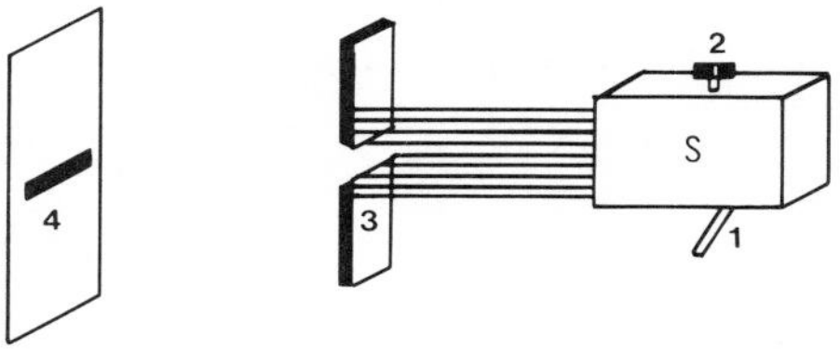

FIGURE 4.1

The experimental setup. (S) A source of light or electrons. (1) A switch to determine which type of beam. (2) A knob to control the beam intensity. (3) A slit of variable width. (4) A phosphorescent screen.

We begin with a large slit width and an intense beam of electrons. A clearly defined glowing stripe appears on the phosphorescent screen where the beam of electrons strike it.

We next switch to an intense beam of ultraviolet light. Again we see a well defined bright stripe on the phosphorescent screen. So far we cannot distinguish between the electrons and the light.

In order to do so we switch the source to electrons at very low intensity. Sure enough we now see little flashes on the phosphorescent screen as individual electrons strike it. We keep track of exactly where each of the flashes occurs and find that they are randomly distributed over the area of the phosphorescent screen that had been occupied by the bright stripe.

We now switch the source back to an intense beam of light and then make the slit very narrow. As expected, we see a single-slit diffraction pattern on the screen. The pattern consists of a series of alternating bright and dark bands. The total width of the pattern is much larger than the slit width. From the details of the pattern, and a knowledge of the slit width, we can calculate the wavelength of the light. All this is just what we would have predicted from classical theory; the electrons behave as particles, and the light behaves as waves.

Now we open the slit wide and switch to a low intensity light source. We see little localized flashes as "light particles" hit the phosphorescent screen. The light seems to be a wave in one case and particles in another.

To further compare light with electrons we now switch to a high intensity electron source and narrow the slit. In this case we see a single slit diffraction pattern on the phosphorescent screen. The electron beam is behaving like waves. Both the electron beam and the light beam behaved as waves for small slit width and as particles for low intensity. But the slit width and the intensity can be set independently. What will happen if we make the slit width small *and* the intensity low?

We switch to a low intensity electron beam and a small slit width. On the phosphorescent screen we see flashes as individual electrons strike the surface. At first sight it seems that the electrons are behaving as particles. The electrons seem to be striking the screen in random positions, but when we keep careful account of exactly where the electrons hit, we notice that no electrons ever hit the screen where the electron diffraction pattern we obtained before was completely dark. In fact, if we average over a very long time, we find that the rate at which electrons hit any small area on the screen is proportional to the intensity of the single slit diffraction pattern at that point. If we replace the phosphorescent screen with a piece of photographic film and expose the film for a long time so that very many electrons have hit it, each one producing a microscopic dot, then upon developing the film we get an excellent photograph of the single-slit diffraction pattern. The electrons have devised a way of being particle-like and wave-like simultaneously. Our whole classical dichotomy of particles and fields has been outflanked. As expected we get exactly the same phenomenon when we use a low intensity light source and a narrow slit width (See Figure 4.2).

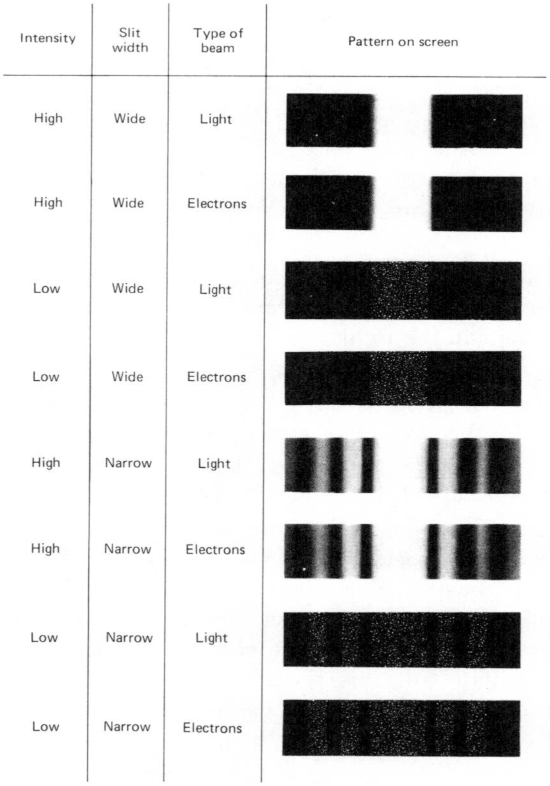
Intensity
Slit width
Type of beam
Pattern on screen
High
Wide
Light
High
Wide
Electrons
Low
Wide
Light
Low
Wide
Electrons
High
Narrow
Light
High
Narrow
Electrons
Low
Narrow
Light
Low
Narrow
Electrons

FIGURE 4.2

Nature does not consist of certain elements that are particles and other elements that are fields, but rather each entity is both particle and field. The theory that quantitatively describes these particle-field entities is called *quantum theory* (or sometimes *quantum mechanics*). There are two fundamental elements in quantum theory.

1. There is a wave equation, which may have different forms for different kinds of particles. The wave equation for electrons, called Schroedinger's equation after the man who first suggested it, we shall introduce shortly. The wave equation for photons is simply the ordinary wave equation which can be derived from Maxwell's equations.

2 There is a rule for interpreting the wave function in terms of the observed particle flux or the observed particle density. That is, if we have found the solution of the electron wave equation (the Schroedinger equation) that is appropriate for a particular physical setup, we can use the rule for predicting the electron flux at any point in space.

Before we develop the mathematical structure of the theory, we can help communicate the physical content of the quantum theory, if we describe how a working physicist uses the theory to make calculations and how the results of the calculations are interpreted in terms of experimental predictions.

In carrying out the calculations we simply ignore the particle properties of the electrons or photons. Suppose we want to analyze the scattering of electrons of momentum $\mathbf{p}$ by stationary protons. We use the wave equataion, namely Schroedinger's equation, which is appropriate for electron waves. We consider the case of an incoming plane electron wave with wave vector given by the de Broglie relation $\mathbf{p} = \hbar\mathbf{k}$. In other words, we represent the incoming electron beam by an incoming beam of electron waves with the appropriate wave vector. By using Schroedinger's equation it is possible to calculate the electron wave radiation given off by the proton (See Figure 4.3). Just as there is a formula for the energy flux due to an electromagnetic wave, there is a corresponding formula giving the electron flux due to a traveling electron wave. One can use that formula to calculate the electron flux in any direction due to the scattered electron wave. Once we have made that calculation we are in a position to interpret the result in terms of experimental observations. The interpretation is as follows. If we place an electron detector at a particular position, then the rate at which electrons enter the detector will be given by the product of the area of the input port of the detector and the calculated electron flux at that point. The exact times at which electrons enter the detector cannot be predicted by the quantum theory in any way. Only the average rate is calculable.

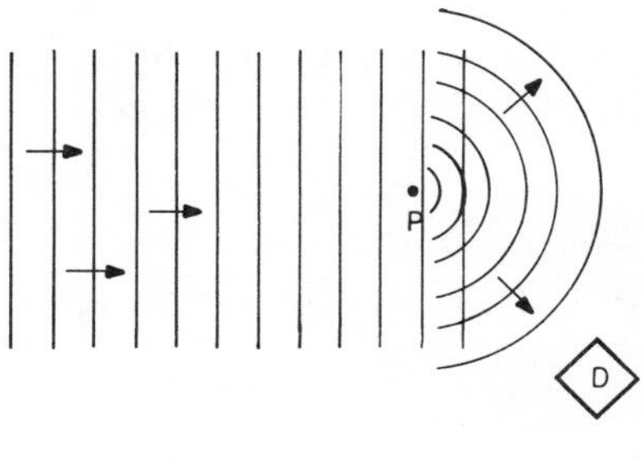

FIGURE 4.3

The quantum picture of electron scattering by a proton

4.2. THE RELATIONSHIP BETWEEN QUANTUM AND CLASSICAL PHYSICS

Since the method of making and interpreting calculations in quantum theory is so totally different from classical theory, one is immediately faced with the question of why the classical theory ever gives the correct results, if in fact the quantum theory is the true description of nature. In preparation for answering that question, let us consider how one would analyze the scattering of electrons by a proton using the classical theory. One would consider a single electron of momentum **p** which had a definite position in the electron beam. (The electron beam in the classical theory is simply a stream of electrons, all of momentum **p**, but distributed at random positions within the beam). For that particular electron one would use Newton's Second Law and Coulomb's Law to calculate the trajectory and in particular to determine the direction of travel of the electron after it is scattered by the proton (See Figure 4.4). One would carry out such a calculation for all possible initial positions of the electron in the beam. Having done that, one would then assume that the electrons are randomly distributed throughout the beam and calculate how many electrons per unit time will be scattered in a particular direction by the proton. That would allow you to predict the flux of electrons at the detector for any position of the detector. The result would agree precisely with the quantum theory calculation! Actually the case of electron-proton scattering is an exceptional case in that respect. Usually the quantum theory calculation of scattered electron flux agrees with the classical theory calculation only for large values of the electron's momentum; that is, for very small values of the electron wavelength. For small values of the momentum the two calculational methods make different predictions, and only the quantum theory agrees with the observations.

FIGURE 4.4

The scattering of electrons by a proton as viewed by the classical theory.

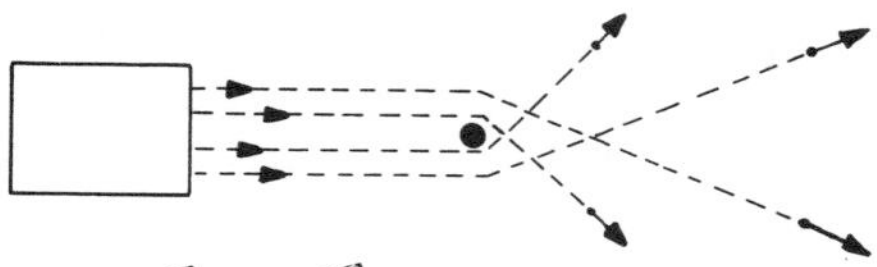

FIGURE 4.5

Electrons are accelerated in the field between the plates.

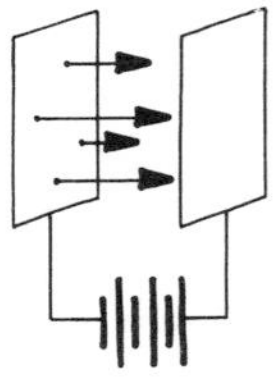

The reason why the wave aspects of electrons went unnoticed for so much longer than the wave aspects of photons is that the wavelength of typical electrons is extremely small, even in comparison with the very short wavelengths of visible photons. Consider the case of an electron that has been accelerated from rest by moving between two metal plates which have a potential difference of 100 volts (See Figure 4.5). The energy of the electron when it strikes the positive plate is

$$E = eV = 1.6 \times 10^{-17}\,J$$

The momentum of the electron is related to its energy by $E = p^2/2m$, which gives

$$p = \sqrt{2mE} = \sqrt{(2)(9.1 \times 10^{-31}\,kg)(1.6 \times 10^{-17}\,J)} = 5.4 \times 10^{-24}\,kg\,m/s$$

The corresponding wavelength is

$$\lambda = \frac{h}{p} = 1.2 \times 10^{-10} m = 1.2\,\text{Å}$$

This distance is close to the diameter of a hydrogen atom. In comparison, the wavelength of yellow light is about 6000 Å When electron waves are sent through devices such as single or double slits (which give rise to noticeable diffraction patterns for light waves) the wavelength is so much smaller than the slit width that it is extremely difficult to detect the diffraction pattern. It is only when electron waves are scattered by crystal specimens in which the typical lattice spacing is a few angstroms, that electron wave interference effects are noticed.

Having seen how Newtonian mechanics and quantum theory can make the same predictions in some situations, we shall turn our attention to the related question of how quantum theory can agree with the field theory of electromagnetism. The two theories do not always agree. The quantum theory says that the electromagnetic field gives the photon flux at any point and that any charged particle near that point will absorb energy from the field by absorbing photons, each of energy $h\nu$. The classical theory says that the electromagnetic field gives the force on a charged particle and that the charged particle absorbs energy from the field continuously as that force does work on the particle. The greatest disagreement occurs when the frequency of the electromagnetic wave is very large (so that $h\nu$ is large), but the amplitude of the wave is very small. The classical theory says that if such an electromagnetic wave falls on a collection of charged particles, the particles will all absorb energy very slowly because of the small amplitude of the electric field. In contrast, the quantum theory says that whenever one of the charged particles absorbs a photon, it will receive a large amount of energy due to the large value of $h\nu$, but that such absorptions will be very infrequent, because the photon flux, which is proportional to the square of the amplitude, is very small. In a process such as the photoelectric effect, where a large amount of energy is needed in a very small locality in order to drive the electron out of the metal, the classical theory gives completely false predictions, while the quantum theory adequately explains the observed phenomena. Although the two theories give completely different descriptions of the details of the energy absorption process, they agree in many cases on the average rate of energy absorption. Infrequent absorptions of large amounts of energy can give the same net absorption as continuous absorption of small amounts of energy. Thus the two theories would agree on how much heat would be generated by sunlight falling on a black surface.

The predictions of the two theories agree in detail concerning the motion of macroscopic charged bodies in slowly varying electric and magnetic fields. As an example, let us consider the effect on the motion of a small charged body whose mass is 10^{-3} gm due to a passing electromagnetic wave of frequency 10^3 Hz. If the body is initially at rest and absorbs one photon, its energy will be

$$E = h\nu = 6.6 \times 10^{-31}\,\mathrm{J}$$

The velocity of the body will then be

$$v = \sqrt{2E/m} = \sqrt{(1.32 \times 10^{-30}\,\mathrm{J})/(10^{-6}\,\mathrm{kg})} = 1.15 \times 10^{-12}\,\mathrm{m/s}$$

Since this velocity is obviously undetectable, we would not notice the absorption of individual photons but would detect only the resultant effect of very large numbers of photons. It can be shown that, if we average over large numbers of photons, then the resultant momentum imparted to the body will agree with the predictions of the classical theory.

Before going deeper into the physical interpretation of the theory, we must introduce more of its mathematical structure. In particular, we shall introduce the fundamental equation satisfied by the electron field wave function.

4.3. THE SCHROEDINGER EQUATION

The electron wave effects we have discussed are roughly comparable to what was known to Erwin Schroedinger at the time he began his major work on quantum theory. Electron wave interference effects had been experimentally verified, and shown to satisfy the de Broglie relations.

The question Schroedinger asked was: "What is the wave equation satisfied by the electron wave function?" We shall present an imaginary reconstruction of the train of thought that led Schroedinger to postulate the equation that bears his name. What is presented here should not be interpreted as a description of historical reality. The arguments that were actually presented by Schroedinger in support of his wave equation involve advanced mathematical formulations of mechanics and cannot be meaningfully repeated here.

We shall assume that the wave number, k, and angular frequency, $\omega = 2\pi\nu$, of the plane wave solutions of the wave equation are related to the observed momentum and energy by the de Broglie Relations:

$$p = \hbar k$$

and

$$E = \hbar\omega$$

The values of p and E must satisfy the well established relationship between momentum and energy for a free particle of mass m

$$E = p^2/2m$$

Therefore we need a wave equation whose plane wave solutions have

$$\hbar\omega = \hbar^2 k^2/2m \tag{4.1}$$

We shall begin with those aspects of wave theory with which we are familiar, namely the wave equation and its plane wave solutions, and try to find the simplest modifications of the theory that will yield plane waves satisfying Equation 4.1.

The wave equations for string waves, sound waves, and electromagnetic waves are all of the form

$$\frac{\partial^2\psi(x,t)}{\partial x^2} - \alpha\,\frac{\partial^2\psi(x,t)}{\partial t^2} = 0 \tag{4.2}$$

where the constant α is different for the three different types of waves. We shall first see whether we can get an appropriate equation for electron waves simply by adjusting the constant α.

The plane wave solutions of Equation 4.2 are of the form

$$\psi(x,t) = A\cos(kx-\omega t) + B\sin(kx-\omega t) \tag{4.3}$$

Substituting Equation 4.3 into Equation 4.2 and using the relations

$$\frac{\partial\psi}{\partial t} = \omega A\sin(kx-\omega t) - \omega B\cos(kx-\omega t), \tag{4.4}$$

$$\frac{\partial^2\psi}{\partial t^2} = -\omega^2 A\cos(kx-\omega t) - \omega^2 B\sin(kx-\omega t), \tag{4.5}$$

and

$$\frac{\partial^2\psi}{\partial x^2} = -k^2 A\cos(kx-\omega t) - k^2 B\sin(kx-\omega t) \tag{4.6}$$

gives

$$(k^2 - \alpha\,\omega^2)(A\cos(k\,x - \omega\,t) + B\sin(k\,x - \omega\,t)) = 0$$

which will be satisfied if

$$\alpha\,\omega^2 = k^2.$$

This relation is obviously different from Equation 4.1. The simplest way to obtain a first power of ω rather than a second power is to take only a first time derivative rather than a second. Therefore let us try a wave equation of the form

$$\frac{\partial^2\psi}{\partial\,x^2} - \alpha\,\frac{\partial\,\psi}{\partial\,t} = 0. \tag{4.7}$$

If we try the same type of plane wave solution in this equation, we get

$$-\,k^2A\cos(k\,x - \omega\,t) - k^2B\sin(k\,x - \omega\,t) - \alpha\,\omega\,A\sin(k\,x - \omega\,t) + \alpha\,\omega\,B\cos(k\,x - \omega\,t) = 0.$$

This can be satisfied if and only if

$$\alpha\,\omega\,B = k^2A \tag{4.8}$$

and

$$\alpha\,\omega\,A = -\,k^2B \tag{4.9}$$

Dividing the first equation by the second gives

$$\frac{B}{A} = -\,\frac{A}{B}$$

or

$$B^2 = -\,A^2$$

with the solution

$$B = i\,A$$

Putting this into Equation 4.9 and cancelling a factor of A from both sides yields a relation between ω and k.

$$\alpha\,\omega = -i\,k^2.$$

This relation will be identical with Equation 4.1 if we choose

$$\alpha = -i\;2m/\hbar$$

In order to obtain an equation that gives the desired relationship between frequency and wave number, we are forced to introduce the imaginary number i into both the wave equation and the basic plane wave solution! Putting this value of α into Equation 4.7 and multiplying through by $\hbar/2m$ gives the form of the Schroedinger equation that is commonly used.

$$i\,\hbar\frac{\partial\,\psi}{\partial\,t} = -\,\frac{\hbar^2}{2m}\,\frac{\partial^2\psi}{\partial\,x^2} \tag{4.11}$$

Using the relation $B = i\,A$ we see that our basic plane wave solution has the form

$$\psi = A\,(\cos\theta + i\sin\theta)$$

where $\theta \equiv k\,x - \omega\,t$. The combination $\cos\theta + i\sin\theta$ appears frequently in the study of complex numbers. It is equal to $e^{i\,\theta}$. In case the reader is unfamiliar with complex exponentials, the following derivation may help. The basic definition of the exponential function is

$$e^x = 1 + x + x^2/2! + x^3/3! + ...$$

Since $(i\theta)^n$ is easy to evaluate, we can substitute $i\theta$ for x in this power series definition of e^x to obtain

$$e^{i\theta} = 1 + i\theta - \theta^2/2! - i\theta^3/3! + \theta^4/4! + i\theta^5/5! + ...$$
$$= (1 - \theta^2/2! + \theta^4/4! - ...) + i(\theta - \theta^3/3! + \theta^5/5! - ...)$$
$$= \cos\theta + i\sin\theta$$

It is assumed that the reader recognizes the power series expansions of $\cos\theta$ and $\sin\theta$. Thus the Schroedinger equation has plane wave solutions of the form

$$\psi(x,t) = A e^{i(kx - \omega t)} \tag{4.12}$$

The *complex conjugate* of any complex number $C = A + iB$ is defined as the complex number $C^* = A - iB$. The *magnitude* of a complex number $C = A + iB$ is written $|C|$ and is defined as $|C| = \sqrt{C^* \cdot C} = \sqrt{A^2 + B^2}$. A fact that we shall frequently use in subsequent sections is that, for any real number θ, the magnitude of $e^{i\theta}$ is one.

$$(e^{i\theta})^* (e^{i\theta}) = (\cos\theta - i\sin\theta)(\cos\theta + i\sin\theta) = \cos^2\theta + \sin^2\theta = 1$$

4.4. THE PHYSICAL MEANING OF THE WAVE FUNCTION

We are now in the situation in which Schroedinger was half a century ago. We have a wave equation without knowing what the physical meaning is of the solutions of the equation. If we solve the equation, we shall end up with a complex electron wave function $\psi(x,t)$. But what does an "electron wave function" mean? We noted before, in the description of the single slit diffraction of electrons and light, that the detectable electron flux exhibited a pattern similar to the energy flux in a classical sound wave. For those classical waves the energy flux is proportional to the square of the wave function. Experiments show that, in order to obtain agreement between the solutions of Schroedinger's equation and the observed electron flux, we must assume that the particle flux associated with a plane wave of wave vector k is given by

$$\text{Electron flux} = \frac{\hbar k}{m} \psi^* \psi \tag{4.13}$$

where ψ^* is the complex conjugate of the complex number $\psi(x,t)$.

When Equation 4.13 is combined with the de Broglie relation $p = \hbar k$, one is naturally led to an even simpler physical interpretation of the wave function ψ. The quantity

$$\frac{\hbar k}{m} = \frac{p}{m} = v$$

is the classical velocity of the electrons in the beam that is described by the plane wave function. If the flux of electrons in the beam is $v\psi^*\psi$, then the density of electrons (i.e. the number of electrons per unit volume) must be given by

$$\text{Electron density} = \psi^* \psi.$$

There is yet one more form in which we shall present the physical interpretation of the electron wave function. Consider the case of an electron beam of very low intensity; such low intensity that we see a flash on the phosphorescent screen on an average of once every few minutes. It would sound silly to talk about the electron density in such a beam; something like saying that the number of pet cobras per person is 10^{-6}. What we really mean in the latter example is that the probability of a person chosen at random owning a pet cobra is 10^{-6}. In the same way what we really mean by the electron density in a very low intensity beam is that if we choose a volume ΔV in the beam and we simultaneously search everywhere within ΔV for an electron, then the probability of finding one in ΔV is equal to ΔV times the electron density. Thus we can say that the

$$\text{Probability of finding an electron in } \Delta V = \psi^* \psi \times \Delta V \qquad (4.14)$$

The advantages of the last formulation are twofold.

1. We can use it also for high intensity beams simply by making ΔV extremely small.
2. We can use it also for other types of solutions of the Schroedinger equation. In particular we can use it for standing wave solutions of the equation for which the electron flux interpretation of the wave function has no meaning.

The factor $\psi^* \psi$ in Equation 4.14 is called the probability density or, speaking less precisely, the electron density.

QUESTION: Is it possible to measure the electron field ψ directly rather than detecting its existence by means of interactions with the associated electrons?

ANSWER: No. It is a basic tenet of the quantum theory that the only way that one can physically interact with the electron field is by interacting with whole electrons. For instance, if light is passed through a region in which an electron field exists, then either a number of photons are scattered or absorbed by electrons, or else the light is completely unaffected.

QUESTION: Since it is impossible to measure the electron field except by interacting with electrons, why not formulate the theory in terms of the motion of electrons and drop any reference to the seemingly fictitious electron field?

ANSWER: During the last fifty years no one has been able to formulate any theory (which agrees with experiment) in terms of a deterministic motion of electrons, rather than an electron field with a probabilistic interpretation. But feel free to try yourself. In order to appreciate the difficulties you would face in formulating a theory without an electron field, consider the following experimental effect (See Figure 4.6).

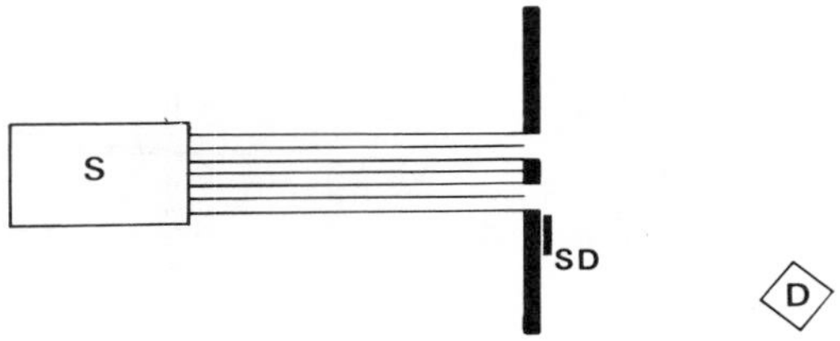

FIGURE 4.6 While the sliding door (SD) is open electrons from the source (S) can reach the detector (D) through either slit.

A low intensity electron beam is incident upon a metal sheet in which there are two slits. At a distance from the slits a small electron detector is located at a position at which Schroedinger's equation predicts complete destructive interference. No electrons are ever detected in the detector no matter how long we wait. We now close one of the slits by means of a little door. Immediately we begin to detect electrons. Any deterministic electron theory is going to have to explain how one gets more electrons to the detector by closing off one of the paths between the source and the detector.

4.5. STANDING WAVE STATES AND DISCRETE ENERGY VALUES

In addition to the plane wave solutions we have been considering, the Schroedinger equation has solutions in the form of standing waves. The standing wave solutions are applicable to situations in which the electron field oscillations are confined to a finite region of space. The simplest of those situations is a one dimensional electron system confined to a bounded region of space. It is an unrealistic example in that real electron systems are three dimensional, but it shows all of the same properties as the real system, and the mathematics is much easier. We look for wave functions $\psi(x,t)$ which are solutions of the Schroedinger equation in the presence of electron-repelling "walls" at $x = 0$ and $x = L$. Due to the walls no electrons will be found outside the interval of length L. Since the electron density is zero outside that interval, the wave function must be zero there. Thus we must look for a solution of the equation that is zero at $x = 0$ and $x = L$. This is reminiscent of the situation one has for a string of length L with both ends fixed. In that case the solutions of the wave equation are of the form

$$u(x,t) = A \sin kx \sin \omega t.$$

As it stands, this function does not satisfy the Schroedinger equation. It must be modified in the same way as the plane wave solution was. As a problem, the reader will be asked to verify that the following function is a solution of the Schroedinger equation.

$$\psi = A \sin kx\, e^{-i\omega t}$$

where

$$\omega = \hbar k^2/2m$$

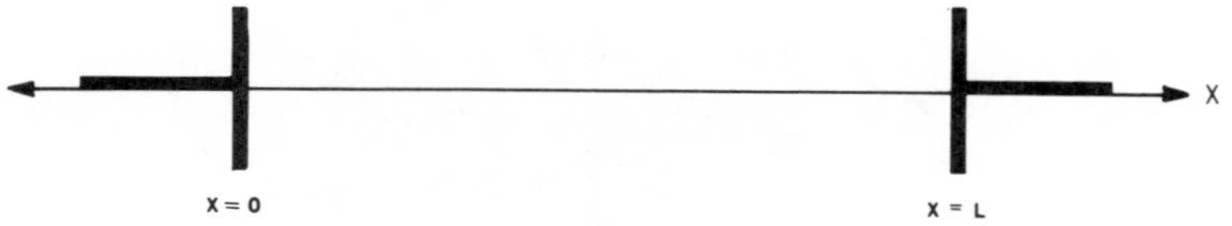

FIGURE 4.7 A one dimensional box of length L.

Since the sine of zero is zero, the above solution is zero, as it should be, at $x = 0$. The solution must also be zero at $x = L$. This will only be the case if $kL = n\pi$, where n is any integer. Thus, for each integer n there is a standing wave solution of the form

$$\psi = A \sin k_n x\, e^{-i\omega_n t}$$

with

$$k_n = n\pi/L$$

and

$$\omega_n = \hbar k_n^2/2m$$

Using the fact that $(e^{-i\omega_n t})^* (e^{-i\omega_n t}) = 1$, we can easily evaluate the probability density for these standing wave solutions.

$$\psi^*\psi = A^2 \sin^2 k_n x$$

The probability densities for the cases n = 1, 2, and 3 are plotted in Figure 4.8 Although the wave function for the standing wave state varies periodically with t because of the factor $e^{-i\omega_n t}$ the probability density remains constant. For that reason such solutions are called *stationary states.* They are also called *quantum states.* In general any solution of Schroedinger's equation of the form $\psi(x,t) = u(x)e^{-i\omega t}$ gives a stationary state, as can be seen by computing the probability density, $\psi^*\psi$. As we stated before, the angular frequency of a quantum state is related to the observed energy of the system by the de Broglie relation. Thus the only possible energies of a particle in a one dimensional box of length L are the discrete energies

$$E_n = \hbar\omega_n = \hbar^2 k_n^2/2m$$

Using the facts that $k_n = n\pi/L$, and $\hbar = h/2\pi$ we get

$$E_n = \frac{n^2 h^2}{8mL^2} \tag{4.15}$$

There is no solution to the wave equation that has a frequency other than one of the normal mode frequencies, and thus there is no solution that has a corresponding energy other than one of the normal mode energies. This effect occurs in any system in which the electrons remain localized in space. The energy of the electrons is related to the frequency of the electron wave. The only

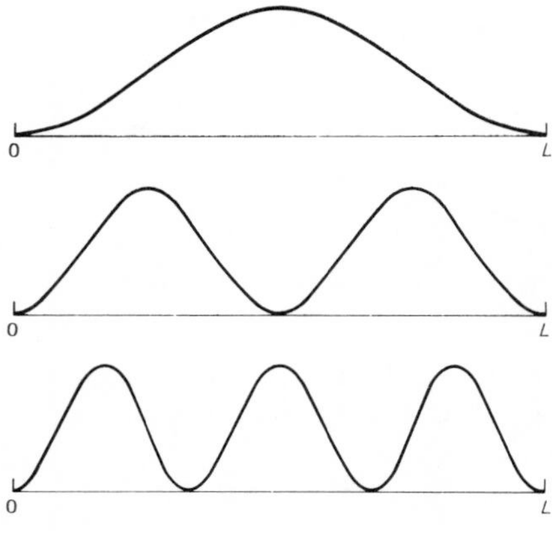

FIGURE 4.8

Plots of the probability density for the standing wave solutions representing electrons confined to a box of length L.

solutions for which the electron field (and thus the electron density) remains localized are standing wave solutions which have discrete frequencies and hence discrete energies. For a given system the set of allowed energy values, E_1, E_2, E_3, $\cdots$ is the energy spectrum of that system. The relationship between the energy spectrum of a system such as a hydrogen atom and the frequencies of electromagnetic radiation given off or absorbed by such a system has been described in the last chapter. The lowest allowed energy value is called the *ground state energy*, and the corresponding quantum state is called the *ground state*.

Let us briefly review the elements of the quantum theory which have been presented.

1. There is an equation (the Schroedinger equation) for the electron field wave function. The equation has both traveling wave solutions and standing wave solutions.
2. The traveling wave solutions are associated with electron beams or fluxes. The average rate at which electrons cross a unit area is given by a formula that is similar to the formulas which give the energy flux in light or sound waves. The experimentally observable energy and momentum of the electrons is related to the frequency and wavelength of the electron waves by the de Broglie relations.
3. The standing wave solutions are associated with bound systems of electrons such as an electron in a closed box or the electron bound to the nucleus of a hydrogen atom. The square of the magnitude of the wave function at a particular location gives the probability density for finding an electron near that location. The energy of the electron is related to the frequency of the standing wave by the de Broglie relation.

QUESTION: As we shall see later, the hydrogen atom is a somewhat fuzzy entity according to the quantum theory. Its size is therefore not well defined. A reasonable estimate of the outer diameter of a hydrogen atom is about 1.5 Å. In its lowest energy state the kinetic energy of the electron of a hydrogen atom is about 2.2×10^{-18} J. How well does this compare with the energy of an electron trapped in a one dimensional box of length L = 1.5Å?

ANSWER: The lowest energy state of an electron in a one dimensional box of length $L = 1.5\times10^{-10}$ m is given by Equation 4.15 with n equal to one.

$$E = \frac{h^2}{8mL^2}$$

The value of h is about 6.6×10^{-34} J·s and, for an electron, $m = 9.1\times10^{-31}$ kg. These values give

$$E = 2.4\times10^{-18}\,J$$

Just knowing the size of the region in which an electron is confined allows us to estimate its kinetic energy.

QUESTION: If a particle which is confined to a finite region of space can have only certain discrete energies, why was this discreteness of the energy not noticed in macroscopic phenomena long ago?

ANSWER: For macroscopic systems the spacing of the allowed energies is so close that the set of possible energies is effectively continuous. For example, the allowed energies for a one gram particle in a one centimeter box are

$$E_n = \frac{n^2h^2}{8mL^2}$$

In this form the equation is hard to interpret in terms of classical ideas. Let us set E_n equal to $\frac{1}{2}mv_n^2$ and see what the change in velocity would be in going from the nth allowed energy to the (n+1)st. Since

$$E_n = \frac{n^2h^2}{8mL^2} = \frac{1}{2}mv_n^2$$

we get

$$v_n = \frac{nh}{2mL}$$

This gives

$$v_{n+1} - v_n = \frac{h}{2mL} = 3.3 \times 10^{-29} m/s$$

Such minute changes in velocity could not be detected.

4.6. NORMALIZATION OF THE WAVE FUNCTION

The ground state wave function for an electron in a one dimensional box of length L was shown to be

$$\psi(x,t) = A \sin kx\, e^{-i\omega t}$$

where $k = \pi/L$ and $\omega = \hbar k^2/2m$. We can make the standing wave of any amplitude, A, without affecting the wavelength or the frequency. How are we to choose the appropriate amplitude? At first sight, due to the fact that the electron field is not directly measurable, there does not seem to be any physical quantity that corresponds to the amplitude. As we shall now see, the probabilistic interpretation of the wave function is enough to determine the amplitude. Suppose we know by prior measurement that there is one electron in the box. The probability that the electron would be found in a small interval of length dx centered at x, is

$$(A^2 \sin^2 kx) \cdot dx$$

If we search for the electron within the finite interval $a \leqslant x \leqslant b$, the probability of detecting it there is

$$P_{ab} = \int_a^b A^2 \sin^2 kx\, dx$$

If we choose $a = 0$ and $b = L$, then the interval becomes the complete box within which we know the electron will be found somewhere. Thus, for that case, the probability P_{ab} must be equal to one.

$$1 = \int_o^L A^2 \sin^2 kx\, dx$$

This equation determines the amplitude A (See Problem 4.6)

$$A = \sqrt{2/L} \tag{4.16}$$

In general the wave function $\psi(x,t)$, describing any system which contains just one electron must satisfy a *probability normalization condition.*

$$\int_a^b \psi^* \psi \, dx = 1 \tag{4.17}$$

where the interval $a \leqslant x \leqslant b$ is the space available to the system. If there are no definite bounds on the system, then $a = -\infty$ and $b = +\infty$.

4.7. THE HEISENBERG UNCERTAINTY RELATION

When we try to apply the normalization condition (Equation 4.17) to the wave function of an electron of momentum p, we encounter a paradox. Such a state of the electron field is described by a plane wave

$$\psi(x,t) = A e^{i(kx - \omega t)}$$

There are no bounds on the system. The normalization condition is therefore

$$\int_{-\infty}^{\infty} \psi^* \psi \, dx = 1$$

But $(e^{i(kx-\omega t)})^*(e^{i(kx-\omega t)}) = 1$. Thus

$$\int_{-\infty}^{\infty} \psi^* \psi \, dx = A^2 \int_{-\infty}^{\infty} 1 \cdot dx = A^2 \cdot \infty$$

An exact plane wave state is spread out uniformly over an infinite interval and cannot be normalized. An exact plane wave state is simply not a physically realizable state. A realizable state would have a wave function similar to the one depicted in Figure 4.9. Such a wave function is spread out over a large but finite

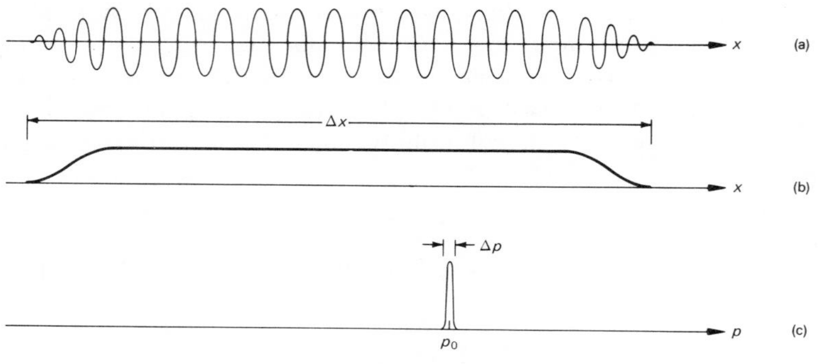

FIGURE 4.9 (A) A normalizable state that is almost a plane wave. The wave function is actually complex, so only the real part of it is plotted here.
(B) The position probability density $\psi^*\psi$ associated with the wave function shown in part (A).
(C) The momentum probability density associated with the wave function shown in part (A).

region of space. At any instant of time there is some interval of length Δx within which the electron would almost surely be detected if it were searched for. We shall call Δx the *uncertainty in position* of the electron. An electron whose wave function was an exact plane wave of wave number k would, if its momentum were measured, always be found to have momentum $p = \hbar k$ exactly. In other words, the probability of finding the electron with momentum p would be one and the probability of finding the electron with any momentum other than p would be zero. If the wave function is not an exact plane wave, that is no longer true. There is a probability distribution associated with the momentum just like that associated with the position of the electron. As shown in Figure 4.9 it is very narrow for a state which is almost a plane wave with large position uncertainty. There is a formula for computing the momentum probability density from the wave function. We shall not present the formula until later, but at this point we merely state without proof a very important theorem called the *Heisenberg uncertainty principle.*

For any wave function whatsoever the spread in the position probability density, Δx, and the spread in the momentum probability density, Δp, are related by the inequality

$$\Delta x \cdot \Delta p \geqslant \hbar$$

There is no state in quantum theory which corresponds to the classical idea of a particle with a definite position x and a definite momentum p. This is true for electrons, photons, and all other particles.

For particles in three dimensions (i.e. real particles) the uncertainty principle relates the uncertainty in a given coordinate x, y, or z to the uncertainty in the corresponding component of momentum p_x, p_y, or p_z. That is

$$\Delta x \cdot \Delta p_x \geqslant \hbar \qquad (4.18)$$

$$\Delta y \cdot \Delta p_y \geqslant \hbar$$

$$\Delta z \cdot \Delta p_z \geqslant \hbar$$

We have approached the uncertainty principle from a mathematical point of view. To fully appreciate its meaning we must now consider what will happen if we experimentally attempt to thwart the principle by constructing a device that will produce a beam of photons whose y component of position and y component of momentum are better defined than Equation 4.18 says is possible. We begin with a device that produces an electromagnetic plane wave which is extended over a large region and has very well defined wave fronts that propagate in the x direction. The photons associated with such a beam would have $p_y \approx 0$ with a vary small uncertainty (See Figure 4.10). The photon beam is, however,

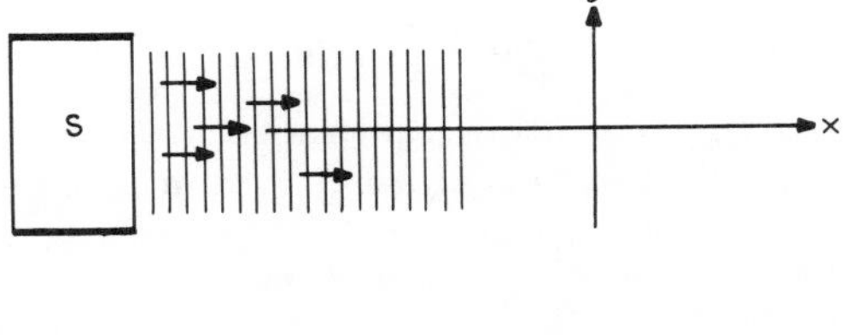

FIGURE 4.10 A source of photons of very small momentum uncertainty.

extensively spread out in the y direction. The y coordinate uncertainty is correspondingly large. Since Δp_y is already acceptably small, we attempt to reduce Δy by placing in the path of the beam a screen which will only allow photons whose y coordinate lies in a narrow range to pass through. We shall then have produced a beam with simultaneously small values of Δp_y and Δy. What actually happens is that the electromagnetic wave, in passing through the narrow slit, is diffracted into a range of directions $\theta \approx \lambda/\Delta y$. The associated photons will no longer be found to have very small values of p_y. In fact, if p is the magnitude of the momentum, then a typical value of p_y would be

$$P_y = p\sin\theta \approx p\,\theta = (h/\lambda)(\lambda/\Delta y) \qquad (4.19)$$

The value of p_y in Equation 4.19 can be taken as an estimate of Δp_y for the beam after it passes through the slit. We see that

$$\Delta y \cdot \Delta p_y \approx h > \hbar$$

Our attempt to reduce Δy has caused a corresponding increase in Δp_y. Our failure to violate the uncertainty principle could have been predicted from the outset. Since there is a mathematical theorem that says that no wave function exists that simultaneously has a position and momentum uncertainty whose product is smaller than $\hbar$, it is obvious that no device is going to produce an electromagnetic beam with a wave function that has such properties.

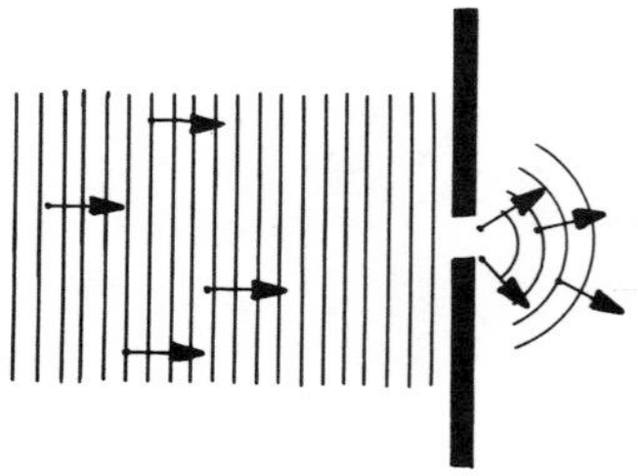

FIGURE 4.11
In passing through the narrow slit the electromagnetic wave is diffracted and this causes the photons on the right to have a wider range of p_y's.

QUESTIONS: If there is no state in which a particle has definite momentum and position, then how can classical mechanics, which assumes particles have definite momentum and position, ever give correct predictions?

ANSWER: The minimum uncertainty allowable by the uncertainty principle is too small to be observable in the motion of macroscopic objects. As an example let us suppose we have an object of mass one gram moving with a velocity of 100 m/s. Suppose that the velocity is known to within one part in a trillion. The uncertainty in v is then

$$\Delta v = 10^{-10} \text{m/s}$$

This means that

$$\Delta p = m\,\Delta v = 10^{-13}\,\text{kg m/s}$$

With that uncertainty in momentum the uncertainty principle states that the uncertainty in the position of the object cannot be less than

$$\Delta x = \frac{\hbar}{\Delta p} = \frac{10^{-34}}{10^{-13}} = 10^{-21}\,\text{m}$$

This distance is about one hundred billionth the diameter of a hydrogen atom. The uncertainty principle is not a significant constraint on the accuracy of a classical position determination.

4.8. THE SCHROEDINGER EQUATION WITH A POTENTIAL

A one-particle state in which the particle has fairly well defined momentum and fairly well defined position is described by a wave function of the form shown in Figure 4.9. Such a wave function is referred to as a *wave packet.* As we saw in the above question, for a particle of one gram mass it is possible to have a quite well defined momentum and yet have a wave packet of very small spatial extent. According to the quantum theory, this little wave packet moves through space in accordance with the Schroedinger equation. The particle will always be found within the wave packet. If we can show that the wave packet moves in such a way as to satisfy the laws of Newtonian mechanics, then we shall have shown that Newtonian mechanics is a logical consequence of the Schroedinger equation. However, before we can do this we must modify the form of the Schroedinger equation somewhat. The form of the Schroedinger equation we have been using was designed to insure that the de Broglie relations led to the classical energy equation

$$E = p^2/2m$$

This classical equation is valid only if no external forces are applied to the particle. If the particle is subject to a conservative force, then the above energy equation must be replaced by the equation

$$E = \frac{p^2}{2m} + V(x) \tag{4.20}$$

where $V(x)$ is the potential function associated with the force.

We now address ourselves to the question of what is the appropriate wave equation for electrons subject to a force? The force is assumed to be a conservative one with an associated potential function $V(x)$. Newtonian mechanics can be derived from the Schroedinger equation under appropriate conditions, but one cannot logically derive the Schroedinger equation from the laws of Newtonian mechanics. In spite of the fact that the more general theory (quantum theory) cannot be derived from the less general theory (Newtonian mechanics), it is still true that the demand that quantum theory should reduce to Newtonian mechanics in the appropriate limit so constrains our freedom to choose a wave equation for the electron field that there seems to be only one simple modification of Equation 4.11 (the Schroedinger equation for electrons without forces) that will lead to Equation 4.10 in the classical limit.

Let us consider a case in which classical Newtonian mechanics should certainly be valid. We consider a particle, whose mass m is about one gram, moving in a potential $V(x)$ that varies appreciably only over distances of the order of a meter. The wave function for the particle is some moving wave packet. As we have seen above, the value of $\hbar$ is so small that we may assume that within the wave packet the wave function differs very little from a pure plane wave and at the same time assume that the total length of the wave packet is so small that the potential is essentially constant over the packet. Then Equation 4.20 can be replaced by the equation

$$E = \frac{p^2}{2m} + V_0$$

where V_0 is a constant (over the wave packet). The value of the "constant" will actually vary gradually as the wave packet moves from one place to another, but at any one time the variation of $V(x)$ over the packet will be completely negligible. If we assume that the de Broglie relations are still valid we must search for an equation that, for a constant potential, has a plane wave solution

$$\psi(x,t) = A e^{i(kx - \omega t)} \tag{4.21}$$

with

$$\hbar\omega = \frac{\hbar^2 k^2}{2m} + V_o \tag{4.22}$$

When applied to a plane wave solution such as Equation 4.21, the terms we now have in the Schroedinger equation give

$$i\hbar\frac{\partial \psi}{\partial t} = \hbar\omega\, \psi(x,t)$$

and

$$-\frac{\hbar^2}{2m}\frac{\partial^2 \psi}{\partial x^2} = \frac{\hbar^2 k^2}{2m}\psi$$

It is clear that the simplest possible term we might add to the Schroedinger equation in order to obtain the relation given in Equation 4.22 rather than the relation $\hbar\omega = \hbar^2 k^2/2m$ is a term of the form $V_o\psi(x,t)$. That is, if we assume that the equation is of the form

$$i\hbar\frac{\partial \psi}{\partial t} = -\frac{\hbar^2}{2m}\frac{\partial^2 \psi}{\partial x^2} + V_o\psi$$

then it will have plane wave solutions which satisfy Equation 4.22. In order for Equation 4.22 to be satisfied wherever the wave packet happens to be, we must have V_o be appropriate to that location. This can be done by replacing V_o by $V(x)$. Thus, if we assume that the fundamental equation of quantum theory is

$$i\hbar\frac{\partial \psi}{\partial t} = -\frac{\hbar^2}{2m}\frac{\partial^2 \psi}{\partial x^2} + V(x)\psi \tag{4.23}$$

we are assured of obtaining Newtonian mechanics in those situations in which we know that Newtonian mechanics is valid. Equation 4.23 was first proposed by Erwin Schroedinger. That the Schroedinger equation is also valid in those cases in which one cannot use Newtonian mechanics can only be confirmed by finding solutions to the equation in those cases and comparing the predictions of the theory with experimental observations. During the last fifty years this has been done in great detail for an extremely wide variety of systems with uniformly positive results. The quantum theory may, sometime in the future, become a limiting case of a more general physical theory, just as Newtonian mechanics is the low velocity limit of relativistic mechanics and the short wavelength limit of quantum mechanics. However, within the wide realm of atomic, molecular, and solid state phenomena, the quantum theory has been so thoroughly confirmed that its essential correctness is beyond reasonable doubt.

4.9. QUANTUM STATES OF A PARTICLE IN A POTENTIAL

The possible energy values of a particle of mass m in a one dimensional potential $V(x)$ are $\hbar\omega_n$ where the numbers ω_1, ω_2, $\cdots$ are the angular frequencies of the standing wave solutions of the Schroedinger equation. A standing wave has a wave function of the form

$$\psi(x,t) = u(x)\, e^{-i\omega t} \tag{4.24}$$

The function u(x) describes the wave pattern of the normal mode. The wave function, ψ, oscillates with the angular frequency ω. If we put this form of solution into Equation 4.23 and use the fact that

$$i\hbar\frac{\partial(u e^{-i\omega t})}{\partial t} = E u e^{-i\omega t}$$

we can cancel the factor $e^{-i\omega t}$ from both sides of the equation and obtain an equation for the function u(x) alone.

$$-\frac{\hbar^2}{2m}\frac{d^2u(x)}{dx^2} + V(x)\,u(x) = E\,u(x) \qquad (4.25)$$

This is called the *time independent Schroedinger equation.* As it stands it is a fairly complicated equation. However, the real source of trouble in using this equation is that *most of the solutions of the time independent Schroedinger equation are spurious.* That is, they do not represent any possible physical state of the system. The reader has probably seen some simple cases of equations with spurious solutions before. For instance, one might have a classical particle moving in two dimensions along the path $y = a^3 - x^3$ and wish to calculate the point at which it hits the ground. That is, the value of x at which $y = 0$. This problem leads to the cubic equation

$$a^3 - x^3 = 0$$

whose solutions are $x = a$, $x = a(-1+i\sqrt{3})/2$, and $x = a(-1-i\sqrt{3})/2$. It is obvious that the complex solutions have no physical interpretation and should simply be ignored. Unfortunately, the spurious solutions of the Schroedinger equation cannot be eliminated quite so easily. Before we look at the problem of identifying the physically meaningful solutions of the Schroedinger equation, let us look at a classical system, described by an equation with spurious solutions that is more closely related to the quantum mechanical problem than is the simple cubic equation we have just considered.

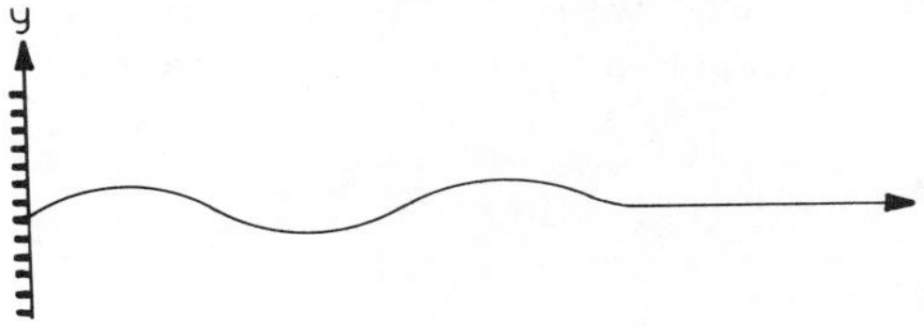

FIGURE 4.12 Waves propagating on a semi-infinite string.

Wave propagation on a uniform string under tension is described by the wave equation

$$\frac{\partial^2 y(x,t)}{\partial x^2} = \frac{1}{v^2}\frac{\partial^2 y(x,t)}{\partial t^2} \qquad (4.26)$$

where v is the wave speed, and y(x,t) is the y coordinate of the string at position x and time t. There are two reasons for which a function y(x,t) that satisfies the wave equation may be rejected as unphysical. We can illustrate them both by assuming that the system we are trying to describe is a semi-infinite string, fixed at $x = 0$ and stretching in the positive x direction.

For this system a wave function may be considered spurious if:

1. The wave function does not satisfy the *boundary condition*, $y(0,t) = 0$. For instance, the function

$$y(x,t) = A \cos k(x - vt)$$

is a solution to Equation 4.26, but it has no meaning as a wave function for a string that is fixed at $x = 0$.

or

2. The wave function diverges as $x \rightarrow +\infty$. An example of a spurious wave function of this type is

$$y(x,t) = (x - vt)^3 + (x + vt)^3$$

This function satisfies the wave equation *and* the boundary condition at $x = 0$, but does not represent any wave motion of the string as can easily be seen by plotting it for $t = 0$.

It is clear from this example that the nature of the *supplemental conditions* that an acceptable wave function must satisfy depends upon the system one is trying to describe by the wave equation. If our string had been infinite in both directions, there would have been no reason to demand that $y(0,t) = 0$ for an acceptable wave function. If the string had been fixed at two places, $x = 0$ and $x = L$, then the wave function would only be defined within the interval $0 \leqslant x \leqslant L$ and no question of divergence at infinity could arise.

Let us now return to the task of defining physically acceptable solutions of the time independent Schroedinger equation. For mathematical convenience we have restricted our study to one dimensional systems. As a visualization of such a system we can take one or more electrons moving within a long, extremely thin tube with smooth hard walls. Another picture of such a system is a quantum mechanical bead on a frictionless wire. We shall look at such obviously artificial systems only while we are developing the basic principles. In later chapters those basic principles will be applied to atoms, molecules, nuclei, and crystals.

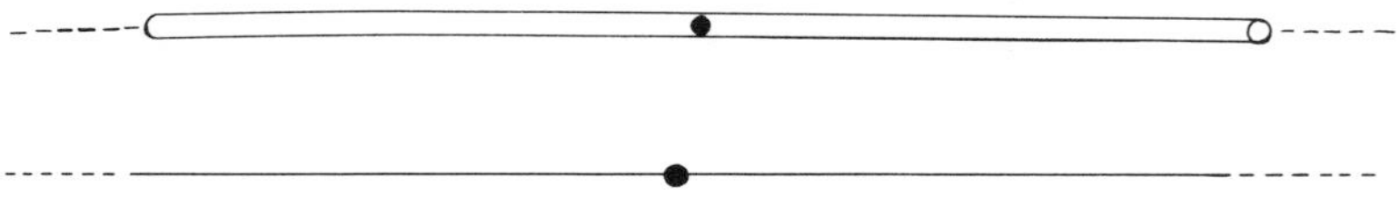

FIGURE 4.13 Two models of a one dimensional system.

The mathematical problem that must be solved in order to determine the wave function is the following one. We must find all the solutions of the time independent Schroedinger equation that satisfy certain supplemental conditions whose detailed nature depends on the system we are investigating. A few representative systems, the supplemental conditions associated with them, and the physical nature of the solutions produced by those supplemental conditions are presented below. Very few mathematical details will be given here and certainly no mathematical proofs. We are just trying to give the reader an overall picture of the kinds of physical phenomena predicted by quantum theory and the relation of the physical phenomena to the mathematical theory. First let us introduce some terminology.

A system is called *unbounded* if the particles in it are free to move from $-\infty$ to ∞. If they are restricted to some finite interval $a \leqslant x \leqslant b$ it is called *bounded.*

A *localized, repulsive* potential function is a function that has a single maximum and steadily decreases to zero on both sides of the maximum. It is called repulsive because a particle in such a potential would always experience a force pushing it away from the location of the potential maximum. The potential is localized in the sense that it is significant only in some finite region and is negligible at very large distances in either direction.

A *localized, attractive* potential function is simply the negative of a localized, repulsive potential. That is, it is a function with a single negative minimum that increases to zero on both sides of the minimum.

A *positively divergent* potential function is one that approaches $+\infty$ at large distances in both directions. That is

$$V(x) \rightarrow +\infty \quad \text{as} \quad |x| \rightarrow \infty$$

We shall consider four different physical systems that we will call cases 1, 2, 3, and 4.

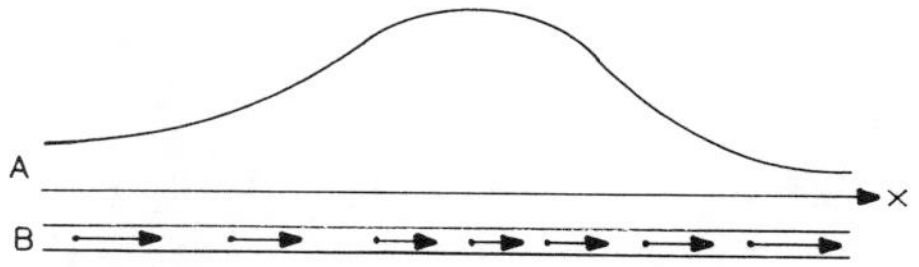

FIGURE 4.14 (A) Graph of a localized repulsive potential.
(B) A stream of classical particles would slow down as they entered the potential region, but, if E were greater than V_{max}, no particles would reverse their direction.

CASE 1 An unbounded system with a localized, repulsive potential and a steady stream of particles coming in from the left.

A classical system such as that shown in Figure 4.14 would have two types of solutions depending on the energy of the incoming particles. If E were less than the maximum value of V(x), then any particle would slow down as it came in until it reached a point at which $V(x) = E$ where its kinetic energy would be zero. It would then retrace its steps in the opposite direction and end up moving back out toward $x = -\infty$ with the same speed with which it came. On the other hand, if E were greater than the maximum of V(x), the particles would never come to rest and would all end up moving toward $x = +\infty$. Thus at a large distance from $x = 0$ we would have an outgoing stream of reflected particles moving toward the left in one case and an outgoing stream of transmitted particles moving toward the right in the other case.

In the quantum mechanical problem we must introduce three supplemental conditions. The first is that the wave function u(x) must remain finite as x goes to plus or minus infinity. It turns out that this condition eliminates all those solutions of the time independent Schroedinger equation (Equation 4.25) for which the energy parameter in the equation, E, is negative. Thus the equation has acceptable solutions only for positive values of E. For negative values of E the equation has solutions but the solutions diverge as $|x| \to \infty$. That one only gets solutions of the equation for positive energy values is certainly reasonable, since the potential energy is positive everywhere (it is a repulsive potential) and one would expect to get only positive values of the kinetic energy.

At very large values of $|x|$, the potential V(x) goes to zero. (It is a localized potential.) Therefore, far from the origin, on both sides, the potential term in the Schroedinger equation can be neglected and the equation becomes the same as the Schroedinger equation for a free particle. That equation is

$$-\frac{\hbar^2}{2m}\frac{d^2u}{dx^2} = Eu$$

The general solution of this equation has the form of two plane waves of wave vectors k and −k, where $\hbar^2k^2/2m = E$

$$u(x) = Ae^{ikx} + Be^{-ikx}$$

(Recall that u(x) is just the space part of the wave function $\psi(x,t)$. It must be multiplied by $e^{-i\omega t}$ to give the full wave function for plane waves that we have used before.) These two parts of the wave function, valid far from the origin where the potential is negligible, represent steady streams of particles moving in opposite directions with the same speed $v = \hbar|k|/m$. The two additional supplemental conditions that we need are:

a. For large positive values of x the amplitude of the left-going plane wave should be zero because we are looking for a solution that represents a system with particles approaching the origin only from the left.

b. For large negative values of x the square of the amplitude of the right-going plane wave is equal to the known particle density in the incoming stream of particles.

It turns out that adding these two supplemental conditions is just enough to determine a unique solution of the Schroedinger equation. That unique solution has the following characteristics:

1. For large negative x the wave function consists of an incoming plane wave whose amplitude is determined by the density of particles in the incoming stream, and an outgoing reflected wave whose amplitude depends on the details of the potential function V(x) and the value of E. However, for a given potential function, the particle flux in the reflected wave is always proportional to the particle flux in the incoming wave.
2. For intermediate values of x, where V(x) is not negligible, the wave function is not simply a combination of two plane waves, but is a solution of the Schroedinger equation.
3. As x approaches $+\infty$, V(x) goes to zero and the wave function u(x) approaches an outgoing plane wave. The outgoing plane wave at large positive x represents a stream of particles that have gotten through the potential. It is called the transmitted wave. The particle flux in the transmitted wave plus the particle flux in the reflected wave is always equal to the particle flux in the incoming wave.

Thus, instead of the incoming particle stream being *either* completely reflected *or* completely transmitted, it is partially reflected and partially transmitted. This does not mean that any particular particle in the incoming stream is partly reflected and partly transmitted so that the outgoing beams contain pieces of electrons rather than complete particles. It means that any particular particle in the incoming beam may, in a random, unpredictable way, be *either* transmitted *or* reflected with the relative probabilities of the two alternatives being determined by the amplitude of the transmitted and reflected waves. Even if the maximum value of V(x) is larger than E, so that according to classical mechanics the incoming particles would all be reflected, the quantum mechanical calculation gives a finite probability that the particle will "tunnel through" this potential and appear in the transmitted stream. This tunneling phenomena has been confirmed in many experiments.

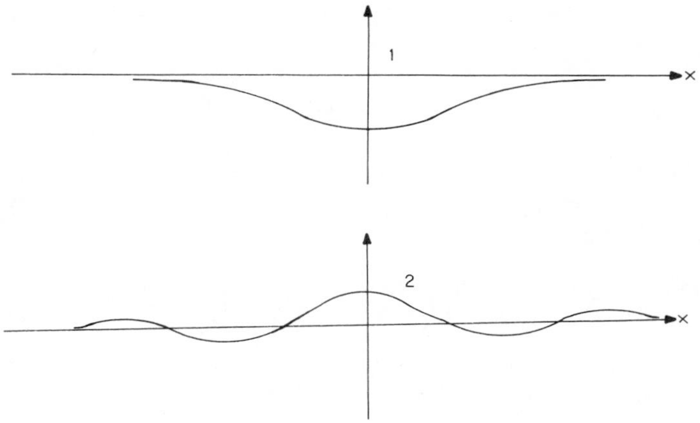

FIGURE 4.15 1. Graph of a localized attractive potential.
2. For particular values of the energy the Schroedinger energy equation has solutions that approach zero as $x \rightarrow \pm\infty$. One such function is shown here.

CASE 2 An unbounded system with a localized attractive potential

In this case there are two types of solutions depending on the value of the energy parameter E in the Schroedinger equation. For any positive E there is a solution that is qualitatively the same as the one we described above for a repulsive potential. At large distances from the origin it approaches an incoming plane wave and reflected and transmitted outgoing waves whose amplitudes depend on the details of V(x).

For most negative values of E the Schroedinger equation has only divergent solutions which must be rejected by the supplemental condition that the wave function not diverge as x approaches plus or minus infinity. However, for particular negative energies E_1, E_2, $\cdots$ the equation has solutions that approach zero as $|x| \rightarrow \infty$. That is $|u(x)| \rightarrow 0$. Those particular negative energies for which the Schroedinger equation has normalizable solutions are called *energy eigenvalues* The collection of eigenvalues is called the *discrete energy spectrum.* The continuous set of positive allowed energy values is called the *continuous energy spectrum.* For each discrete eigenvalue, E_n, there is a corresponding convergent solution of the Schroedinger equation, $u_n(x)$. That is

$$-\frac{\hbar^2}{2m}\frac{d^2u_n}{dx^2} + V u_n = E_n u_n \tag{4.27}$$

The full time-dependent wave function $\psi(x,t)$ is related to $u_n(x)$ by Equation 4.24

$$\psi(x,t) = u_n(x)\, e^{-i\omega_n t}$$

where

$$\omega_n = E_n/\hbar$$

Since $|\psi(x,t)|^2 \rightarrow 0$ as $|x| \rightarrow \infty$ this wave function has no part that represents a flux of particles into or out of the system. When properly normalized these standing wave solutions of the time-independent Schroedinger equation represent the possible quantum states of a single particle in the potential V(x). The physical interpretation of the wave function, ψ, is that

$$\int_a^b \psi^*(x,t)\psi(x,t)dx$$

is the probability of finding the particle within the interval $a \leqslant x \leqslant b$ at time t. Because of the special form of the standing wave solution, that probability can be written more simply in terms of $u_n(x)$.

$$\begin{matrix}\text{The probability of finding the particle} \\ \text{within the interval } a < x < b\end{matrix} = \int_a^b |u_n(x)|^2 dx$$

where $|u_n^2| \equiv u_n^* u_n$. This interpretation of the normal mode wave function makes sense only if $u_n(x)$ satisfies the normalization condition

$$\int_{-\infty}^{\infty} |u_n(x)|^2 dx = 1 \qquad (4.28)$$

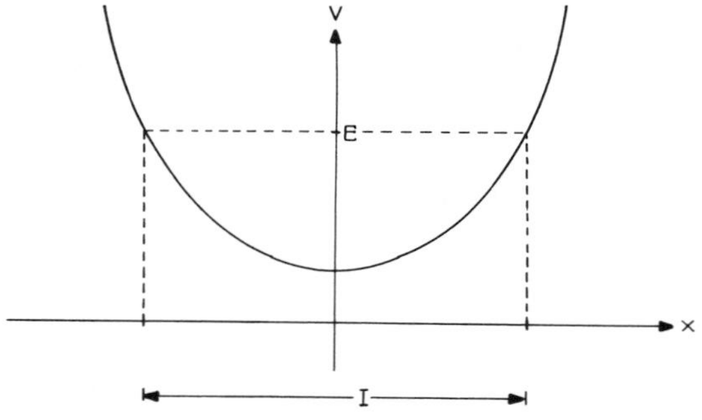

FIGURE 4.16 A divergent potential approaches $+\infty$ in both directions (i.e., $x \rightarrow \pm\infty$). A classical particle in such a potential is restricted to an interval I that depends on its energy.

CASE 3 An unbounded system with a positively divergent potential.

A classical particle in a divergent potential, such as the potential associated with a Hooke's law force, $V(x) = \frac{1}{2}kx^2$, could never escape to infinity. Whatever its energy, it would be forever confined to that finite region of space in which $V(x) \leqslant E$. It could never have an energy less than the minimum value of V(x). In the corresponding quantum system one also obtains only bound states in which no particles come from or go to infinity. The energy spectrum for a divergent potential is purely discrete. The ground state energy (the lowest of the discrete energy eigenvalues) is larger than the minimum value of V(x). There are an infinite number of energy eigenvalues

$$V_{min} < E_1 < E_2 < E_3 < \cdots$$

The set of discrete energies is unbounded. That is $E_n \rightarrow \infty$ as $n \rightarrow \infty$. For each eigenvalue, E_n, there is a corresponding eigenfunction, $u_n(x)$ that satisfies the normalization condition (Equation 4.28).

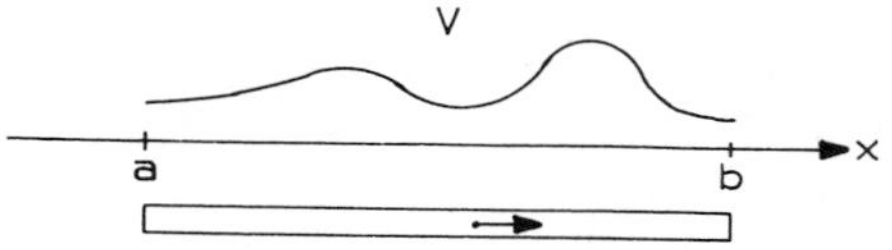

FIGURE 4.17
The particle is confined to an interval $a < x < b$ in which a potential field V(x) exists.

CASE 4 A bounded system with an arbitrary potential.

In this case the wave function is defined only within the interval $a \leqslant x \leqslant b$. The supplemental condition that must be used at $x = a$ and $x = b$ is that u(x) be equal to zero there. That is

$$u(a) = 0 = u(b)$$

We have already looked at one example of such a system, namely the particle in a box. In that case V(x) was equal to zero.

For a bounded system one always obtains an energy spectrum that is *qualitatively* like the spectrum of a particle in a box. That is, it is a purely discrete spectrum extending to $+\infty$.

$$E_1 < E_2 < E_3 < \cdots$$

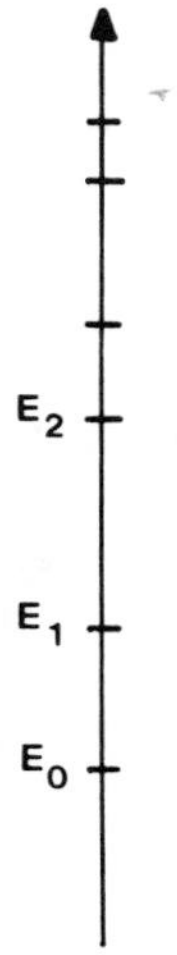

FIGURE 4.18
A particle confined to a finite region always has a purely discrete energy spectrum that extends from some ground state energy, E_o, to plus infinity.

For each eigenvalue E_n there is an associated eigenfunction $u_n(x)$ such that

$$-\frac{\hbar^2}{2m}\frac{d^2u_n}{dx^2} + V u_n = E_n u_n$$

and

$$u_n(a) = 0 = u_n(b)$$

The somewhat difficult problem of actually solving the Schroedinger energy equation for a few representative cases will be left to another chapter.

QUESTION: Is there any way, without going through the complete quantum theory of radiation, to get an idea of why an electron in an energy eigenstate does not radiate even though the wave function oscillates with angular frequency ω_n, and why, when the electron makes a transition, the angular frequency of the radiation is $\omega_{nm} = \omega_n - \omega_m$? In other words, is there any way of identifying anything that oscillates at that frequency and can be considered as the source of the electromagnetic radiation?

ANSWER: We have identified $\psi^*\psi\,\Delta V$ as the probability of finding an electron in the volume element ΔV. If we are a little bit careless about our definitions we can interpret $\rho(r,t) = -e\,\psi^*(\mathbf{r},t)\psi(\mathbf{r},t)$ as the electric charge density associated with the electron density $\psi^*\psi$. If the electron is in an energy eigenstate, then $\psi(\mathbf{r},t)$ is of the form

$$\psi(\mathbf{r},t) = \psi_n(\mathbf{r},t) = u_n(\mathbf{r})\,e^{-i\omega_n t}$$

For simplicity we shall assume that $u_n(\mathbf{r})$ is a real function. This gives an electric charge density of

$$\rho(\mathbf{r},t) = -e\,u_n^2(\mathbf{r})$$

which is constant in time and would therefore not be expected to create any electromagnetic radiation.

To answer the second part of the question, let us consider a wave function that describes an electron making a slow transition from the quantum state $\psi_n(\mathbf{r},t)$ to the quantum state $\psi_m(\mathbf{r},t)$. Such a wave function might be written in the form

$$\psi(\mathbf{r},t) = a(t)\,\psi_n(\mathbf{r},t) + b(t)\,\psi_m(\mathbf{r},t)$$

where a and b are real functions which gradually change from $a = 1$, $b = 0$ to $a = 0$, $b = 1$ (See Figure 4.19). When $a = 1$ and $b = 0$ the electron is in quantum state ψ_n. When $a = 0$ and $b = 1$, the electron has made a transition to quantum state ψ_m. The electric charge density associated with this state is

$$\rho(\mathbf{r},t) = -e\,[a^2\psi_n^*\psi_n + b^2\psi_m^*\psi_m + a\,b(\psi_n^*\psi_m + \psi_m^*\psi_n)]$$

If we now use the facts that $\psi_n = u_n(\mathbf{r})\,e^{-i\omega_n t}$ and $\psi_m = u_m(\mathbf{r})\,e^{-i\omega_m t}$ we obtain

$$\rho(\mathbf{r},t) = -e\,[a^2u_n^2 + b^2u_m^2 + 2\,a\,b\,u_m u_n \cos(\omega_n - \omega_m)t]$$

By assumption, a(t) and b(t) vary slowly in time. Therefore the major time dependence is in the term containing the factor $\cos(\omega_n - \omega_m)t$ which represents a charge density that is oscillating at the angular frequency ω_{nm} and would thus be expected to produce radiation of that frequency.

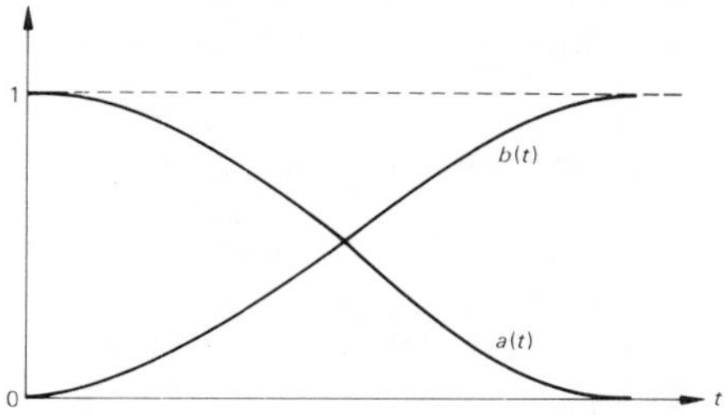

FIGURE 4.19 The functions a(t) and b(t).

QUESTION: The plane wave solutions of the Schroedinger equation for a free particle are $\psi = Ae^{i(kx-\omega t)}$ where $\omega = \hbar k^2/2m$.
This wave could be written as

$$\psi = Ae^{ik(x-ut)} = f(x-ut)$$

where the wave velocity u is given by $u = \omega/k$. But $\omega/k = \hbar k/2m = p/2m$. This is not the correct velocity for free particles. Free particles should travel at velocity $v = p/m$. Why does the wave velocity not come out equal to the velocity of the particles?

ANSWER: The wave function is not directly measurable and so there is no good reason to equate the wave velocity with the particle velocity. The directly measurable quantity, namely the particle density $\psi^*\psi$ is simply a constant, and thus one cannot tell how fast the particles in the beam are traveling. If we take the ratio of the particle flux to the particle density (See Equations 4.13 and 4.14), we get the expected particle velocity $v = p/m$. But this is a very unconvincing argument, since we used the particle velocity in deriving Equation 4.14. What we need to do to get a believable formula for the particle velocity from Schroedinger's equation is to choose a solution for which the electron density is not so smooth. If the electron density is lumpy, then we can watch how the lumps of electron density move and say with reasonable confidence that the only way a lump of electrons can move with velocity v is for the electrons in the lump to move with that velocity. This is not really as hard to do as it sounds. If we add two plane waves of slightly different momenta, we obtain a lumpy electron density for which we can easily calculate the velocity of the lumps. We take one plane wave with wave number $k+\Delta k$ and angular frequency $\omega+\Delta\omega$ and another with wave number $k-\Delta k$ and angular frequency $\omega-\Delta\omega$ where Δk and $\Delta\omega$ are very small. In order for both wave functions to satisfy the Schroedinger equation we need to have

$$\omega+\Delta\omega = \hbar(k+\Delta k)^2/2m \approx \hbar k^2/2m + \hbar k\,\Delta k/m$$

and

$$\omega-\Delta\omega = \hbar(k-\Delta k)^2/2m \approx \hbar k^2/2m - \hbar k\,\Delta k/m$$

Subtracting the bottom equation from the top, we get

$$\Delta\omega = \hbar k\,\Delta k/m$$

a fact we shall later use. If we assume each plane wave has an amplitude $\frac{1}{2}A$, then the wave function for the addition of the two plane waves is

$$\begin{aligned}\psi(x,t) &= \tfrac{1}{2}A\,e^{i(kx+\Delta kx-\omega t-\Delta\omega t)} + \tfrac{1}{2}A\,e^{i(kx-\Delta kx-\omega t+\Delta\omega t)} \\ &= \tfrac{1}{2}A\,e^{i(kx-\omega t)}\left(e^{i(\Delta kx-\Delta\omega t)} + e^{-i(\Delta kx-\Delta\omega t)}\right) \\ &= A\,e^{i(kx-\omega t)}\cos(\Delta k\,x-\Delta\omega\,t)\end{aligned}$$

Looking at the electron density we get

$$\psi^*\psi = A^2\cos^2(\Delta k\,x-\Delta\omega\,t)$$

The measurable electron density now comes in a series of clumps that move to the right with velocity given by

$$v = \frac{\Delta\omega}{\Delta k} = \hbar k/m = p/m \tag{4.29}$$

When we make a series of wave packets by adding more than one plane wave, we find that the wave packets move with the expected classical velocity. This analysis supplies the missing element in our discussion of the classical limit of quantum mechanics. We showed that the energy was related to the momentum in the proper way ($E = p^2/2m + V(x)$). We did not show that the wave packet with momentum p actually moved at the classical velocity $v = p/m$. We can now do so simply by noting that the equation

$$\hbar\omega = \hbar^2k^2/2m + V_o$$

also leads to the relation $\Delta\omega = \hbar k\,\Delta k/m$. Therefore the rest of our analysis can be used without change for the case of a particle in a potential, and thus Equation 4.29 which says that $p = mv$ will be valid for a particle moving in a slowly varying potential.

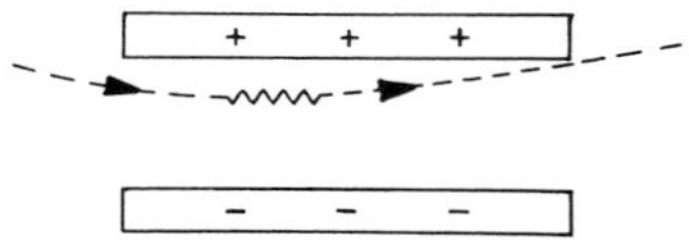

FIGURE 4.20

If the potential changes very little over distances comparabale to the wave packet size, then the packet moves like a classical particle.

SUMMARY

The classical picture of nature in which certain things such as electrons are purely particles while other things such as electromagnetism and gravitation are purely fields does not agree with experiment. Instead there are electron particles associated with an electron field; proton particles associated with a proton field; photon particles associated with an electromagnetic field, etc....

There is a wave equation satisfied by each one of the fields. For traveling waves there is a particle flux associated with the waves. The wavelength and frequency of the field are related to the momentum and energy of the associated particles by the de Broglie relations

$$p = h/\lambda = \hbar k \qquad \text{and} \qquad E = h/T = \hbar\omega$$

For photons the additional relation $c = \lambda/T$ gives $E = cp$, which indicates that photons must be particles of zero mass.

For electrons, not in a potential field, the known relationship between energy and momentum is $E = p^2/2m$ which leads, via the de Broglie relations to a relation between ω and k.

$$\hbar\omega = \hbar^2 k^2/2m$$

The simplest wave equation that has solutions satisfying this relation is the Schroedinger equation

$$i\hbar\frac{\partial\psi}{\partial t} = -\frac{\hbar^2}{2m}\frac{\partial^2\psi}{\partial x^2}$$

It has complex plane wave solutions

$$\psi = Ae^{i(kx-\omega t)} = A(\cos(kx-\omega t)+i\sin(kx-\omega t))$$

The physical interpretation of the wave function is that the positive real number $\psi^*(x,t)\psi(x,t)\cdot\Delta x$ gives the probability for finding an electron at time t in the interval Δx.

The Schroedinger equation also has standing wave solutions which are associated with particles that remain localized in space. The standing wave solutions have certain discrete possible frequencies, and therefore the bounded system can have only certain discrete energies E_1, E_2, $\cdots$. For a particle in a one dimensional box of length L the quantum states are

$$\psi_n = A\sin k_n x\, e^{-i\omega_n t}$$

where $k_n = n\pi/L$ and $\omega_n = \hbar k_n^2/2m$.

The allowed energy levels are $E_n = \hbar^2 k_n^2/2m$.

When an electron system goes from a quantum state of energy E_n to one of energy E_m, where $E_n > E_m$, the excess energy is usually given off as a single photon of frequency

$$\nu_{nm} = (E_n - E_m)/h$$

If the system initially has energy E_m, it can absorb a photon of energy $E_n - E_m$ and shift to energy level E_n.

Because of its interpretation in terms of probability density, the wave function of a bound state must satisfy the probability normalization condition

$$\int_A^B \psi^* \psi \, dx = 1$$

where the interval $A < x < B$ is the space available to the particle.

The Heisenberg uncertainty principle shows that it is impossible to have an electron in a quantum state in which its uncertainty in position, Δx, and its uncertainty in momentum, Δp_x, do not satisfy the inequality

$$\Delta x \cdot \Delta p_x \geqslant \hbar$$

For systems of macroscopic size there exist solutions of the Schroedinger equation in the form of moving wave packets which are well localized in space. The motion of the packet satisfies the equations of classical dynamics.

For a particle in a potential $V(x)$ the correct form of the wave equation is

$$i\hbar \frac{\partial \psi}{\partial t} = -\frac{\hbar^2}{2m} \frac{\partial^2 \psi}{\partial x^2} + V(x)\psi$$

The stationary states (or quantum states) are solutions of Schroedinger's equation of the form

$$\psi(x,t) = u(x)\, e^{-i\omega t}$$

The function $u(x)$ is a solution of the time-independent Schroedinger equation (also called the Schroedinger energy equation).

$$-\frac{\hbar^2}{2m} \frac{d^2 u}{dx^2} + V(x)u = Eu$$

This equation has solutions that do *not* diverge at infinity only for certain values of the energy parameter E. A value of E for which the Schroedinger energy equation has a nondivergent solution is called an energy eigenvalue. If the parameter E in the equation is set equal to an energy eigenvalue and the equation is solved, the nondivergent solution obtained, $u(x)$, is called an *energy eigenfunction.* If the potential is localized ($V(x) \rightarrow 0$ as $|x| \rightarrow \infty$) and the system is unbounded, the eigenfunctions are of two types. For any positive value of E there is a solution $u(x)$ that describes a steady incoming stream of particles that are, in a random and unpredictable way, either reflected by or transmitted through the potential region. If the localized potential is repulsive, then there are no acceptable solutions for any negative energy. If the localized potential is attractive, then for certain particular negative energies E_1, E_2, $\cdots$ the Schroedinger energy equation has bound state solutions $u_1(x)$, $u_2(x)$, $\cdots$ that satisfy the probability normalization condition

$$\int_{-\infty}^{\infty} |u_n|^2 dx = 1$$

If the potential is positively divergent at infinity or if the particles are constrained to a finite region of space, then the energy spectrum is purely discrete and all the energy eigenfunctions represent bound states.

PROBLEMS

4.1* If a photograph of a still life is taken using very low intensity light by leaving the shutter open for a very long time, which one of the statements below would be true?

A. The film will remain unexposed because the photons of very low intensity light do not have sufficient energy to carry out the chemical reaction in the film.

B. The image obtained will be simply a random scatter of dots where individual photons hit the film.

C. The image obtained will be similar to the picture that would be obtained with normal intensity, but much more "grainy" due to the particle-like properties of low intensity light.

D. The picture obtained will be the same as would be obtained by using normal intensity light and a fast shutter speed.

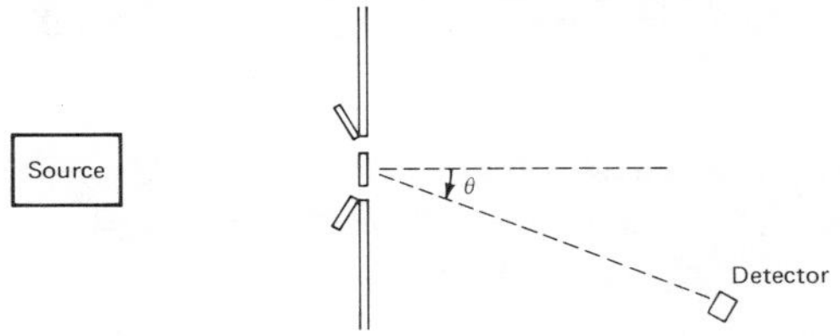

FIGURE P.4.2

4.2* Shown in Figure P.4.2 is a light source, a screen with two holes that can be separately opened or closed, and a photon detector at a large distance from the screen. When only the upper hole is open, the rate at which photons are detected is R_1. When only the lower hole is open, the rate is R_2. When both holes are open, the rate at which photons are detected is R_{12}. According to quantum theory, which of the following is true?

A. $R_{12} = R_1 + R_2$.

B. R_{12} may be smaller than, the same as, or larger than $R_1 + R_2$, depending upon the location of the detector.

C. R_{12} is always less than $R_1 + R_2$ due to interference between the two holes.

D. $R_{12} = (R_1 + R_2)\sin\theta$.

4.3 What is the de Broglie wavelength of a one gram ping pong ball traveling at three meters per second?

4.4 What is the wavelength and frequency associated with an electron that has been accelerated through a potential difference of 40,000 V?

4.5 Calculate the first two energy levels of an electron in a one dimensional box of length 1 Å.

4.6* By evaluating the integral $\int_0^L \sin^2 kx\,dx$ where $k = n\pi/L$, show that the amplitude of the normalized wave function for a particle in a box must be $\sqrt{2/L}$.

4.7** Consider a quantum system of a particle of mass m confined to a one dimensional box of length L. The system is in its ground state. Assume that the right wall at $x = L$ is movable. If it is moved rapidly the particle may make a transition to a different quantum state. However, it can be shown that if the wall is moved very slowly, then the system will always remain in its ground state. Suppose the length of the box is decreased very slowly from L to $L - \Delta L$ (Assume $\Delta L << L$). The energy of the particle will then increase from $h^2/8mL^2$ to $h^2/8m(L-\Delta L)^2$. The increased energy of the particle must come from the work done by the outside force needed to move the wall.

(a) Use this fact to determine the force exerted by the particle on the wall.

(b) Compare the answer you got in (a) with the average force that a classical particle of the same energy, bouncing back and forth between the two walls, would exert on the wall.

4.8* (a) By taking the required partial derivatives, verify that the wave function

$$\psi(x,t) = A \sin kx\, e^{-i\omega t}$$

satisfies Equation 4.11, if $\omega = \dfrac{\hbar k^2}{2m}$.

(b) Show that $\psi(L,t) = 0$ only if k has one of the values $k_n = n\pi/L$.

4.9* As was mentioned before, the diameter of a hydrogen atom in its ground state is about 1.5 Å.

(a) Calculate the minimum uncertainty in p_x of an electron whose position uncertainty Δx is 1.5 Å

(b) Calculate the kinetic energy of an electron with momentum equal to the value of Δp_x you obtained in (a).

4.10* Show that one obtains a standing wave solution of Equation 4.11 that is equivalent to the one given in Problem 4.8 by adding two plane wave solutions of opposite momenta, p and −p. The solution obtained is not identical to that given in Problem 4.8. Why is the difference of no physical importance?

4.11** Find two energy eigenvalues and normalized eigenfunctions for a particle of mass m in a potential of the form $V(x) = -\alpha/x$ if we impose the extra condition that the wave function must be zero for all negative values of x.

Hint: try wave functions of the form $u_1 = x e^{-ax}$ and $u_2 = (x^2 + cx)e^{-bx}$.

4.12* Find the value of c for which the wave function

$$u(x) = c/\sqrt{x^2+1}$$

is normalized within the region $-\infty < x < \infty$.

4.13** By defining two quantities,

$$v(x) = \frac{2m}{\hbar^2} V(x)$$

and

$$\epsilon = \frac{2m}{\hbar^2} E$$

one can write the Schroedinger energy equation as

$$-u'' + vu = \epsilon u$$

Find the potential v(x) and the energy eigenvalue ϵ given that $u(x) = x e^{-(x/\lambda)^2}$ is the energy eigenfunction and that $v(0) = 0$ and $\lambda = 2Å$.

4.14* Match the three energy spectra shown in Figure P.4.14 with the three potential functions listed below. (Both constants, α and β are positive.)

1. $V(x) = \alpha |x|$
2. $V(x) = \alpha/(1 + x^2/\beta^2)$
3. $V(x) = -\alpha e^{-(x/\beta)^2}$

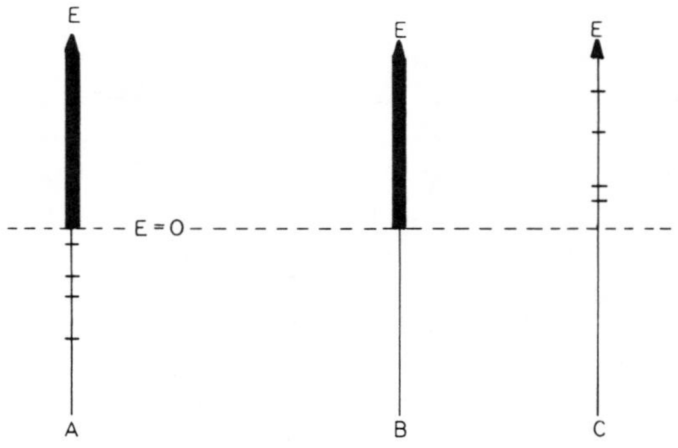

FIGURE P.4.14

CHAPTER FIVE

SOLVING THE SCHROEDINGER EQUATION

5.1. STATIONARY STATES

A stationary state is a solution of the Schroedinger equation that has the form

$$\psi(x,t) = u(x)\, e^{-i\omega t}$$

In the last chapter we showed that the space wave function, u(x), which from now on we shall simply call the wave function, must satisfy the time-independent Schroedinger equation

$$-\frac{\hbar^2}{2m}u''(x) + V(x)\,u(x) = E\,u(x)$$

The energy of a particle in the stationary state is related to the angular frequency of the time-dependent wave function, ψ, by

$$E = \hbar\omega$$

As we shall see in this and subsequent chapters, it is the time-independent Schroedinger equation that appears in almost all applications of quantum theory to real physical systems. It is the central mathematical element in any analysis of nature on an atomic scale. Finding and interpreting solutions of that equation constitutes the major daily activity of theoretical physicists and chemists doing research in the fields of atomic physics, molecular physics, solid state physics, and nuclear physics. Since these fields of fundamental physics have very wide application in most branches of engineering it is important that engineers who do not want to narrowly restrict the kinds of problems they are competent to handle be able to read and understand the published results of research in such areas of basic physics. The basic mathematical concepts used in describing research results in those fields often come directly from particular solutions of the time-independent Schroedinger equation. In the last chapter we described, in a qualitative way, some of the solutions of that equation and their physical interpretation. However, we actually wrote down detailed solutions for only two cases; the free particle, for which the solutions are plane waves, and the particle in a box, for which the solutions are sine functions and the possible energy values are discrete. In this chapter we shall increase the student's library of detailed solutions. It is only by seeing the equation solved and the solutions interpreted for a variety of physical situations that he or she will develop confidence in using the mathematical techniques needed and develop some physical intuition regarding quantum phenomena.

In working with the time-independent Schroedinger equation it is convenient to simplify the equation by introducing the following two definitions.

$$\epsilon \equiv \frac{2m}{\hbar^2}E \tag{5.1}$$

and

$$v(x) \equiv \frac{2m}{\hbar^2}V(x) \tag{5.2}$$

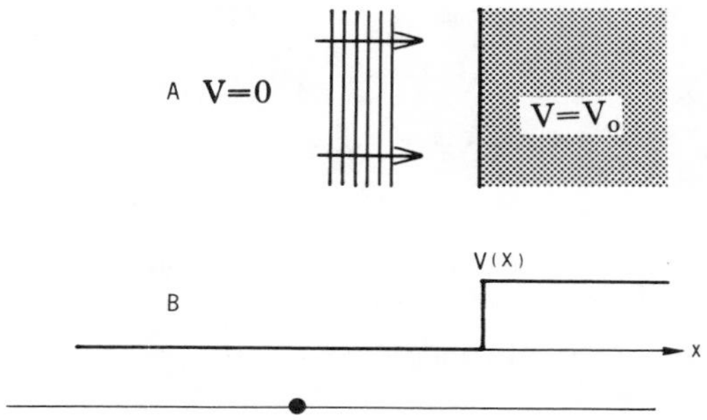

FIGURE 5.1 Two visualizations of a system described by a step potential.

The equation then has the form

$$-u''(x) = (\epsilon - v(x))\,u(x) \tag{5.3}$$

5.2. THE STEP POTENTIAL

When a sound wave or an electromagnetic wave meets the boundary between two different uniform media, such as air and water, it is partly transmitted into the new medium and partly reflected back into the medium in which it was traveling. The quantum equivalent of this phenomenon is the partial reflection of a plane wave solution of the Schroedinger equation at a surface on which the potential changes abruptly from one constant value to another. We can construct such a situation by choosing a potential function v(x) that is equal to zero for negative values of x and equal to a constant, v_0, for positive values of x. The potential jump, v_0, may be either positive or negative. The wave function will also be assumed to depend only on the variable x. This system may be pictured as one in which a three dimension plane wave is incident normally on a flat plane that defines the boundary between two regions of constant potential or it may be pictured as a truly one dimensional system in which there is a potential discontinuity at the point x = 0. (See Figure 5.1) Written explicitly for negative and positive x, Equation 5.3 takes the form

$$u'' = -\epsilon u \qquad x < 0 \tag{5.4}$$

and

$$u'' = -(\epsilon - v_0)u \qquad x > 0 \tag{5.5}$$

Since we expect partial reflection of the incoming wave at the discontinuity in v(x) we can shorten our analysis by assuming a wave function on the left which contains an incoming wave of wave vector k and a reflected wave of wave vector −k. The amplitude of the reflected wave is not necessarily equal to the amplitude of the incoming wave.

$$u(x) = A_{in}e^{ikx} + A_{refl}e^{-ikx} \quad (\text{for } x < 0) \tag{5.6}$$

Substituting this into Equation 5.4 we see that

$$\epsilon = k^2 \tag{5.7}$$

In the region to the right of the potential jump we expect to find only a transmitted wave moving to the right. Its amplitude and wavelength are in general different from those of the incoming wave. Thus

$$u(x) = A_{tran}e^{iqx} \quad (\text{for } x > 0) \tag{5.8}$$

This wave function satisfies Equation 5.5 if

$$\epsilon = q^2 + v_o \tag{5.9}$$

We shall assume that $\epsilon > v_o$ so that Equation 5.9 has a real solution for q. Later we shall consider what happens when $\epsilon < v_o$.

We now have separate solutions for negative and positive x. In order for the two solutions to agree at x = 0 we must have

$$A_{in} + A_{refl} = A_{tran} \tag{5.10}$$

It can be shown (see Problem 5.2) that, for any finite potential the derivative of the wave function, u'(x), must also be continuous at x = 0. If we differentiate Equations 5.6 and 5.8 and then set x = 0 in the two expressions, we get

$$i\,k\,A_{in} - i\,k\,A_{refl} = i\,q\,A_{tran} \tag{5.11}$$

We define a *reflection coefficient*, R, and a *transmission coefficient*, T, by

$$R = A_{refl}/A_{in}$$

and

$$T = A_{tran}/A_{in}$$

Dividing Equations 5.10 and 5.11 by A_{in} gives two simultaneous equations for R and T

$$T - R = 1$$

$$q\,T + k\,R = k$$

These equations can be solved without difficulty, giving

$$R = \frac{k-q}{k+q} \tag{5.12}$$

and

$$T = \frac{2k}{k+q} \tag{5.13}$$

5.3. PARTICLE CONSERVATION

The value of A_{in} fixes the particle density in the incoming beam. It can be chosen arbitrarily. The flux of particles in the incoming beam is $v|A_{in}|^2$ where $v = \hbar k/m$. In order for our interpretation of the plane wave solutions in terms of particle fluxes to be logically consistent we must have the particle flux in the incoming beam equal to the sum of the particle flux in the reflected beam plus the particle flux in the transmitted beam. This gives

$$\frac{\hbar k}{m}|A_{in}|^2 = \frac{\hbar k}{m}|A_{refl}|^2 + \frac{\hbar q}{m}|A_{tran}|^2 \tag{5.14}$$

Multiplying Equation 5.14 by $m/\hbar|A_{in}|^2$ gives a relation between R and T that must be satisfied if the theory is to give a conserved particle flux.

$$k|R|^2 + q|T|^2 = k \tag{5.15}$$

Substituting R and T from Equations 5.12 and 5.13 gives the equation

$$k\left(\frac{k-q}{k+q}\right)^2 + q\left(\frac{2k}{k+q}\right)^2 = k$$

which is an identity in k and q. Thus the Schroedinger equation *does* give a conserved particle flux.

The probability that a particular particle is reflected at the potential jump is equal to the ratio of the reflected beam flux to the incoming beam flux. That ratio is equal to $|R|^2$

$$\text{Probability of reflection} = |R|^2 \tag{5.16}$$

Since a particle must be either reflected or transmitted, it is clear that the

$$\text{Probability of transmission} = 1 - |R|^2 \tag{5.17}$$

5.4. BARRIER PENETRATION

In Section 5.1 we assumed that the energy ϵ was larger than the potential step v_o. In this section we shall see that the case in which $0 < \epsilon < v_o$ introduces an interesting new quantum phenomenon, called *barrier penetration.* In this case we can define two positive real parameters, k and λ, by

$$k = \sqrt{\epsilon} \tag{5.18}$$

and

$$\lambda = \sqrt{v_o - \epsilon} \tag{5.19}$$

Equations 5.4 and 5.5 can then be written as

$$u'' = -k^2 u \qquad (x < 0)$$

and

$$u'' = \lambda^2 u \qquad (x > 0)$$

for $x < 0$ the general solution of the equation contains incoming and reflected plane waves as before.

$$u(x) = A_{in}e^{ikx} + A_{refl}e^{-ikx} \quad (\text{for } x < 0) \tag{5.20}$$

The general solution in the region $x > 0$ is made up of two real exponentials

$$u(x) = A_+e^{\lambda x} + A_-e^{-\lambda x}$$

In order to prevent the wave function from diverging as $x \rightarrow +\infty$ we must set A_+ equal to zero. Therefore

$$u(x) = A_-e^{-\lambda x} \qquad (\text{for } x > 0) \tag{5.21}$$

The conditions that u(x) and u'(x) be continuous at x = 0 give two equations relating the three constants A_{in}, A_{refl}, and A_-.

$$A_{in} + A_{refl} = A_- \tag{5.22}$$

and

$$ik(A_{in} - A_{refl}) = -\lambda A_- \tag{5.23}$$

Defining $\alpha \equiv A_-/A_{in}$ and $R \equiv A_{refl}/A_{in}$ we can write these equations as

$$\alpha - R = 1$$

and

$$\lambda\alpha - ikR = -ik$$

These equations have the solution

$$R = -\frac{\lambda + ik}{\lambda - ik} \tag{5.24}$$

and

$$\alpha = \frac{2k}{\lambda - ik} \tag{5.25}$$

Let us first show that in this case, in which the potential jump is larger than the kinetic energy of the incoming particles, the particle beam is completely reflected. In using Equation 5.16 to calculate the probability that an incoming particle is reflected at the potential jump we must keep in mind that the reflection coefficient, R, is now a complex number.

$$\text{Probability of reflection} = R^*R$$
$$= \left(-\frac{\lambda - ik}{\lambda + ik}\right)\left(-\frac{\lambda + ik}{\lambda - ik}\right) = 1$$

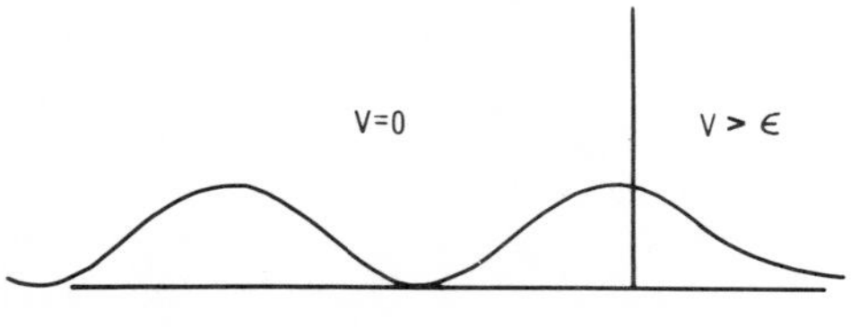

FIGURE 5.2 A graph of the probability density, $u^*(x)u(x)$, near the potential jump. The periodic pattern on the left is due to interference between the incoming and reflected waves. Within the classically forbidden region the probability density goes to zero exponentially with increasing x.

Complete reflection of a particle beam at a potential jump that is larger than the kinetic energy of the particles is exactly what we would expect according to Newtonian mechanics. Since, in classical mechanics, the kinetic energy can never be negative, it would be impossible for a particle of total energy E to enter a region in which the potential energy is greater than E. The significant difference between the quantum mechanical and classical results is that, according to quantum theory, although the particles are all eventually reflected, they do partially penetrate the classically forbidden region. That is, the probability of finding a particle in the region $x > 0$ is not zero. This can easily be seen by looking at Figure 5.2 in which the spatial probability density $|u(x)|^2$ is plotted. In the classically forbidden region $u(x) = A_- e^{-\lambda x}$ and therefore the probability density $|u|^2$ is proportional to $e^{-2\lambda x}$. This phenomenon of the wave function extending past a discontinuity at which there is perfect reflection is well known in classical wave theory. The most important case of it is associated with the perfect internal reflection of light. When a light wave approaches, from the inside, the surface of a medium of index of refraction n it will be completely reflected if the angle between the wave fronts and the surface is larger than the critical angle θ_0 defined by $n \sin \theta_0 = 1$

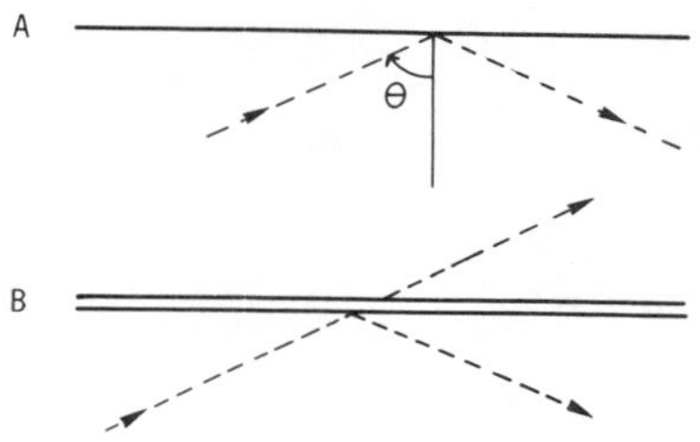

FIGURE 5.3 A. If $\sin \theta > 1/n$ the light ray is completely reflected at the surface.
B. If another piece of glass is brought close, without touching, the wave partially jumps the gap and a transmitted wave appears whose amplitude depends on the width of the gap and decreases exponentially with increasing gap width. The amplitude of the reflected wave is correspondingly diminished so that the reflected plus transmitted waves carry off the incident wave energy.

In this case, although there is no propagating wave transmitted into the vacuum, the electromagnetic fields there are not zero. Above the interface in Figure 5.3(A) there is an oscillating field whose amplitude decreases exponentially with distance from the interface. This field can be detected by placing another flat piece of the same medium at a small distance from the surface. The propagating wave will partially "jump the gap" and a transmitted wave will appear whose amplitude decreases exponentially as the gap width is increased. In a similar way the penetration of the Schroedinger wave function into the classically forbidden region can be detected by setting up a situation in which the potential function v(x) returns to zero at a finite value of x. In that case a transmitted wave appears to the right of the potential barrier. If the barrier width, a, is larger than λ^{-1}, then the amplitude of the transmitted wave is, to a good approximation, proportional to $e^{-\lambda a}$. This transmitted wave represents a flux of particles that have passed through the potential barrier. Thus, when a particle strikes a

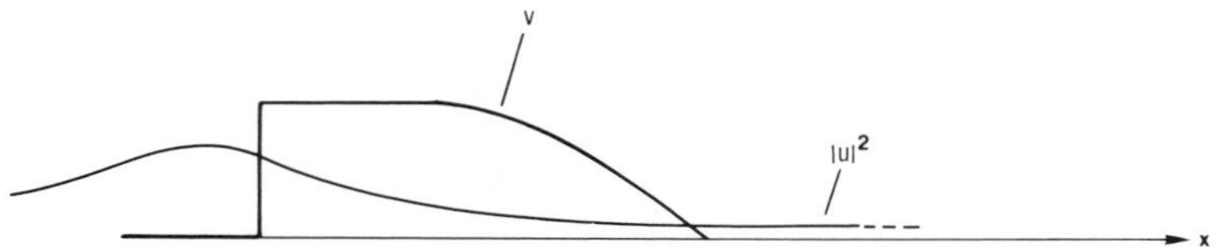

FIGURE 5.4 If the potential returns to zero in a finite distance a small amplitude transmitted wave appears on the right. This wave represents a small but finite probability that any incident particle will pass through the barrier rather than be reflected. It is not possible to predict which thing any particular particle will do.

potential barrier of finite width, Schroedinger's equation predicts that there is a finite probability that the particle will be transmitted through the barrier rather than being reflected, even if the classical energy conservation law would prohibit the particle from ever entering the barrier.

QUESTION: An electron with a kinetic energy of .5 eV approaches a region 1 mm wide in which its potential energy would be 1 eV. What is a rough estimate of its probability of passing through the potential barrier?

ANSWER: First let us calculate the quantity λa. We know that

$$v_o = \frac{2m}{\hbar^2} V_o \quad \text{and} \quad \epsilon = \frac{2m}{\hbar^2} E$$

where

$$V_o = 1 \text{ eV} = e \text{ Joules}$$

and

$$E = .5\text{eV} = .5e \text{ Joules}$$

(The conversion factor in going from electron volts to Joules is the electronic charge e). Therefore,

$$\lambda = \left[\frac{2me}{\hbar^2}(V_o - E)\right]^{1/2}$$

$$= \sqrt{me}/\hbar$$

$$= \frac{\sqrt{(9.1 \times 10^{-31})(1.6 \times 10^{-19})}}{1.05 \times 10^{-34}}$$

$$= 3.6 \times 10^9 \text{m}^{-1}$$

The barrier width, a, is 10^{-3}m. Therefore

$$\lambda a = 3.6 \times 10^6$$

Since this is much larger than one, we can assume that the transmitted wave amplitude is proportional to $e^{-\lambda a}$ and, therefore, the probability of transmission is proportional to $e^{-2\lambda a}$. If we neglect the proportionality constant, which would be of order one, we can estimate the transmission probability as

$$P_{tran} \approx e^{-2\lambda a} = e^{-7.2\times 10^6} \approx 1/10^{3000000}$$

This is such a ridiculously small number that we can say that a potential barrier of macroscopic dimension such as this one allows no transmission at all.

QUESTION: A potential energy of 1 eV is typical of an electron in an atom. However, a typical electronic length is 1 Å. What happens to the estimated transmission probability in the last question when we make the barrier width of atomic dimensions?

ANSWER: In that case

$$\lambda a = .36$$

and

$$e^{-2\lambda a} = e^{-.72} \approx .5$$

which shows that, on an atomic scale, barrier penetration is likely to be an important effect.

QUESTION: If a particle hits a potential step that is much smaller than its kinetic energy, is the probability of reflection very small?

ANSWER: Yes. If we define a parameter β as the ratio of the potential step to the kinetic energy,

$$\beta = \frac{v_0}{\epsilon}$$

then we can write the reflection coefficient in terms of β. For $\beta \leqslant 1$, Equation 5.12 gives

$$R = \frac{k-q}{k+q} = \frac{\sqrt{\epsilon}-\sqrt{\epsilon - v_0}}{\sqrt{\epsilon}+\sqrt{\epsilon - v_0}} = \frac{1-\sqrt{1-\beta}}{1+\sqrt{1-\beta}}$$

The reflection probability is equal to $|R|^2$. for $\beta \geqslant 1$ we know that $|R|^2 = 1$. In Figure 5.5 we have plotted $|R|^2$ as a function of β. It is clear that $|R|^2$ becomes very small when $v_0 << \epsilon$.

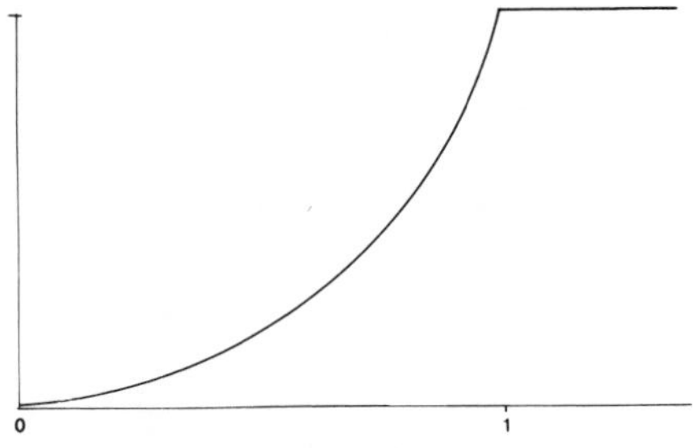

FIGURE 5.5 The reflection probability, $|R|^2$, as a function of $\beta = v_0/\epsilon$.

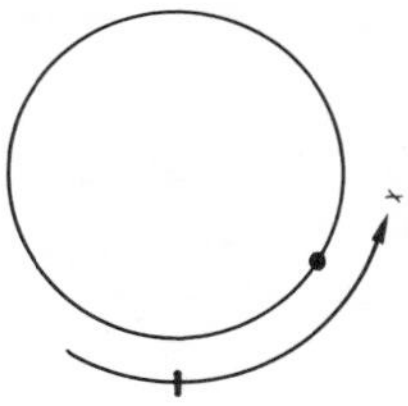

FIGURE 5.6

A quantized bead on a frictionless circular wire of length L.

5.5. THE QUANTIZED BEAD ON A CIRCULAR WIRE

In Figure 5.6 we see a bead of mass m on a frictionless wire loop of total length L. We assume that no potential field is present. If x is a variable that defines displacement along the wire from some arbitrary point, the Schroedinger equation is

$$-u''(x) = \epsilon\, u(x) \qquad (5.26)$$

The only way in which this system differs from a single free particle in one dimension is that, for any value of x, the point specified by the coordinate $x+L$ is identical to the point specified by the coordinate x. If these two values of the coordinate specify the same point on the loop, then the wave function at $x+L$ must equal the wave function at x, or else we would have two different values of the wave function at one point. Thus any acceptable wave function for this system must satisfy the *periodic boundary condition*

$$u(x+L) = u(x) \qquad \text{(for any x)} \qquad (5.27)$$

The solutions of Equation 5.26 are plane waves

$$u(x) = A\,e^{ikx}$$

where $k = \pm\sqrt{\epsilon}$. In order for this solution to satisfy the periodic boundary condition we must have

$$A\,e^{ik(x+L)} = A\,e^{ikx}$$

which gives

$$e^{ikL} = 1 \qquad (5.28)$$

In the complex plane the number $e^{i\theta}$ is a point on the unit circle that makes an angle θ with respect to the real axis. Thus $e^{i\theta} = 1$ if $\theta = 0,\ \pm 2\pi,\ \pm 4\pi,\ \cdots$ or in general if $\theta = 2\pi n$ where n is any integer. Thus the only solutions of Equation 5.28 are

$$k_n = \frac{2\pi}{L} n \qquad (n = \text{integer}) \qquad (5.29)$$

This quantization of the wave vector leads to a discrete energy spectrum.

$$E_n = \frac{\hbar^2}{2m}\epsilon_n = \frac{\hbar^2}{2m}\left(\frac{2\pi n}{L}\right)^2 = \frac{h^2 n^2}{2m\,L^2}$$

This system is sometimes referred to as a *particle in a periodic box.* Its energy spectrum differs in a number of significant ways from the energy spectrum of the ordinary particle in a box.

a. The integer $n = 0$ which leads to the energy value $E_0=0$ is allowed for the periodic box but not for the box with hard walls.

b. for $n > 0$ the periodic box energy level is four times as large as the corresponding hard-wall box energy level.

c. for any energy level, E_n, other than the ground state energy there are two distinct quantum states ($n = \pm |n|$) with that energy. When there are a number of states with the same energy, E_n, that energy level is said to be *degenerate* and the *degeneracy* of the energy level is an integer equal to the number of distinct states with the given energy. for the particle in a periodic box the degeneracy of every energy level other than the ground state energy is two.

The energy eigenfunctions are of the form

$$u_n(x) = A e^{ik_n x}$$

The normalization constant, A, is fixed by the demand that

$$\int_{-L/2}^{L/2} |u_n|^2 dx = 1$$

This gives the following normalized energy eigenfunction.

$$u_n(x) = \frac{1}{\sqrt{L}} e^{ik_n x} \tag{5.31}$$

with k_n given by Equation 5.29.

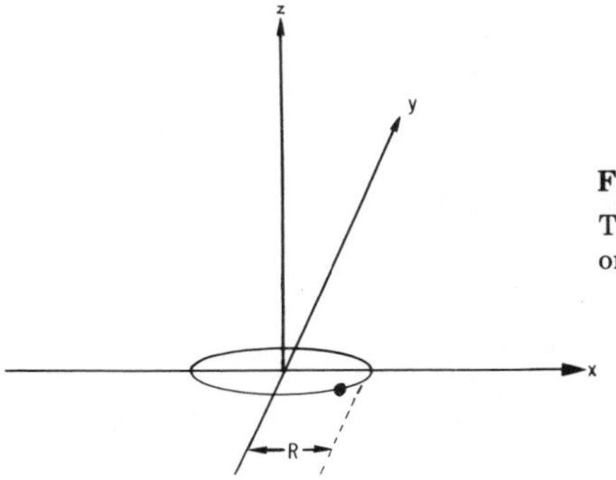

FIGURE 5.7

The z component of angular momentum about the origin is $L_z = Rp$.

5.6. QUANTIZATION OF ANGULAR MOMENTUM

The energy eigenfunctions are obviously plane waves of momentum $p_n = \hbar k_n$. It is useful, as an introduction to the quantum theory of angular momentum, to calculate the angular momentum, about the center of the circular loop, of a particle in the quantum state u_n. If we place the loop in the x-y plane with its center at the origin, then the angular momentum component we are investigating is L_z. From the definition of angular momentum it is clear that

$$L_z = Rp$$

where R is the radius of the loop and p is the momentum of the particle along the loop. In state u_n the value of L_z is

$$L_z = R p_n = R\left(\hbar \frac{2\pi}{L} n\right) = n\hbar \quad \text{(note: } L = 2\pi R\text{)} \tag{5.32}$$

Thus, the possible values of L_z are

$$L_z = 0, \pm \hbar, \pm 2\hbar, \cdots$$

Notice that the allowed values of angular momentum do not depend on the radius of the loop or the mass of the bead. They are the same as the values assumed in the Bohr theory of hydrogen except for the fact that $L_z = 0$ is an allowed value here but was not in the Bohr theory. Later, when we take up the proper quantum theory of hydrogen, we shall find that the Bohr angular momentum values are in error by one unit and that the correct angular momentum of the hydrogen ground state is zero rather than $\hbar$ as was assumed in the Bohr picture.

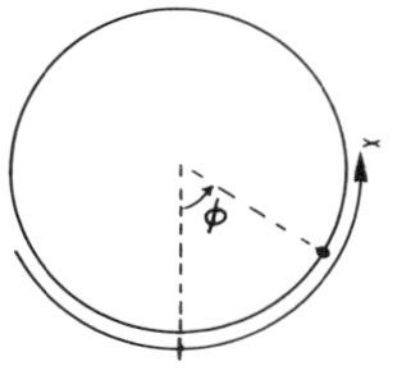

FIGURE 5.8

The angular variable, ϕ, is related to x by $x = R\phi$, where R is the radius of the loop.

The eigenfunctions take a particularly simple form if we express them in terms of a variable ϕ, which is an angular variable measured from the center of the loop. ϕ is related to the variable x we used before by

$$x = R\phi$$

Thus

$$k_n x = (\frac{2\pi}{L}n)(\frac{L}{2\pi}\phi) = n\phi$$

and the eigenfunctions are

$$u_n(\phi) = A e^{in\phi}$$

In order to normalize u_n in terms of the variable ϕ, we must choose A so that

$$\int_{-\pi}^{\pi} |u_n(\phi)|^2 d\phi = 1$$

This gives

$$u_n = \frac{1}{\sqrt{2\pi}} e^{in\phi} \tag{5.33}$$

Later, in the chapter on the hydrogen atom, this eigenfunction will reappear where it will describe the circulation of the electron about the proton in a state with $L_z = n\hbar$.

5.7. THE HARMONIC OSCILLATOR

One of the most important examples of the Schroedinger energy equation is that in which the potential V(x) is a harmonic oscillator potential

$$V(x) = \tfrac{1}{2} k x^2 \tag{5.34}$$

The quantum theory of the harmonic oscillator should not be viewed as being relevant to a macroscopic mass attached to a spring. Although the theory is technically applicable to such a system, the distinctly quantum mechanical aspects of the solutions, such as the discrete energy levels, would be completely undetectable for a system of macroscopic size. Rather, the theory is appropriate and leads to observable consequences for the many microscopic systems in which a particle of atomic dimensions is bound to other particles by a force which can be well approximated by a harmonic oscillator force over a finite range of distances. An excellent example of such a system is a diatomic molecule.

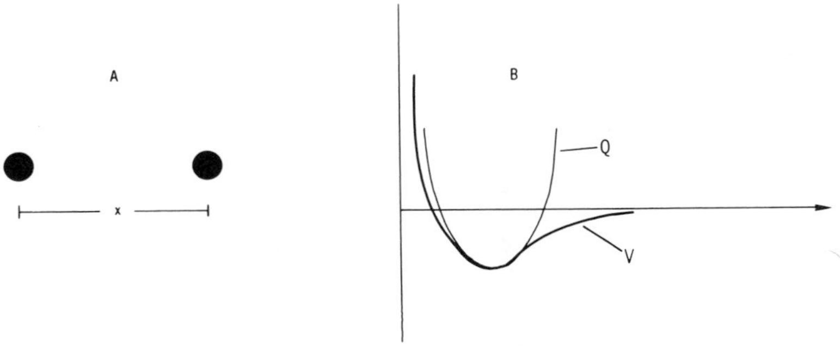

FIGURE 5.9 A. The potential energy of a diatomic molecule, made of two interacting atoms, depends on the separation of the atoms, x.
B. Near the equilibrium distance the potential, V(x), can be approximated by a quadratic function, giving a Hooke's law force.

In such a molecule the two atoms would ordinarily be found close to their equilibrium distance from one another. The equilibrium distance is that distance at which the potential energy due to the interaction between the atoms is a minimum. If the separation between the atoms is made larger or smaller than the equilibrium distance, a restoring force will be encountered which, over a distance which is different for different diatomic molecules, can be approximated by a "Hooke's Law" force. Such a force leads to a potential energy function of the harmonic oscillator form (See Figure 5.9).

The classical harmonic oscillator (in one dimension) is a particle of mass m subject to a force,

$$F = -k x$$

The equation of motion is therefore

$$m \frac{d^2x}{dt^2} = -k x$$

This equation has the solution

$$x = x_o \sin(\omega_o t - \phi)$$

where $\omega_o = \sqrt{k/m}$ and x_o is the amplitude of vibration of the oscillator.

The first thing we want to determine is when it is permissible to use the classical rather than the quantum theory. As we have seen before, classical mechanics is applicable in those cases in which the uncertainties in position and momentum, required by the uncertainty principle, are negligible in comparison to the distances and momenta that occur in the motion of the particle. for a classical oscillator the natural measure of the distances involved in the motion is just the amplitude of vibration, x_o. The momentum also varies periodically with an amplitude, p_o, given by

$$p(t) = m\frac{dx}{dt} = m\,\omega_o x_o \cos(\omega_o t - \phi) = p_o \cos(\omega_o t - \phi)$$

That is

$$p_o = m\,\omega_o\,x_o$$

Quantum effects will be important when $x_o p_o$ is comparable to the minimum uncertainty

$$(\Delta\,x)(\Delta\,p) = \hbar$$

This gives the criterion for the applicability of classical mechanics.

$$x_o p_o = m\,\omega_o x_o^2 >> \hbar$$

If we define a *quantum length*, d, by

$$d = \sqrt{\hbar/m\,\omega_o} \tag{5.35}$$

then classical mechanics may be used whenever the amplitude of the motion is much larger than d. For a macroscopic oscillator, with m = l kg and ω_o= 1 rad/s, d is about 10^{-17}m, which is very much smaller than atomic dimensions and thus quantum effects are always negligible. For a typical diatomic molecule m is about 10^{-26} kg and ω_o is about 10^{14} rad/s. This gives $d \approx 10^{-11}\,m = .1\,Å$. Since the typical size of such a molecule is 1 Å, an uncertainty of .1 Å is quite noticeable and thus the system must be treated by quantum mechanics.

The Schroedinger energy equation for a harmonic oscillator is

$$-\frac{\hbar^2}{2m}\frac{d^2u}{dx^2} + \tfrac{1}{2}k\,x^2 u = E\,u \tag{5.36}$$

This equation can be simplified by introducing a variable s whose physical meaning is simply the distance measured in quantum lengths rather than meters.

$$s \equiv x/d \tag{5.37}$$

We assume that

$$u(x) = f(s) \tag{5.38}$$

Differentiating Equation 5.38 twice and using Equation 5.37 gives

$$\frac{d^2u}{dx^2} = \frac{1}{d^2}\frac{d^2f}{ds^2}$$

When this is substituted into the Schroedinger equation, it takes the form

$$-f''(s) + s^2 f(s) = 2\lambda f(s) \tag{5.39}$$

where

$$\lambda \equiv \frac{E}{\hbar \omega_0} \tag{5.40}$$

Our task is now to find a function f(s) and a number λ such that the pair will satisfy Equation 5.39. The simplest function that yields a solution is

$$f_0(s) = c_0 e^{-s^2/2}$$

If this function is tried in Equation 5.39 a solution is obtained if the eigenvalue λ is chosen to be

$$\lambda_0 = \frac{1}{2}$$

The value of the normalization constant, c_0, is determined by the condition

$$\int_{-\infty}^{\infty} u_0^2(x)\, dx = d \int_{-\infty}^{\infty} f_0^2(s)\, ds = d c_0^2 \int_{-\infty}^{\infty} e^{-s^2} ds = 1$$

As shown in the Table of Integrals

$$\int_{-\infty}^{\infty} e^{-s^2} ds = \sqrt{\pi}$$

This gives the normalized eigenfunction

$$f_0(s) = d^{-1/2} \pi^{-1/4} e^{-s^2/2} \tag{5.41}$$

A graph of $f_0(s)$, which can be shown to be the ground state energy eigenfunction, is given in Figure 5.10. One can see from the graph that, in the ground state, the uncertainty in the position of the particle is about one quantum length.

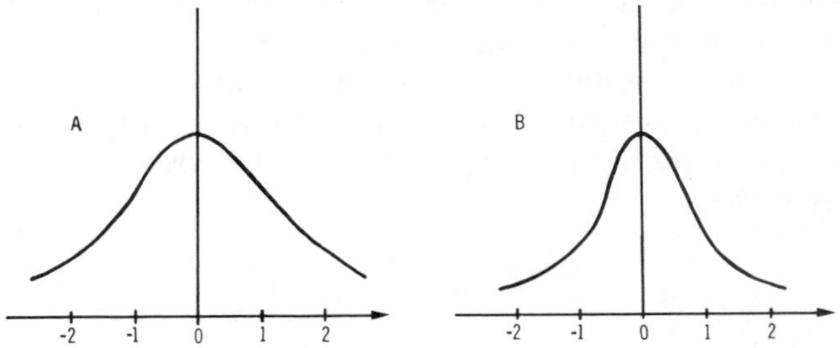

FIGURE 5.10 A. A graph of $f_0(s)$.
B. A graph of the probability density, $f_0^2(s)$.

Now that we have the ground state wave function, we would like to determine the energy eigenvalues and eigenfunctions associated with the excited states. The key to doing this is given in two simple theorems whose proofs are left as problems. The first theorem is

THEOREM I

If f(s) is a solution of the equation

$$f'' = (s^2 - 2\lambda)f \tag{5.42}$$

and

$$g \equiv sf - f' \tag{5.43}$$

then g(s) is a solution of the equation

$$g'' = (s^2 - 2(\lambda + 1))g \tag{5.44}$$

Equation 5.42 is just the harmonic oscillator eigenvalue equation. (Compare with Equation 5.39.) This theorem tells us that if f is an eigenfunction with an associated eigenvalue λ, then the function g defined in Equation 5.43 is another eigenfunction with the eigenvalue $\lambda + 1$. Thus, starting from the ground state we can generate an infinite sequence of new eigenfunctions and eigenvalues. The eigenvalue λ_0 is equal to ½. The sequence of eigenvalues will therefore be

$$\lambda_0 = 1/2$$

$$\lambda_1 = \lambda_0 + 1 = 3/2$$

$$\lambda_2 = \lambda_1 + 1 = 5/2$$

$$\lambda_n = n + 1/2$$

Using Equation 5.40 which relates the eigenvalue λ to the energy eigenvalue we get the following simple formula for the energy levels of a quantized oscillator.

$$E_n = (n + \tfrac{1}{2})\hbar\omega_0 \tag{5.45}$$

where $\omega_0 = \sqrt{k/m}$ is the angular frequency of vibration of an equivalent classical oscillator.

For a harmonic oscillator the spacing between adjacent energy levels is constant and equal to $\hbar\omega_0$ (See Figure 5.11). This fact has the interesting consequence that if such a quantum oscillator is initially in a quantum state with energy E_n and makes a transition to the next lower energy state, emitting a photon with the excess energy in the process, then the frequency of the emitted photon will be given by

$$\hbar\omega_{photon} = E_n - E_{n-1} = \hbar\omega_0$$

or

$$\omega_{photon} = \omega_0$$

Thus a quantum harmonic oscillator would emit electromagnetic radiation whose frequency was equal to the expected classical frequency of vibration. Because of this fact many classical theories of the electromagnetic properties of solids which assumed that the particles were subject to Hooke's law forces were at least partially successful. However, their degree of success was quite limited, and they can hardly be compared with the very detailed and accurate calculations presently made with the quantum theory of solids.

If we begin with a *normalized* eigenfunction f(s) and use Equation 5.43 to generate a new eigenfunction, then the new eigenfunction, g(s), will unfortunately not be correctly normalized. The next theorem, which concerns

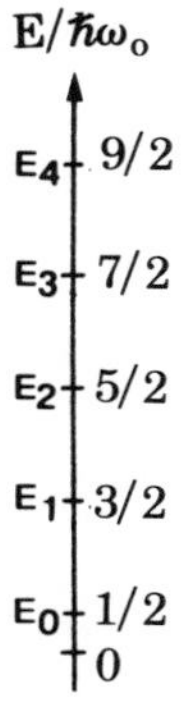

FIGURE 5.11
The energy spectrum of the quantum harmonic oscillator.

the normalization of g(s) will show us how to make a slight change in the formula for generating g(s) so that we always get a new *normalized* eigenfunction.

THEOREM II

If f(s) satisfies Equation 5.42 and g(s) is related to f(s) by Equation 5.43, then

$$\int_{-\infty}^{\infty} g^2 ds = (2\lambda + 1) \int_{-\infty}^{\infty} f^2 ds$$

If we divide the right hand side of Equation 5.43 by $\sqrt{2\lambda + 1}$ we shall then get a normalized eigenfunction, assuming the eigenfunction f was normalized. Thus, starting with the nth normalized eigenfunction, $f_n(s)$, we can generate the $(n+1)$st normalized eigenfunction $f_{n+1}(s)$ by the formula

$$f_{n+1} = \frac{1}{\sqrt{2\lambda_n + 1}} (s f_n - f'_n)$$

Using the fact that $\lambda_n = n + \frac{1}{2}$ we can write this as

$$f_{n+1} = \frac{1}{\sqrt{2n+2}} (s f_n - f'_n) \tag{5.46}$$

Using this formula we can easily generate the following series of energy eigenfunctions.

$$f_0 = d^{-1/2} \pi^{-1/4} e^{-s^2/2}$$

$$f_1 = \frac{1}{\sqrt{2}} (s f_0 - f'_0) = \sqrt{2/d}\, \pi^{-1/4} s\, e^{-s^2/2}$$

$$f_2 = \frac{1}{\sqrt{4}} (s f_1 - f'_1) = \frac{\pi^{-1/4}}{\sqrt{2d}} (2s^2 - 1) e^{-s^2/2}$$

etc.

It is clear that every eigenfunction will have the form of a polynomial times $e^{-s^2/2}$

$$f_n = P_n(s) e^{-s^2/2}$$

The polynomials, P_0, P_1, P_2, $\cdots$ are called the *Hermite polynomials* and the function $e^{-s^2/2}$ is called the *Gaussian function.*

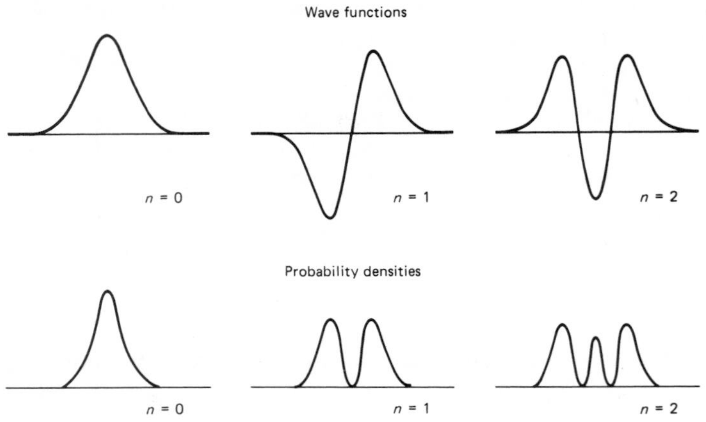

FIGURE 5.12 A. Graphs of the first three eigenfunctions for the harmonic oscillator.
B. Graphs of the associated probability densities, $f_n^2(s)$.

5.8. THE WAVE PACKET SOLUTION

The quantum harmonic oscillator has a certain property that nicely reveals the connection between quantum and classical mechanics. We shall simply state the property without proving it. (The proof is very complicated.)

We consider the time dependent Schroedinger equation (Equation 4.23) which determines how an arbitrary initial wave function changes in time.

$$i\hbar\frac{\partial \psi(x,t)}{\partial t} = -\frac{\hbar^2}{2m}\frac{\partial^2\psi}{\partial x^2} + \tfrac{1}{2}kx^2\psi \qquad (5.47)$$

We look for a solution which, at time t = 0, is just the harmonic oscillator ground state displaced a distance, a, from the origin. That is

$$\psi(x,o) = u_0(x-a)$$

The probability density at time t = 0 is

$$P(x,o) = u_0^2(x-a)$$

What is remarkable is that, as $\psi(x,t)$ evolves in time according to Equation 5.47, the probability density keeps exactly the same shape but simply oscillates back and forth, at angular frequency ω_0, just like a classical oscillator. That is (See Figure 5.13)

$$P(x,t) = u_0^2(x - a\cos\omega_0 t) \qquad (5.48)$$

The only real difference between this solution and the predictions of classical mechanics is that, according to Equation 5.48, the position of the oscillator at any instant is indefinite by a distance of the order of the quantum length d. This example illustrates the fact that it is not appropriate to compare the solutions of classical mechanics with energy eigenstate solutions of the Schroedinger equation. Rather, they should be compared with wave packet solutions of the time dependent Schroedinger equation.

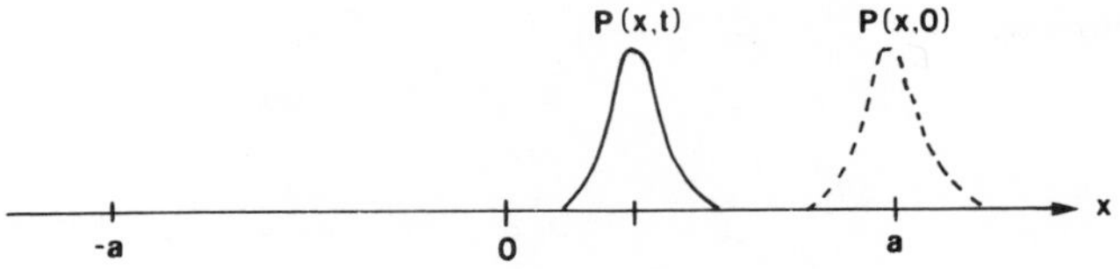

FIGURE 5.13 The shape of the probability density remains constant but the location of the maximum moves like a classical oscillator.

SUMMARY

A *stationary state* solution of the Schroedinger equation has the form

$$\psi(x,t) = u(x)\, e^{-i\omega t}$$

where the *space function* u(x) is a solution of the *Schroedinger energy equation.*

$$-\frac{\hbar^2}{2m}u''(x) + V(x)\, u(x) = E\, u(x)$$

With the definitions

$$E = \frac{\hbar^2}{2m}\epsilon$$

and

$$V(x) = \frac{\hbar^2}{2m}v(x)$$

the energy equation can be written

$$-u'' + v\,u = \epsilon\, u$$

A *step potential* is of the form

$$v(x) = \begin{Bmatrix} 0 & x < 0 \\ v_o & x > 0 \end{Bmatrix}$$

A plane wave of wave vector $k = \epsilon^{1/2}$ and amplitude A_{in}, coming in from the left, will be partially reflected and partially transmitted. The reflected wave will have wave vector $-k$ and amplitude $A_{refl} = R\, A_{in}$. The transmitted wave will have wave vector $q = (\epsilon - v_o)^{1/2}$ and amplitude $A_{trans} = T\, A_{in}$, where

$$R = \frac{k-q}{k+q}$$

and

$$T = \frac{2k}{k+q}$$

The probability that a particle will be reflected at the potential discontinuity is $P_{refl} = |R|^2$. The probability that it will be transmitted is $P_{trans} = 1 - P_{refl}$.

If a stream of particles hits a potential step that is greater than the particle energies, the particles are completely reflected. However, the wave function extends into the *classically forbidden region* where $v(x) > \epsilon$ and thus there would be a finite probability of finding a particle in that region.

If the potential v(x) returns to zero after a finite distance, a, so that the particles are incident on a finite width barrier, then they will no longer be completely reflected. If $\lambda a > 1$, where $\lambda = (v_o - \epsilon)^{1/2}$, then an approximate formula for the *transmission coefficient* of the barrier is

$$T = e^{-\lambda a}$$

The probability that any given particle will pass through the barrier rather than be reflected is equal to $T^2 = e^{-2\lambda a}$.

The wave function of a quantized bead on a circular wire of length L must satisfy the *periodic boundary condition*

$$u(x+L) = u(x)$$

The energy eigenfunctions of a particle in such a "periodic box" are of the form

$$u_n = \frac{1}{\sqrt{L}} e^{ik_n x}$$

where $k_n = \frac{2\pi}{L} n$ with $n = 0, \pm 1, \pm 2, \cdots$

The energy eigenvalues are

$$E_n = \frac{\hbar^2 k_n^2}{2m}$$

All of the energy levels, except the ground state energy ($E_0 = 0$), are degenerate, with a degeneracy equal to two. This means that there are two distinct eigenfunctions with the same energy eigenvalue for each energy eigenvalue except E_0. If the x-y plane is chosen as the plane of the wire loop with the coordinate origin at the loop center, then one finds that the z-component of angular momentum of the system is quantized with the possible values of L_z being given by

$$L_z = n\hbar \qquad n = 0, \pm 1, \pm 2, \cdots$$

The eigenfunctions for these quantized angular momentum states have a very simple form if they are written in terms of the angular variable ϕ which defines the angular position of the bead on the loop. They are

$$u_n = \frac{1}{\sqrt{2\pi}} e^{in\phi}$$

A harmonic oscillator is a one dimensional particle subject to a potential $V(x) = \frac{1}{2} k x^2$.

A harmonic oscillator may be treated by classical mechanics if its amplitude of vibration is much larger than its *quantum length*, d, given by

$$d = \sqrt{\hbar / m \omega_0}$$

where $\omega_0 = \sqrt{k/m}$ is the classical angular frequency of the oscillator.

The energy eigenvalues of a quantized harmonic oscillator are

$$E_n = (n + \tfrac{1}{2})\hbar \omega_0 \qquad n = 0, 1, 2, \cdots$$

The ground state wave function is

$$f_0 = d^{-1/2} \pi^{-1/4} e^{-s^2/2}$$

where $s = x/d$.

Further normalized energy eigenfunctions can be generated using the formula

$$f_{n+1} = \frac{1}{\sqrt{2n+2}} (s f_n(s) - f_n'(s))$$

They are of the form

$$f_n = P_n(s) e^{-s^2/2}$$

where the *Hermite polynomial*, P_n, is an nth order polynomial in s.

The time-dependent Schroedinger equation for a harmonic oscillator has a solution in the form of a moving wave packet. for that solution the probability density at time t, P(x,t), maintains a fixed shape, but the location of its maximum oscillates back and forth like a classical particle. If the position uncertainty associated with $P(x,t)$ is very small in comparison to the amplitude of vibration, then one can simply say that the "location of the particle" is oscillating back and forth. This is exactly the classical description of the motion of a harmonic oscillator.

PROBLEMS

5.1* Suppose the complex-valued function $u_1(x)$ is a solution of the Schroedinger energy equation with the energy E.

(a) Show that the function $u_2 \equiv u_1^*$ is also a solution of the equation with the same energy. (* means complex conjugate.)

(b) A function of the form

$$u(x) = a_1 u_1(x) + a_2 u_2(x) \quad (a_1 \text{ and } a_2 \text{ constants})$$

is called a linear combination of u_1 and u_2. Show that u(x) is a solution of the energy equation with energy E.

(c) Find two linear combinations that are real functions.

5.2** Suppose that u(x) is a solution of the Schroedinger energy equation with the potential V(x). Assume that $|V(x)| \leqslant V_0$ for some constant V_0 and that u(x) is continuous. By considering the integral

$$I = \int_{-\Delta}^{\Delta} u''(x)\,dx$$

and using Equation 5.3 prove that

$$\lim_{\Delta \to 0} (u'(\Delta) - u'(-\Delta)) = 0$$

which states that u′(x) is continuous at x = 0.

5.3 A beam of electrons of energy 10 eV is incident upon a potential step of magnitude 5 eV. (That is $V(x) = 0$ for $x < 0$ and $V(x) = 5\,\text{eV}$ for $x > 0$.) What are the reflection and transmission coefficients?

5.4** (a) Consider a one dimensional step potential of the form

$$v(x) = \begin{Bmatrix} v_o \\ 0 \end{Bmatrix} \text{for} \begin{Bmatrix} x < 0 \\ x > 0 \end{Bmatrix} \quad (v_0 > 0)$$

A flux of particles of energy $\epsilon > v_0$ is coming in from the left. Introducing reflected and transmitted waves calculate the reflection and transmission coefficients.

(b) Compare the values of R and T obtained in part (a) with those that would be obtained from a potential

$$v(x) = \begin{Bmatrix} 0 \\ -v_o \end{Bmatrix} \text{for} \begin{Bmatrix} x < 0 \\ x > 0 \end{Bmatrix}$$

using the equation of Section 5.1. Comment upon the comparison.

5.5 (a) Use the method of Section 5.3 to estimate the probability that a 10 eV electron will pass through a barrier of height 12 eV and width 2 Å.

(b) Do the same thing for a 10 eV proton.

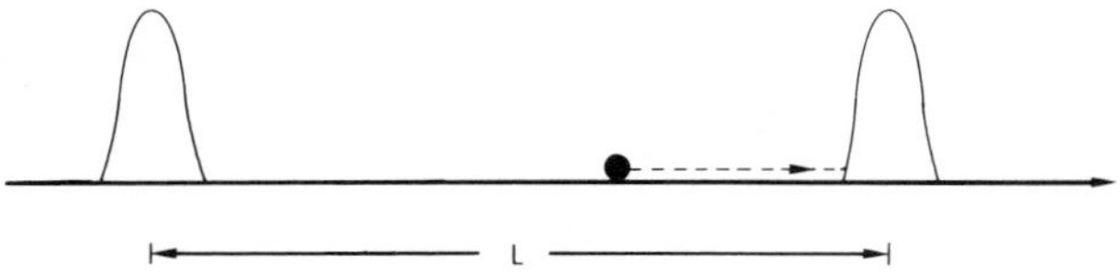

FIGURE P.5.6

5.6** A particle of speed v bounces back and forth between two barriers separated by a distance l. (See Figure P.5.6). Each time the particle hits one of the barriers it has a probability $P_T << 1$ of tunneling through and escaping. Show that, after a long period of time, t, the probability of finding the particle still trapped between the two barriers is $\exp[-(v P_T/l)t]$.

5.7 (a) A bead of mass .1 g is on a frictionless ring of diameter 1 cm. It has one unit of angular momentum. Using a classical picture of the bead's motion, estimate how many times it goes around the ring in one billion years.

(b) An electron with one unit of angular momentum is in an atomic orbit of radius a_0 (one Bohr radius). Estimate how many times it circles the nucleus during one second.

5.8* Carry out the detailed operation leading to Equations 5.39 and 5.40.

5.9* By substituting the suggested wave functions into the Schroedinger equation derive Equations 5.7 and 5.9.

5.10** This is the proof of Theorem I of Section 5.6. Assume that the function f(s) satisfies the equation $f'' = (s^2 - 2\lambda)f$. Define g(s) by $g = sf - f'$.

(a) Evaluate g' in terms of f and f' by using the differential equation satisfied by f to eliminate f''.

(b) Differentiate the result you got in part (a) and again eliminate f'' to obtain g'' in terms of f and f'.

(c) Show that the result of (b) can be written in the form of Equation 5.44.

5.11** This is the proof of Theorem II of Section 5.6. Assume the function f(s) is a solution of the equation $f'' = (s^2 - 2\lambda)f$ and that $\int f^2 ds = 1$. Let $g = sf - f'$. Using partial integration and the property that $f(-\infty) = f(\infty) = 0$, show that

(a) $\int s f f' ds = -\tfrac{1}{2}$

(b) $\int f'^2 ds = 2\lambda - \int s^2 f^2 ds$

(c) $\int g^2 ds = 2\lambda + 1$

5.12* The probability that a quantum harmonic oscillator in state f_n will be found in the interval x to x+dx is $f_n^2(x)\,dx$. If we multiply this by the potential energy at location x (namely $V(x) = \frac{1}{2}k\,x^2$) and integrate over x, we obtain the average of V(x) in a statistical sense.

$$\overline{V}_n = \int_{-\infty}^{\infty} f_n^2 V(x)\,dx$$

For the ground state show that $\overline{V}_n = \frac{1}{2}E_n$. The result is actually true for any harmonic oscillator energy state. (Hint: Use Equation 5.41 and the Table of Integrals).

CHAPTER SIX

STATES, OPERATORS, AND MEASUREMENTS

6.1. THE NEED FOR MORE DETAIL

The combination of the Schroedinger equation and the probabilistic interpretation given to its solution, $\psi(x,t)$, is in many respects not adequate as a complete physical theory. We are not saying that anything presented in the last two chapters is wrong. It is simply not enough. To illustrate the inadequacy of our present description of quantum theory, let us suppose we have a system that is well described by the harmonic oscillator wave equation and we know that the system is in its ground state. According to our present interpretation of the theory we would have two pieces of information about the system. We would know its energy exactly, and we would know the probability of finding the particle within any distance interval. But suppose we measure the momentum of the particle. What does the theory predict about the results of such a measurement? Since the wave function is not a plane wave, there is no wave vector that we can use in the de Broglie relation to predict the momentum. The same can be said about a measurement of the kinetic energy alone, the particle velocity, or many other potentially measurable characteristics of the system. Does the particle have a well defined kinetic energy, momentum, etc., and if it does, then what mathematical elements in the theory represent those properties? In this chapter we want to complete the quantum theory so that such questions can be given a precise answer. When we do, we shall find that all the statements made in the last chapter regarding the energy and probability density associated with any quantum state are left unchanged. The laws of quantum theory will be presented in the form of five basic postulates. Their ultimate justification is that they agree with experimental observations.

6.2. STATES

Although we eventually want to state the laws of quantum theory in a completely general way, so they can be applied to all systems, it will help to avoid confusion if we first restrict ourselves to a definite simple system, namely a system of one particle in one dimension. We assume that the particle is subject to a force which depends on its location and that the force can be derived from a potential function $V(x)$.

In classical mechanics, in order to completely specify the physical state of such a single particle system (at a given instant of time) we would only need two numbers, x and v, the particle's position and velocity. Given those two numbers we could then calculate any other physically observable characteristic of the system, such as its kinetic energy, its momentum, or its potential energy.

Our first postulate of quantum theory concerns the question of how we mathematically describe the physical state of a quantum system.

POSTULATE 1

Any physical state of a quantum system, at one instant in time, is described by a normalized wave function. Also, every normalized wave function describes some possible physical state.

A normalized wave function is one for which*

$$\int_{-\infty}^{\infty} \psi^*(x)\psi(x)\, dx = 1$$

QUESTION: Is normalization the only mathematical requirement on a wave function in order that it represent a physical state?

ANSWER: Not really. In order for the Schroedinger equation to make any sense the wave function has to have a second derivative (which automatically implies that it is continuous). But we are going to avoid such mathematical technicalities. If we included all the mathematical details associated with quantum theory, this book would resemble a text in functional analysis.

Specifying the wave function defines the state of the system as precisely as is possible according to the quantum theory. For example, the statement that a harmonic oscillator is in its ground state *and* has momentum p is meaningless in the quantum theory just as the statement that a particle has zero velocity *and* positive kinetic energy is meaningless in classical mechanics. Once you have given the wave function of the system, you cannot add further constraints on the physical state of the system. If the system had momentum p, then its wave function would be a plane wave of momentum p and it could not simultaneously be a harmonic oscillator ground state. These two requirements are logically contradictory.

The wave function, ψ, gives us as much information about the physical state of the system as it is possible to obtain according to the quantum theory. We have not yet seen how to extract all that information from the wave function. In order to do so we must introduce a new mathematical idea.

*Since the state of the system usually changes in time, the wave function, ψ, is generally a function of both x and t. We are not indicating its t dependence because we shall be considering the state of the system only at a fixed instant. Later we shall discuss the time dependence.

6.3. OPERATORS

The notion of an *operator* in mathematics is a simple and general one. Anything that will convert one function into another one is called an operator. If we denote the operator as $\hat{O}$ (we write operators with a caret ^ to distinguish them from functions), then the statement that $\hat{O}$ converts the function f(x) into the function g(x) would be written as the equation

$$\hat{O}f(x) = g(x)$$

The following four examples of simple operators should make the notion clear. In each case we show what happens when the operator operates on the function, f(x) = sin x.

(1) Multiplication by a constant. $\hat{O} = c$

$$\hat{O}f(x) = c\,f(x). \qquad [\hat{O}\sin x = c\sin x]$$

(2) Multiplication by a function. $\hat{O} = u(x)$

$$\hat{O}f(x) = u(x)\,f(x) \qquad [\hat{O}\sin x = u(x)\sin x]$$

(3) The first derivative. $\hat{D}_1 = \frac{d}{dx}$

$$\hat{D}_1 f(x) = \frac{df(x)}{dx} \qquad [\hat{D}_1 \sin x = \cos x]$$

(4) The second derivative. $\hat{D}_2 = \frac{d^2}{dx}$

$$\hat{D}_2 f(x) = \frac{d^2 f(x)}{dx^2} \qquad [\hat{D}_2 \sin x = -\sin x]$$

Notice that $\hat{D}_2$ has the same effect on any function as $\hat{D}_1$ acting twice. That is

$$\hat{D}_2 f(x) = \hat{D}_1\,\hat{D}_1 f(x)$$

This fact is usually expressed as an operator equation.

$$\hat{D}_2 = \hat{D}_1 \cdot \hat{D}_1 = \hat{D}_1^2$$

6.4. OPERATOR EIGENVALUES

One operator we have already used, without calling it such, is the *energy operator*

$$\hat{E} = -\frac{\hbar^2}{2m}\frac{d^2}{dx^2} + V(x)$$

The effect of the energy operator on any function, f(x) is

$$\hat{E}f(x) = -\frac{\hbar^2}{2m}\frac{d^2 f(x)}{dx^2} + V(x)\,f(x)$$

Comparing this with Equation 4.25 we see that Schroedinger energy equation can be written in the form

$$\hat{E}u(x) = E\,u(x) \tag{6.1}$$

(Remember $\hat{E}$ is an operator but E is a number.)

As we mentioned before, this equation will have normalized solutions only if the constant, E, is one of the energy eigenvalues, E_1, E_2, E_3, $\cdots$. The numerical values of the E_n's depend on the form of the potential, V(x). For each one of the energy eigenvalues, E_n, there is a normalized energy eigenfunction, $u_n(x)$, which satisfies Equation 6.1

$$\hat{E}\,u_n(x) = E_n\,u_n(x)$$

If we make a measurement of the energy of a one particle quantum system with potential V(x), the result of the energy measurement will always be one of the energy eigenvalues. We shall never find that the system has some energy in between two energy eigenvalues.

Also, if the system has a definite energy, E_n, then its wave function is proportional to the corresponding energy eigenfunction. The proportionality factor is $e^{-i\omega_n t}$.

$$\psi = e^{-i\omega_n t}\,u_n(x) \tag{6.2}$$

6.5. THE MOMENTUM OPERATOR

Now let us see that we can find another operator, $\hat{p}$, that is related to the momentum of the particle in the same way as $\hat{E}$ is related to the energy. We know already what the states of definite momentum are. They are the plane waves

$$\psi_p = e^{i(kx - \omega t)}$$

where the corresponding momentum is $p = \hbar k$. We can write this in the same form as Equation 6.2 if we let

$$u_p(x) = e^{ikx}$$

In order for everything to be analogous to the energy case, this function should be a solution of a momentum eigenvalue equation of the same form as Equation 6.1. That is

$$\hat{p}\,u_p(x) = p\,u_p(x) \tag{6.3}$$

The simplest operator that satisfies Equation 6.3 is

$$\hat{p} = -i\hbar\frac{\partial}{\partial x}$$

Then

$$\hat{p}\,u_p(x) = -i\hbar\frac{\partial\, e^{ikx}}{\partial x} = \hbar k\,u_p(x)$$

Since there is a plane wave with every possible value of k, the eigenvalue of the momentum operator, p, can be any number from $-\infty$ to ∞.

Before we go any further let us notice a certain pleasing internal consistency in what we have already developed. If we take the classical energy formula

$$E = \frac{1}{2m}\,p^2 + V(x)$$

and replace E and p by the corresponding quantum operators, we get the equation

$$\hat{E} = \frac{1}{2m}\,\hat{p}^2 + V(x)$$

But

$$\hat{p}^2 = (-i\hbar\frac{\partial}{\partial x})(-i\hbar\frac{\partial}{\partial x}) = -\hbar^2\frac{\partial^2}{\partial x^2}$$

which gives

$$\hat{E} = -\frac{\hbar^2}{2m}\frac{\partial^2}{\partial x^2} + V(x)$$

in agreement with the Schroedinger equation.

The relationship between the eigenvalues of the energy and momentum operators and the measured values of those physical quantities also holds for other physical characteristics of the system. (We shall give a more precise statement of the relationship shortly.) For the angular momentum of the system there is an angular momentum operator, $\hat{L}$, whose eigenvalues give the results of making an angular momentum measurement. For the kinetic energy there is a kinetic energy operator, $\hat{K}$, whose eigenvalues are the possible results of making a kinetic energy measurement, etc.

For any operator $\hat{O}$ the eigenvalues of the operator are determined by solving an equation of the form of Equation 6.1

$$\hat{O} f_\lambda(x) = \lambda f_\lambda(x)$$

If, for a given constant, λ, a normalized solution, $f_\lambda(x)$, exists, then λ is said to be an eigenvalue of the operator $\hat{O}$ and the function $f_\lambda(x)$ is called the corresponding eigenfunction.

6.6. MEASUREMENTS

We are now in a position to state the next of the basic rules or postulates of quantum theory. (Rule 1 was stated in Section 6.2).

POSTULATE 2

For each observable of a system (each measurable characteristic, such as the kinetic energy, the x-coordinate, the momentum, etc.) there is some operator whose eigenvalues are the only possible results of a measurement of that observable. We say that that observable is represented by the operator.

For instance, the energy of a one dimensional harmonic oscillator is represented by the operator

$$\hat{E} = -\frac{\hbar^2}{2m}\frac{d^2}{dx^2} + \frac{k}{2}x^2 \qquad (6.4)$$

For this system, Postulate 2 states that any exact measurement of the energy will yield one of the eigenvalues of the equation

$$-\frac{\hbar^2}{2m}f''(x) + \frac{k}{2}x^2 f(x) = E f(x)$$

But this is the harmonic oscillator eigenvalue equation which we considered in the last chapter. Its eigenvalues are

$$E_n = (n + \tfrac{1}{2})\hbar\omega_0$$

Therefore, Postulate 2 states that these are the only values we would ever get in a measurement of the energy. It does not say which of the possible energy values one would actually get in any particular measurement. That question is answered by the later postulates.

6.7. THE POSITION OPERATOR

One obvious observable for a single particle system is the location of the particle. What is the operator that represents the x-coordinate of a particle in one dimension? One way of discovering the form of the operator, $\hat{x}$, is to begin with the classical formula for the energy of a harmonic oscillator and replace E, p, and x by the corresponding operators. We then get

$$\hat{E} = \tfrac{1}{2} m \hat{p}^2 + \frac{k}{2} \hat{x}^2$$

If this equation is to agree with Equation 6.4 we must have

$$\hat{x} = x$$

That is, the operator $\hat{x}$ just multiplies any function by the variable x. Because the operator, $\hat{x}$, is the same as the variable, x, we shall not bother to write it with a caret above it.

QUESTION: What are the eigenfunctions of the operator x?

ANSWER: One of the little annoyances encountered in learning quantum theory is the fact that the two most fundamental operators, x, and $\hat{p}$, are both mathematically troublesome. We have already mentioned that a perfect plane wave, which is the only exact solution of the momentum eigenvalue equation, cannot be normalized. However, by using very extended, but still finite, wave packets, we can approach, as closely as we like, to a solution of the momentum eigenvalue equation with a normalized wave function. We have a similar situation with the operator, $\hat{x}$. A solution to the coordinate eigenvalue equation would be the function, $f_{x_0}(x)$, with the property that, for some constant, x_0,

$$x f_{x_0}(x) = x_0 f_{x_0}(x) \tag{6.4}$$

which could be written as

$$(x - x_0) f_{x_0}(x) = 0$$

The only function that has that property is one that is zero everywhere except at $x = x_0$ (See Figure 6.1(a)). For such a function $\int |f_{x_0}|^2 dx = 0$, which means that $f_{x_0}(x)$ cannot be normalized by multiplying it by any finite normalization constant. However, we can approach arbitrarily closely to a solution of Equation 6.4 with a function of the form shown in Figure 6.1(b). Although it is not at all simple, it is in fact possible to make mathematical sense out of a function which is the limit, as Δx goes to zero, of the function shown in Figure 6.1(b). Those limit functions are the eigenfunctions of x.

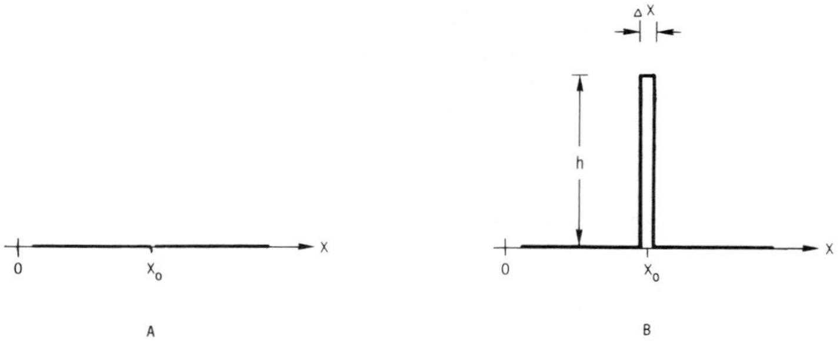

FIGURE 6.1 A. A function for which $(x-x_0)f(x) = 0$ for all x. However, $\int f^2(x)\,dx = 0$ and thus the function cannot be normalized.
B. A function for which $(x-x_0)f(x) \approx 0$ and for which $\int f^2(x)\,dx = 1$ if we choose $h = 1/\sqrt{\Delta x}$.

6.8. OTHER OBSERVABLES

In three dimensions there are three coordinates whose corresponding operators are just the multiplicative operators, x, y, and z. There are also three components of momentum whose operators are the obvious analogues of $\hat{p}_x$.

$$\hat{p}_x = -i\hbar\frac{\partial}{\partial x}$$

$$\hat{p}_y = -i\hbar\frac{\partial}{\partial y}$$

and

$$\hat{p}_z = -i\hbar\frac{\partial}{\partial z}$$

Starting with these six operators we can construct the operator that represents any other observable simply by substituting into the classical formula for the observable. For example, the z-component of angular momentum is given by the classical formula

$$L_z = (\mathbf{r}\times\mathbf{p})_z = x\,p_y - y\,p_x$$

Therefore the quantum operator that represents the z-component of angular momentum is

$$\hat{L}_z = -i\hbar\left(x\frac{\partial}{\partial y} - y\frac{\partial}{\partial x}\right)$$

Operators for the other components of angular momentum could be determined in the same way.

The classical expression for the energy of a single particle in a three-dimensional potential field, V(x,y,z), is

$$E = \frac{p_x^2 + p_y^2 + p_z^2}{2m} + V(x,y,z)$$

Making the above-mentioned operator substitutions gives the energy operator for the corresponding quantum system.

$$\hat{E} = -\frac{\hbar^2}{2m}\left(\frac{\partial^2}{\partial x^2} + \frac{\partial^2}{\partial y^2} + \frac{\partial^2}{\partial z^2}\right) + V(x,y,z) \qquad (6.5)$$

This allows us to determine the Schroedinger energy equation for a three-dimensional system. Its general form is $\hat{E}u = Eu$. Using the energy operator given in Equation 6.5 produces the following differential equation.

$$-\frac{\hbar^2}{2m}\left(\frac{\partial^2 u}{\partial x^2}+\frac{\partial^2 u}{\partial y^2}+\frac{\partial^2 u}{\partial z^2}\right)+V(x,y,z)\,u = Eu \qquad (6.6)$$

In order for this equation to make any sense it is clear that the energy eigenfunction, u, must be a function of all three space coordinates. That is $u = u(x,y,z)$. The physical interpretation of the wave function is that $|u(x,y,z)|^2 dx\,dy\,dz$ is the probability of finding the particle in an infinitesimal volume $dV = dx\,dy\,dz$ located at (x,y,z).

In the next chapter we shall consider this equation for the case of the hydrogen atom in which the potential V(x,y,z) is the Coulomb potential.

$$V(x,y,z) = -\frac{ke^2}{\sqrt{x^2+y^2+z^2}}$$

6.9. THE CALCULATION OF PROBABILITIES

So far we have two rules of quantum theory. (1) The state of a system is described by a wave function. (2) If we measure the value of some observable $\hat{O}$, we shall always get one of the eigenvalues of $\hat{O}$ as a result of our measurement.

But any observable has a great many different eigenvalues. Suppose we know that the system is in state $\psi(x)$ and we make a measurement of the observable $\hat{O}$. What is the probability that we shall get some particular eigenvalue, λ? In phrasing the question this way we are recognizing the fact that quantum theory does not allow us to predict with certainty the results of all possible measurements, even if we know the wave function of the system. We can only predict the probability of getting the different possible values. There is only one operator for which we can presently answer this question; the position operator, x. We know that, if the wave function is $\psi(x)$, then the probability of finding the particle coordinate within a very small interval, Δx, centered at x_0, is $\psi^*(x_0)\,\psi(x_0)\,\Delta x$. Looking at the position eigenfunction given in Figure 6.1(b) it is easy to verify that, in the limit of very small Δx,

$$\psi^*(x_0)\,\psi(x_0)\,\Delta x = \left|\int f_{x_0}(x)\,\psi^*(x)\,dx\right|^2$$

Thus the probability is given by the square of the product integral of the eigenfunction and the state function. This is the clue to the general principle stated in Postulate 3.

POSTULATE 3

If we have a system in a state, $\psi(x)$, and we make a measurement of an observable, $\hat{O}$, then the probability of getting the eigenvalue, λ as a result of the measurement is

$$P_\lambda = \left|\int f_\lambda\,\psi^*\,dx\right|^2$$

where $f_\lambda(x)$ is the eigenfunction of $\hat{O}$ with eigenvalue, λ.

QUESTION: Suppose a particle in a one dimensional box of length $L = \pi$ is in its ground state. If at $t = 0$ a measurement of the momentum of the particle is made, what is the probability of getting the value $p = \hbar k$?

ANSWER: The unnormalized momentum eigenfunction and the unnormalized ground state wave function are respectively $f_p(x) = e^{ikx}$ and $\psi(x) = \sin x$. Since we are using unnormalized functions we will get only the relative probabilities of obtaining different values of p in the measurement. The integral we need to evaluate is

$$I = \int_0^{\pi} f_p(x)\,\psi^*(x)\,dx = \int_0^{\pi} e^{ikx} \sin x\,dx = \frac{1+e^{i\pi k}}{1-k^2}$$

P(p) is equal to I^*I times some normalization constant, c, that is independent of k. This gives

$$P(p) = c\,\frac{1+\cos \pi k}{(1-k^2)^2}$$

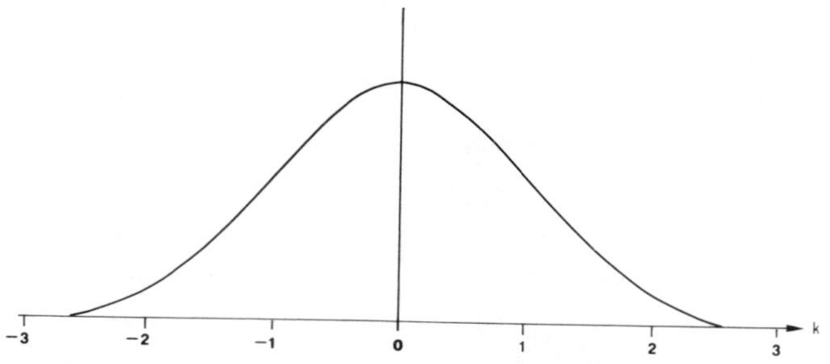

FIGURE 6.2 A graph of the momentum probability function of a particle in a box of length π.

QUESTION: When can we predict with certainty that the result of measuring an observable, $\hat{O}$, will be a particular eigenvalue, λ?

ANSWER: In order for us to be certain of getting the eigenvalue, λ, the probability, P_λ, would have to be equal to one. That is

$$\left| \int_{-\infty}^{\infty} f_\lambda\,\psi^*\,dx \right|^2 = 1$$

It can be shown (See Problem 6.8) that this can be true only if $\psi(x)$ is some constant times $f_\lambda(x)$. Thus, one can be sure of the results of a particular measurement if, and only if, the state of the system at the time of the measurement is an eigenfunction of the measured observable.

The most common example is a measurement of the energy of a system whose wave function is $e^{-i\omega_n t}$ times an energy eigenstate. That is

$$\psi(x,t) = e^{-i\omega_n t}\,u_n(x)$$

In that case we are guaranteed to get the energy value

$$E_n = \hbar\omega_n$$

QUESTION: What exactly does it mean to say that for a system in state ψ we have a probability P_λ of getting the value λ for some observable? In other words, how could we carry out an experimental test to determine whether this law was satisfied? We certainly could not set up the system in state ψ and then make a measurement of the probabilities in order to see if they had the predicted values. Probabilities are just statistical ideas; they are not observables of the system. If we measure the observable, $\hat{O}$, and actually get some value, say λ, does that verify the law? If λ were not one of the eigenvalues of $\hat{O}$, then the experiment would certainly show that Postulate 2 is false. If λ were an eigenvalue but the probability predicted for λ by Postulate 3 were zero, then the experiment would again show that the rules were false. But suppose λ were an eigenvalue and the probability predicted for λ by Postulate 3 were .002. With such a small probability we might be somewhat surprised to actually get the value λ but we cannot really say that we have proven that the theory is wrong. How can we experimentally determine whether the rules are true or false?

ANSWER: Fortunately, the answer is shorter than the question. We must begin with a very large number, N, of identical systems, all of which are in the same state ψ. Make a measurement of $\hat{O}$ on every system. If any of the measurements does not yield an eigenvalue, we have immediately repudiated the theory. Suppose N_1 measurements give λ_1, N_2 measurements give λ_2, etc. Then the theory predicts that, for extremely large N, N_1/N will be extremely close to P_{λ_1}, N_2/N will be extremely close to P_{λ_2}, and so forth. By making N sufficiently large we can test the statistical predictions to any accuracy we choose.

6.10. SETTING UP A QUANTUM STATE

The first three rules of quantum theory allow us to make predictions about the results of measuring any observable of a system which is in a known state $\psi(x)$. We now have to mention how we can prepare a system in a known state and, once we have done so, what happens to the state of the system if it is left alone for a time t. We need two more postulates, numbers (4) and (5), which will then complete all our fundamental laws of quantum theory.

POSTULATE 4

If, at time t_o, a measurement of $\hat{O}$ gives the value, λ, then immediately following the measurement the wave function is f_λ. That is

$$\psi(x,t_o) = f_\lambda(x)$$

This is certainly the strangest of all the laws of quantum theory. To get an idea of the peculiar consequences of this postulate, let us imagine that we have produced a single electron in a very extended plane wave state. Within the region occupied by this large wave packet we have set up a number of very small electron detectors. At time t_o one of our detectors detects the electron. Instantly (faster than the speed of light) the wave function changes to a position eigenfunction at the location of that detector. We cannot use the Schroedinger equation, nor any other equation, to describe the instantaneous change of the

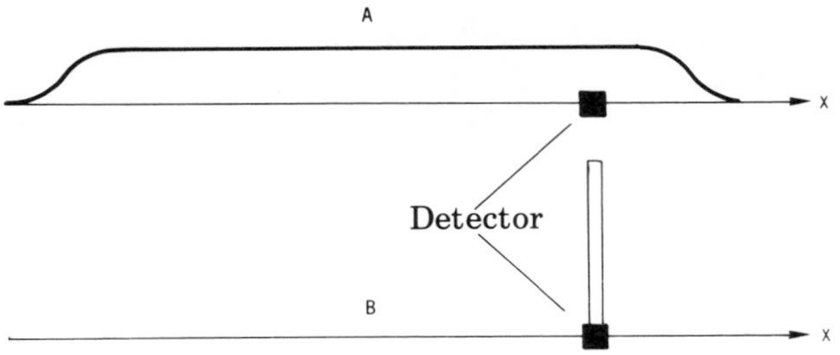

FIGURE 6.3 A. The position probability function before the particle is detected.
B. The position probability function immediately after the particle is detected.

wave function from a momentum eigenfunction to a position eigenfunction. At the time it was proposed, many people, the most influential among them being Albert Einstein, objected vehemently to this postulate on the grounds that it violated the principle of relativity, according to which physical effects cannot propagate from one region to another at any speed greater than c. However, these objections erroneously treat the wave function as a measurable physical field. When the wave function is instead interpreted as a probability function, then these arguments, stemming from the principle of relativity, lose their force because all probability functions have this characteristic of instantaneous propagation as the following example illustrates. Suppose we draw a single card without looking at it, from a well-shuffled deck. We put the card into an envelope and deliver it to an astronaut, who transports it to the moon. The probability That the card on the moon is the ace of spades is then 1/52 and remains constant as long as we do not look at the rest of the deck. However, if here on earth we turn over the top card of the deck and see the ace of spades, then that probability instantly becomes zero. Not because some wave traveled from the deck to the moon in order to affect the card there, but simply because we know that there is only one ace of spades and it cannot be here and there simultaneously. Relating this to the electron experiment that we are discussing, we see that before the electron was detected it had a probability density which could be calculated using the Schroedinger equation and was spread through a large region of space. But, as soon as the electron is detected and therefore known to be at one place, the probability that it can also be any place else, immediately becomes zero. Postulate 4 says that what is true for position measurements is also true for measurements of other observables. If the wave function of a system is not a momentum eigenfunction, then there is a certain calculable probability of getting any particular value of momentum. But, if we measure the momentum and find it to be p_o, then the probability that it is any other value goes instantly to zero. As we have seen in Section 6.9, this would mean that the wave function must be a momentum eigenstate with eigenvalue p_o.

Postulate 4 is valuable in that it tells us how to set up a system so that, at the initial instant we know its wave function. At $t = 0$ we simply make an accurate measurement on the system. Its initial wave function is then the eigenfunction corresponding to the results of our measurement. Postulate 5 below tells us how that wave function changes in time, once it has been set up.

POSTULATE 5

A system which is left undisturbed has a state $\psi(x,t)$ which satisfies the Schroedinger equation

$$i\hbar \frac{\partial \psi}{\partial t} = \hat{E}\psi$$

Being "left undisturbed" means that no measurements are made on the system. It does not mean that there are no external forces acting on the particles of the system. There may be external forces acting on the system, but those forces must be taken into account in determining the energy operator, $\hat{E}$. The most common example is an external force which is derivable from a potential function, V. The potential function of the external force is then one of the terms in the energy operator.

For the rest of this chapter we shall examine how these principles of quantum theory are used in designing and analyzing typical laboratory experiments. We shall first analyze a somewhat abstract general experiment, which we call the *standard quantum mechanical experiment.* We will then look at a typical real experiment.

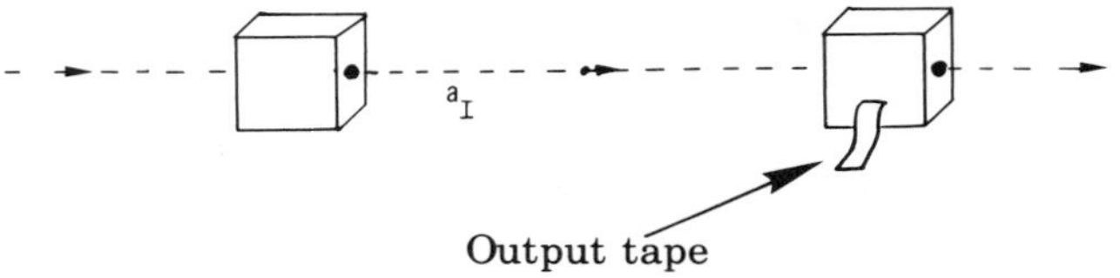

FIGURE 6.4 The setup for a Standard Quantum Mechanical Experiment. The particles are stopped in the first box unless the measured value of $\hat{A}$ is a_I. In the second box the value of $\hat{B}$ is measured and the result is printed on an output tape (OT).

6.11. THE STANDARD QUANTUM MECHANICAL EXPERIMENT

In Figure 6.4 we see a very dilute, low velocity, beam of particles (they may be electrons, atoms, molecules, etc.) coming in from some unspecified apparatus outside the picture on the left. We have chosen to describe a beam experiment simply because, with a beam of moving particles, things that are done to the particles at different times must also happen at different places and therefore we can scan the temporal sequence of the experiment by moving our eye across the diagram from left to right. The density of particles in the beam is assumed to be so low that particles pass through the measuring apparatus one at a time. As any particle passes through the apparatus, the following sequence of events takes place.

(1) The particle passes through the first measuring device at some time we will call $t = 0$. This device makes a measurement of some physical characteristic of the particle, such as its energy or angular momentum. The physical characteristic measured is represented by the operator $\hat{A}$, which has eigenvalues a_1, a_2, a_3 $\cdots$. If the value obtained in the measurement is not equal to some previously decided particular eigenvalue, a_I, the molecule is absorbed in the first device and gets no further. Thus the first device produces a dilute beam of particles all of which are in some known initial state.

(2) Once the first measurement is taken, the particle is not disturbed as it passes from the first measuring device to the second.

(3) When the particle arrives in the second measuring device, at time $t = T$, a measurement is made of some observable represented by an operator $\hat{B}$. A record is kept of which of the eigenvalues, $b_1, b_2, b_3, \cdots$, is actually obtained in the measurement.

After the sequence (1), (2), (3) is completed for a very large number of particles the data is collected in the form of a table in which is listed the number of times each eigenvalue,b_n, has been obtained.

EIGENVALUE	b_1	b_2	b_3	$\cdots$
NO. OF MEAS. IN WHICH IT WAS OBTAINED	784	497	48	$\cdots$
MEASURED PROBABILITY	.261	.165	.016	$\cdots$

Each entry in the table (784, 497, 48, etc.) is then divided by the total number of particles that have passed completely through the apparatus. ($N = 784 + 497 + 48 + \cdots$) in order to get the *measured probabilities* of obtaining the values $b_1, b_2, b_3, \cdots$ in a measurement of $\hat{B}$ at time T when it is known that a measurement of $\hat{A}$ at time $t = 0$ yielded a_I. This set of measured probabilities constitutes the basic experimental data which is to be compared with the results of a theoretical calculation. The calculation is carried out in the following way.

(1) The eigenvalue equation for the operator $\hat{A}$ is solved to obtain the set of eigenvalues $a_1, a_2, \cdots$ and the associated eigenfunctions $f_1(x), f_2(x), \cdots$, where

$$\hat{A} f_n(x) = a_n f_n(x)$$

and

$$\int |f_n|^2 dx = 1$$

(2) The time-dependent Schroedinger equation,

$$i\hbar \frac{\partial \psi(x,t)}{\partial t} = \hat{E}\psi(x,t)$$

is solved, using the *initial condition* that

$$\psi(x,o) = f_I(x)$$

where $f_I(x)$ is the eigenfunction corresponding to the preset eigenvalue a_I. (Remember that particles were allowed to pass through the first measuring device only if their measured value of $\hat{A}$ was a_I.) The state of the system at time $t = T$ (the time of the second measurement) is given by $\psi(x,T)$.

(3) The eigenvalue equation for the operator $\hat{B}$ is solved for the eigenvalues $b_1, b_2, \cdots$, and corresponding eigenfunctions, $g_1(x), g_2(x), \cdots$

(4) The *calculated probabilities* of obtaining the values $b_1, b_2, \cdots$ in the second measurement are given by

$$P_n = |\int g_n(x)\psi^*(x,T)\, dx|^2$$

These calculated probabilities are then compared with the measured probabilities.

In brief, the standard quantum mechanical experiment consists of three parts. (1) The system is prepared in a known quantum state. (2) The system is left undisturbed for a time during which its wave function changes according to Schroedinger's equation. (3) A final measurement is made in order to obtain some information about the wave function of the system.

Let us now see that a realistically described experiment can be interpreted quite directly and naturally within this scheme. As our example we take high energy electron scattering by nuclei. In a series of experiments similar to the one we shall describe, Robert Hofstadter determined the electrical charge density within nuclei.

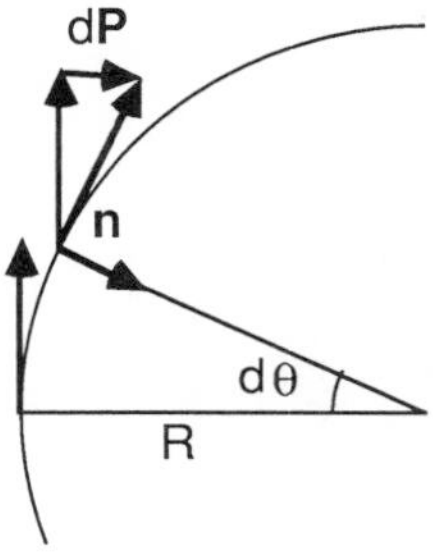

FIGURE 6.5
Definitions of **n** and $d\theta$.

6.12. ELECTRON SCATTERING BY NUCLEI

Electrons are accelerated to high energies within a linear accelerator. The beam of electrons emerging from the accelerator has some distribution of energies and momenta. In order to select a precise initial momentum, the beam is sent through a *magnetic spectrometer*. In this device the beam passes through a region of carefully controlled uniform magnetic field. In computing the motion of high energy electrons in such a large sized device it is appropriate to use classical relativistic dynamics. During a short time interval an electron's momentum changes by an amount **dp** given by

$$d\mathbf{P} = \mathbf{n}\,P\,d\theta$$

where **n** and $d\theta$ are defined in Figure 6.5. Setting the magnetic force, $\mathbf{F} = -e\,\mathbf{v}\times\mathbf{B}$, equal to $\frac{d\mathbf{P}}{dt}$ gives

$$-e\,\mathbf{v}\times\mathbf{B} = \mathbf{n}\,P\frac{d\theta}{dt} = \frac{\mathbf{n}\,P\,v}{R}$$

The direction of the vectors on both ends of this equation are the same, as they should be. The magnitude of the vectors must also be equal. This yields the relation

$$e\,v\,B = \frac{P\,v}{R}$$

or

$$R = \frac{P}{e\,B}$$

The radius of the circular path travelled by any electron is directly proportional to the relativistic momentum of the electron. Thus the electrons which pass through the exit collimator slit of the spectrometer are of precisely known momentum. The width of the exit slit is much too large to cause any noticeable diffraction effects on the very short wavelength electron beam. The magnetic

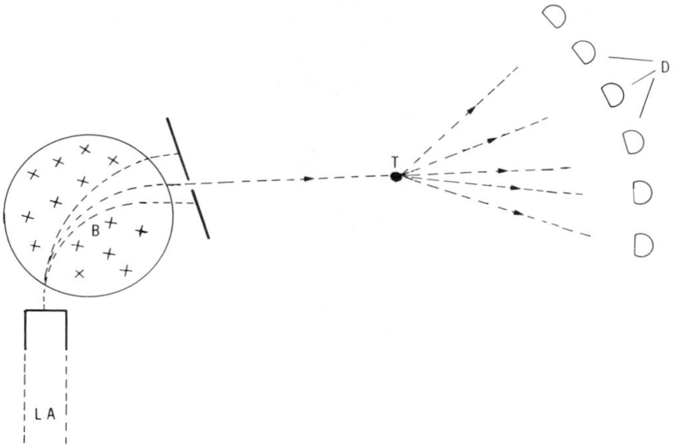

FIGURE 6.6 The setup for studying electron scattering by nuclei. Electrons from a linear accelerator (LA) pass through a magnetic field (**B**) which separates them according to their momentum. Electrons of a certain momentum are allowed to propagate to the target area where they interact with nuclei and are scattered. The outgoing electrons are detected by an array of detectors (D).

spectrometer is obviously a momentum measuring device and, as required by Postulate (4), an electron coming out of the magnetic spectrometer is in a momentum eigenstate. Its wave function is a wave packet whose size is extremely large in comparison to its wavelength. The electron beam is sufficiently dilute that the interactions between electrons in the beam are completely negligible. The target nuclei are also few enough in number and well enough separated that the probability of an electron interacting with more than one nucleus is negligibly small. (This is really a bit of an oversimplification. The effects of multiple interactions are usually small but not negligible. They must be estimated and eliminated from the final data.) We can view the experiment as a large number of repeated experiments in which a single electron interacts with the electric charge of a single nucleus. Our *system* therefore consists of a single electron in the electric potential field created by a single nucleus. We know the initial wave function of the electron. The system is undisturbed while the electron propagates toward the nucleus, while it interacts with the nucleus, and while the scattered electron wave propagates away from the target region in all directions. At a large distance from the target is a large array of electron counters to determine the energy and direction of the scattered electrons. The motive in carrying out such experiments these days is not to test the laws of quantum theory, although that is done as an accidental byproduct. The purpose of such experiments is to determine the form of the energy operator $\hat{E}$ which describes the electron-nucleus system. The energy operator for such a system is much more complicated than the simple operators we have been working with.

Nonetheless, most of the details of the operator could be determined by theoretical analysis before the experiments were performed. Only the values of certain parameters in the operator had to be experimentally determined. Those parameters are determined in the following way. The electron beam is kept on until a large number, N, of electrons have come through the exit slit of the spectrometer. The number of electrons counted in each counter is recorded. Let us say N_i electrons have been detected in the i-th counter. Now some values are assumed for the unknown parameters in the energy operator. A calculation is made of the time dependent wave function ψ by using the Schroedinger equation with the operator $\hat{E}$ which contains the assumed value of the parameters. In the calculation the initial state of the wave function is a momentum eigenfunction coming toward the nucleus from the **collimator** slit. From the Schroedinger equation it is possible to calculate the probability, P_i, that the electron would travel away from the nucleus in the direction of the i-th detector. The set of numbers P_i are compared with the set of numbers N_i/N. The assumed values of the parameters in the calculation are adjusted until the set of calculated probabilities matches the observed number of counts.

The three stages of the standard quantum mechanical experiment, namely (1) selecting the initial state by means of a measuring device, (2) allowing the system to freely propagate, and (3) making a measurement by means of a second measuring device, are easily identified in this experiment. The quantities, $\hat{A}$ and $\hat{B}$, which are observed in this experiment, are the momentum of the electron before and after scattering by the nucleus.

6.13. COMMUTATION OF OPERATORS

If a function, f(x), is multiplied by any two numbers, a and b, the order in which the multiplication is done has no effect on the result. That is

$$a(b\,f(x)) = b(a\,f(x))$$

This property is described by saying that any two numbers *commute.* Unfortunately this convenient commutation property is not shared by operators. For most pairs of operators, $\hat{A}$ and $\hat{B}$

$$\hat{A}(\hat{B}\,f(x)) \neq \hat{B}(\hat{A}\,f(x))$$

An important example of a pair of noncommuting operators is the position and momentum operators. For any function, f(x),

$$x\,\hat{p}\,f(x) = -i\hbar x\,f'(x)$$

but

$$\hat{p}\,x\,f(x) = -i\hbar(x\,f(x))' = -i\hbar x\,f'(x) - i\hbar f(x)$$

Thus

$$\hat{p}\,x\,f(x) = x\,\hat{p}\,f(x) - i\hbar f(x) \qquad (6.7)$$

It is this noncommutability of the two operators x and $\hat{p}$ that is the source of the Heisenberg uncertainty principle involving position and momentum measurements. We can see the relationship between uncertainty and commutability as follows. Suppose there were some wave function, ψ, for which the position had the completely predictable value, x_0, and the momentum also had a completely predictable value, p_o. We have seen before that an observable can have an exactly predictable value only if the wave function of the system is an eigenfunction of the operator representing the observable. Thus ψ would have to be an eigenfunction of both operators x and $\hat{p}$. That is

$$x\psi = x_o\psi \quad \text{and} \quad \hat{p}\psi = p_o\psi$$

But then

$$x\hat{p}\psi = x p_o\psi = p_o x\psi = p_o x_o\psi$$

and

$$\hat{p}x\psi = \hat{p}x_o\psi = x_o\hat{p}\psi = x_o p_o\psi$$

which means that

$$x\hat{p}\psi = \hat{p}x\psi$$

in contradiction to Equation 6.7. Thus it is impossible to find a state in which the values of position and momentum are simultaneously exactly predictable. The following theorem, whose proof is omitted, shows that when two operators commute, there can be *no* uncertainly relation between the corresponding observables.

6.14. THE FUNDAMENTAL COMMUTATOR THEOREM

Suppose a is one of the eigenvalues of an operator, $\hat{A}$ and b is one of the eigenvalues of an operator, $\hat{B}$. Then a function f_{ab} is called a *simultaneous eigenfunction* of $\hat{A}$ and $\hat{B}$ if

$$\hat{A}f_{ab} = a f_{ab} \quad \textit{and} \quad \hat{B}f_{ab} = b f_{ab}$$

The fundamental commutator theorem is:

THEOREM: If $\hat{A}\hat{B}f = \hat{B}\hat{A}f$ for every function, f, then all of the eigenfunctions of $\hat{A}$ can be simultaneously **eigenfunctions** of $\hat{B}$ and vice versa.

Because of this theorem it is possible to simultaneously specify the values of the two physical observables represented by a pair of commuting operators. The most frequent application of this theorem is in cases in which one of the operators involved is the energy operator. We say that the energy is *conserved* for a quantum mechanical system because, if the system is put into an energy eigenstate, ψ_E, at time $t = t_o$ and then left undisturbed for any length of time, it will remain in the energy eigenstate and therefore the energy of the system will remain constant. We now want to show that any observable whose operator commutes with $\hat{E}$ is also conserved. Suppose $\hat{A}$ commutes with $\hat{E}$. Then the energy eigenstates are also eigenfunctions of $\hat{A}$. They can be indexed by the two eigenvalues, E and a

$$\hat{E}\psi_{Ea} = E\psi_{Ea} \quad \text{and} \quad \hat{A}\psi_{Ea} = a\psi_{Ea}$$

But we know that, if the system is initially in state ψ_{Ea} and is not disturbed, then it will remain in that state. Thus the value of the observable, $\hat{A}$, will also remain constant.

All conserved quantities (such as the total momentum of an isolated system, or the angular momentum of a particle in a central force field) are represented by operators that commute with $\hat{E}$.

SUMMARY

Quantum mechanics can be based on five basic postulates.

(1) Any physical state of a quantum system is described by a normalized wave function, and every normalized wave function describes some possible physical state.

(2) For each observable of a system there is a corresponding operator whose eigenvalues are the only possible results of an accurate measurement of that observable. That operator is said to represent the observable.

(3) If a measurement of observable, $\hat{O}$, is made on a system in state, $\psi(x)$, then the probability of getting the eigenvalue, λ, as a result is

$$P_\lambda = \left| \int_{-\infty}^{\infty} f_\lambda \psi^* \, dx \right|^2$$

where $\hat{O} f_\lambda = \lambda f_\lambda$.

(4) If, at time t_o, a measurement of $\hat{O}$ gives the value, λ, then immediately following the measurement the wave function is $f_\lambda(s)$. That is

$$\psi(x,t_o) = f_\lambda(x)$$

(5) During any time interval in which no measurements are made the wave function satisfies Schroedinger's equation.

$$i\hbar \frac{\partial \psi(x,t)}{\partial t} = \hat{E}\, \psi(x,t)$$

The three coordinates of a particle in three dimensions are represented by the multiplicative operators, x, y, and z.

The three components of momentum of a particle are represented by the differential operators

$$\hat{p}_x = -i\hbar \frac{\partial}{\partial x},$$

$$\hat{p}_y = -i\hbar \frac{\partial}{\partial y},$$

$$\hat{p}_z = -i\hbar \frac{\partial}{\partial z}$$

The operators representing other observables may be constructed by substituting the expressions for the six basic operators, x, y, z, $\hat{p}_x$, $\hat{p}_y$, $\hat{p}_z$, into classical formulas for the observables.

If, for some constant, λ, the equation

$$\hat{O} f_\lambda(x) = \lambda f_\lambda(x)$$

has a normalized solution, $f_\lambda(x)$, then λ is called an *eigenvalue* of the operator, $\hat{O}$, and $f_\lambda(x)$ is called the corresponding *eigenfunction.*

The results of making a measurement on a system can be predicted with certainty if, and only if, the wave function of the system is proportional to an eigenfunction of the measured observable.

In a standard quantum mechanical experiment, (1) one measurement is made to set up the system in a known quantum state, (2) the state of the system is allowed to change for a time according to Schroedinger's equation, and then (3) a second measurement is made in order to extract information about how the system has changed during that time interval.

Two operators are said to commute if $\hat{A}\hat{B}f = \hat{B}\hat{A}f$ for every function f.

If two operators commute, then they have a complete set of simultaneous eigenfunctions $f_{a_i b_j}$, where

$$\hat{A} f_{a_i b_j} = a_i f_{a_i b_j}$$

and

$$\hat{B} f_{a_i b_j} = b_j f_{a_i b_j}$$

Operators that commute with $\hat{E}$ represent conserved physical observables.

PROBLEMS

6.1 $\hat{A}$ is a *linear operator* if $\hat{A}(f(x)+g(x)) = \hat{A}f(x)+\hat{A}g(x)$ for any functions, f and g. Which of the following are linear operators?

(a) $\hat{A}f(x) = f'(x)+2f(x)$

(b) $\hat{A}f(x) = f(x)+1$

(c) $\hat{A}f(x) = f(x+a)$

6.2* The wave equation, $\dfrac{\partial^2 u}{\partial x^2} - \dfrac{1}{v^2}\dfrac{\partial^2 u}{\partial t^2} = 0$, contains the *wave operator* $\hat{W} = \dfrac{\partial^2}{\partial x^2} - \dfrac{1}{v^2}\dfrac{\partial^2}{\partial t^2}$

(a) Show that, for any function u(x,t)

$$\hat{W}u = \hat{A}^+\hat{A}^-u = \hat{A}^-\hat{A}^+u$$

where

$$\hat{A}^+ = \frac{\partial}{\partial x} + \frac{1}{v}\frac{\partial}{\partial t} \quad \text{and} \quad \hat{A}^- = \frac{\partial}{\partial x} - \frac{1}{v}\frac{\partial}{\partial t}$$

(b) Show that any function $f(x-vt)$ is a solution of the equation $\hat{A}^+f = O$ and that any function $g(x+vt)$ is a solution of the equation $\hat{A}^-g = O$.

(c) Use the results of parts (a) and (b) to show that any function $u(x,t) = f(x-vt)+g(x+vt)$ is a solution of the wave equation.

6.3* Suppose the operator $\hat{A}$ has eigenvalues $\lambda_1, \lambda_2, \cdots$ and eigenfunctions $f_1, f_2, \cdots$. Find the eigenvalues and eigenfunctions of the operator $\hat{B} = c_0 + c_1\hat{A} + c_2\hat{A}^2$.

6.4** (a) Suppose at a certain instant a particle, in a harmonic oscillator potential, has a wave function which is a position eigenfunction $f_{x_0}(x)$ with a very small uncertainty, Δx. (See Figure 6.1(b) for a description of position eigenfunctions.) If at that instant an energy measurement is made, what is the probability that the energy value will be $E_0 = \frac{1}{2}\hbar\omega_0$?

(b) If a particle in a harmonic oscillator potential is in its ground state and a measurement of its position is made, what is the probability that it will be found within the small interval Δx centered at x_0?

(c) Is there any relationship between the answers in parts (a) and (b) which you can state in a more general way?

6.5 In a region of uniform magnetic field of strength B = 3T a proton moves in a circular path of radius 60 cm. What is the proton's momentum and velocity?

6.6* (a) Construct the operators for the x, y, and z components of the angular momentum.

(b) Show that, for any function f(x,y,z)

$$\hat{L}_x \hat{L}_y f = \hat{L}_y \hat{L}_x f + i\hbar \hat{L}_z f,$$

$$\hat{L}_y \hat{L}_z f = \hat{L}_z \hat{L}_y f + i\hbar \hat{L}_x f,$$

and

$$\hat{L}_z \hat{L}_x f = \hat{L}_x \hat{L}_z f + i\hbar \hat{L}_y f$$

Hint: In part (b) obtain the second and third equations from the first by cyclic permutation of the variables x, y, and z.

6.7** The *parity operator* $\hat{P}$ has the following effect on any function

$$\hat{P} f(x) = f(-x)$$

(a) Show that the eigenvalues of $\hat{P}$ are +1 and −1.

(b) Show that $\hat{P}$ commutes with the energy operator for a harmonic oscillator.

(c) Using (a) and (b) show that the energy eigenfunctions for a harmonic oscillator are either *even* functions (f(x) = f(-x)) or *odd* functions (f(x) = −f(x)).

6.8* Suppose f(x) and g(x) are normalized real functions. That is $\int_{-\infty}^{\infty} f^2 dx = 1$ and $\int_{-\infty}^{\infty} g^2 dx = 1$. Use the fact that $\int_{-\infty}^{\infty} (f(x) - g(x))^2 dx > O$ unless f(x) = g(x) to show that $\int_{-\infty}^{\infty} f(x) g(x)\, dx < 1$ unless f(x) = g(x).

CHAPTER SEVEN

THE STRUCTURE OF ATOMS AND MOLECULES

7.1. THE HYDROGEN ATOM

One of the most satisfactory aspects of the quantum theory was its ability to explain, without additional assumptions, the physical and chemical properties of the elements. Beginning with nothing more than the Schroedinger equation and the Coulomb potential (to describe the force between electrically charged particles) one can predict the layout of the Periodic Table of the Elements, including the lengths of the periods and the valences of the elements. One can predict the geometrical structure of chemical compounds, the hexagonal benzene ring, the tetrahedral methane molecule, the double spiral of DNA. The mathematical problems encountered in trying to apply the quantum theory to systems as complex as atoms and molecules are truly formidable. We shall only be able to sketch the barest essentials and in a number of places we shall be forced to simply quote the results of detailed mathematical calculations.

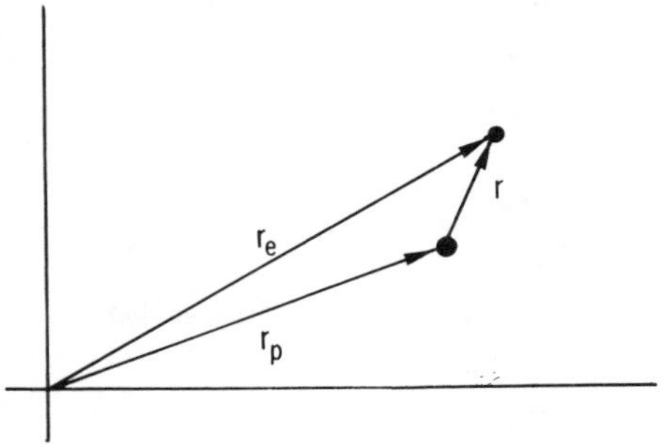

FIGURE 7.1
The relative position vector.

The hydrogen atom is the obvious place to start any study of the quantum theory of atoms. It is the only element for which one can construct the exact solutions of the Schroedinger equation. In fact, when relativistic effects are taken into account and the Schroedinger equation must be replaced by a more complicated equation, called the Dirac equation, that equation can also be solved only for the hydrogen atom. Because of the existence of those exact solutions of the quantum theory, hydrogen is the favorite system of experimentalists studying small and subtle quantum mechanical effects. It is only when you know precisely what the theory predicts that you can detect very small discrepancies between theory and experiment and it is the detection of such small discrepancies that motivates and guides the search for more perfect theories.

The hydrogen atom consists of a proton of mass m_p and charge e and an electron of mass m_e and charge $-e$. If we call the position vectors of the electron and proton $\mathbf{r}_e$ and $\mathbf{r}_p$ respectively, and we define a vector

$$\mathbf{r} = \mathbf{r}_e - \mathbf{r}_p$$

which gives the position of the electron *with respect to the proton* then the attractive electrostatic force on the electron is given by

$$\mathbf{F} = -\frac{k e^2}{r^2}\mathbf{n}$$

The potential energy function of the system is

$$V = -\frac{k e^2}{r}$$

We shall first look at the electron-proton system using classical mechanics. Since the electrostatic interaction causes both the electron and the proton to accelerate it would seem that the problem of predicting the correlated motion of the two particles would be extremely difficult. In reality the system is completely equivalent to a system containing only an electron, of a slightly smaller mass, moving in the Coulomb field of a permanently stationary proton. We can prove this remarkable fact in the following way. The total momentum of the system is

$$\mathbf{P} = m_e \mathbf{v}_e + m_p \mathbf{v}_p$$

If we assume that there are no forces being exerted on the particles by anything outside the system, then the total momentum **P** remains constant. The value of **P** is different in different inertial frames. Let us choose to look at the system in an inertial frame in which the constant vector **P** is zero. This is called the center of mass frame of the two particle system. In the center of mass frame

$$m_e \mathbf{v}_e + m_p \mathbf{v}_p = 0 \tag{7.4}$$

The center of mass vector, **R**, is defined by

$$\mathbf{R} = \frac{m_e \mathbf{r}_e + m_p \mathbf{r}_p}{m_e + m_p} \tag{7.5}$$

If we take the time derivative of both sides of Equation 7.5 and use Equation 7.4 we see that

$$\frac{d\mathbf{R}}{dt} = \frac{m_e \mathbf{v}_e + m_p \mathbf{v}_p}{m_e + m_p} = 0$$

which means that **R** is a constant vector.

The value of the constant vector **R** can easily be determined from Equation 7.5 using the initial positions of the electron and proton. (Naturally we must know the initial positions in order to predict the subsequent motion.) Knowing the center of mass we can easily calculate the positions of both particles if we also know the relative position vector, **r**. We can derive an equation for **r** by first writing the equations of motion of each particle separately.

$$\frac{d^2\mathbf{r}_e}{dt^2} = -\frac{k\,e^2}{m_e r^2}\mathbf{n}$$

and

$$\frac{d^2\mathbf{r}_p}{dt^2} = \frac{k\,e^2}{m_p r^2}\mathbf{n}$$

Subtracting the bottom equation from the top and using the definition of **r** given in Equation 7.1 we get the desired equation of motion

$$\frac{d^2\mathbf{r}}{dt^2} = -\left(\frac{1}{m_e} + \frac{1}{m_p}\right)\frac{k\,e^2}{r^2}\mathbf{n}$$

If we define the *reduced mass*, μ, by

$$\frac{1}{\mu} = \frac{1}{m_e} + \frac{1}{m_p} \tag{7.6}$$

then we can write this equation as

$$\mu\frac{d^2\mathbf{r}}{dt^2} = -\frac{k\,e^2}{r^2}\mathbf{n}$$

which, as we promised, is exactly the same as the equation of an electron of mass μ moving in the field of a proton that is stationary at the origin. It turns out that, because electrons are so much lighter than protons, μ differs only slightly from m_e.

$$m_p = 1.67265 \times 10^{-27}\,\text{kg}$$

$$m_e = 9.1095 \times 10^{-31}\,\text{kg}$$

$$\mu = 9.1045 \times 10^{-31}\,\text{kg}$$

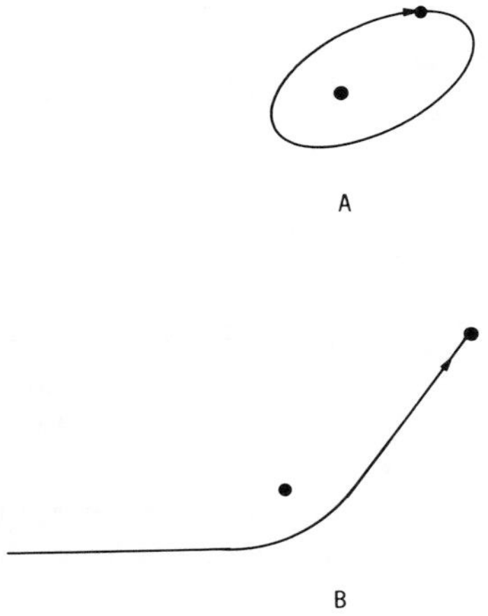

FIGURE 7.2

The orbit of an electron for negative energy (A) is fundamentally different from that for positive energy (B). Only the negative energy orbit describes a bound system.

It was for this reason that we neglected this complication when considering the Bohr theory of hydrogen. We shall now return to the policy of ignoring the proton motion and using the ordinary electron mass, m_e, secure in the knowledge that we can, at any time, correct the small errors this causes in any formula simply by replacing m_e by μ.

7.2. POSITIVE AND NEGATIVE ENERGY SOLUTIONS OF THE EQUATIONS OF MOTION

According to the energy conservation law, the quantity

$$E = \tfrac{1}{2} m v^2 - \frac{k e^2}{r}$$

remains constant along the trajectory of the electron. The kinetic energy, $\frac{1}{2} m v^2$, can have any value from 0 to $+\infty$. The potential energy ranges from $-\infty$ to 0. Therefore the total energy E can, according to classical mechanics, have any value from $-\infty$ to $+\infty$.

However, the classical trajectories with negative energy have a different character from those with positive energy. If the total energy is negative ($E = -|E|$), then, since $\frac{1}{2} m v^2$ cannot be negative,

$$\frac{k e^2}{r} = |E| + \tfrac{1}{2} m v^2 \geqslant |E|$$

This says that $r \leqslant \dfrac{k e^2}{|E|}$. An electron of negative energy can never escape from the proton. It must always be somewhere within a sphere of radius $\dfrac{k e^2}{|E|}$. This is the classical picture of a bound state. It can be shown that the classical orbit of the electron would be an ellipse with one focus at the position of the proton. The exact dimensions of the ellipse, around which the electron repeatedly moves, depend on the values of the energy and angular momentum of the electron. For small values of the angular momentum the orbit is a long thin ellipse. For larger values the shape of the orbit approaches a circle. For a given negative energy the maximum possible angular momentum is obtained in the circular orbit of that energy. In the section on the Bohr theory of hydrogen we saw that, for such an orbit, the angular momentum, L and the energy, E, are related by (see Equation 3.19)

$$L^2 = \frac{m k^2 e^4}{2|E|} \qquad (7.7)$$

This is a fact that we shall later use.

If E is positive the classical trajectory is quite different. Every positive energy solution of the classical equations of motion has the characteristic that the electron comes in from infinity, is accelerated and deflected by the proton, and goes back out to infinity in a different direction. It is clear that the positive energy trajectories do not describe anything that could reasonably be called a hydrogen atom (even a classical hydrogen atom). What they describe is the scattering of an electron by a proton.

The solutions of the Schroedinger equation for an electron in the Coulomb field of a proton have a similar separation according to whether the energy is positive or negative. For any positive energy there are solutions that describe the quantum mechanical scattering of an incoming electron beam (i.e., an incoming Schroedinger plane wave) by the Coulomb field. Since, in this chapter we are concerned with the structure of the hydrogen atom and not the problem of electron scattering, we will ignore those positive energy solutions of the Schroedinger equation. For certain particular negative energy values the Schroedinger equation will be seen to have bound state solutions that describe energy eigenstates of the hydrogen atom. It is those solutions that we will look for.

7.3. THE ATOMIC QUANTUM LENGTH

In Section 5.6 we found that a harmonic oscillator could be analyzed using classical mechanics rather than the more complicated quantum mechanics if its amplitude of vibration was much larger than a certain distance, d, that we called the quantum length of the oscillator. The formula for the quantum length in terms of the particle mass, m, and the classical frequency of the oscillator, ω_0, was $d = \sqrt{\hbar/m\,\omega_0}$. The basic principle that we used in deriving this result was that classical physics is an adequate approximation as long as the minimum uncertainties allowed by the Heisenberg uncertainty principle in the values of x and p can be made simultaneously much smaller than the typical values of x and p during the motion. Using the same idea we would now like to determine when it is appropriate to use classical physics in treating the motion of an orbiting electron. We consider an electron in a circular orbit of radius r. The momentum vector of the electron has a constant magnitude p. If the unavoidable uncertainties in a given coordinate, say x, and its associated momentum component, p_x, are to be ignorably small, then we must have

$$\Delta x << r$$

and

$$\Delta p_x << p$$

But this implies that

$$\hbar \leqslant \Delta x\, \Delta p << r p \tag{7.8}$$

For a circular orbit the "centrifugal force" must balance the Coulomb force

$$\frac{m v^2}{r} = \frac{k e^2}{r^2}$$

which implies that

$$p = \sqrt{m k e^2/r}$$

Using this in Equation 7.8 yields the following condition for the applicability of classical mechanics to electronic orbits.

$$r >> \frac{\hbar^2}{m k e^2} = a_0$$

The natural quantum length for the hydrogen atom is just the Bohr radius, a_0.

7.4. RADIAL SOLUTIONS OF THE SCHROEDINGER EQUATION

In Section 6.8 we showed that the Schroedinger energy equation for a particle in a three-dimensional potential field V(x,y,z) is

$$-\frac{\hbar^2}{2m}\left(\frac{\partial^2 u}{\partial x^2}+\frac{\partial^2 u}{\partial y^2}+\frac{\partial^2 u}{\partial z^2}\right)+Vu = Eu \tag{7.9}$$

The fact that $|u(x,y,z)|^2$ is interpreted as the electron probability density means that u must satisfy the normalization condition

$$\int |u|^2 dx\,dy\,dz = 1$$

In the problem at hand the potential is the Coulomb potential

$$V(x,y,z) = -ke^2/r$$

where

$$r = \sqrt{x^2+y^2+z^2} \tag{7.10}$$

Since V depends only on the distance from the origin one is led to suspect that there may be solutions of the Schroedinger equation that also depend on r alone. Let us see how we may determine those *radial solutions.* We assume that

$$u(x,y,z) = F(r) \tag{7.11}$$

where F(r) is, as indicated, a function of one variable. The partial derivatives needed in Equation 7.9 are computed as follows.

$$\frac{\partial u}{\partial x} = \frac{dF}{dr}\frac{\partial r}{\partial x}$$

and

$$\frac{\partial^2 u}{\partial x^2} = \frac{d^2F}{dr^2}\left(\frac{\partial r}{\partial x}\right)^2 + \frac{dF}{dx}\frac{\partial^2 r}{\partial x^2} \tag{7.12}$$

Using the facts that (see Equation 7.10)

$$\frac{\partial r}{\partial x} = \frac{x}{r} \quad \text{and} \quad \frac{\partial^2 r}{\partial x^2} = \frac{1}{r} - \frac{x^2}{r^3}$$

we can write Equation 7.12 as

$$\frac{\partial^2 u}{\partial x^2} = \frac{x^2}{r^2}F'' + \left(\frac{1}{r} - \frac{x^2}{r^3}\right)F'$$

where $F' \equiv dF/dr$ and $F'' \equiv d^2F/dr^2$.

It is obvious how to write similar equations for $\partial^2 u/\partial y^2$ and $\partial^2 u/\partial z^2$. When that is done, and the three equations are added, one gets

$$\frac{\partial^2 u}{\partial x^2}+\frac{\partial^2 u}{\partial y^2}+\frac{\partial^2 u}{\partial z^2} = F'' + \frac{2}{r}F'$$

This allows us to write the Schroedinger equation as

$$-\frac{\hbar^2}{2m}\left(F''+\frac{2}{r}F'\right) - \frac{ke^2}{r}F = EF \tag{7.13}$$

It will simplify the Schroedinger equation if we introduce the Bohr radius, a_o, as our unit of distance. We do this by using a variable ρ that is equal to the distance from the origin measured in Bohr radii.

$$\rho = r/a_o$$

We can write Equation 7.13 in terms of ρ by assuming that

$$F(r) = f(\rho)$$

The equation we obtain is (The details are assigned as Problem 7.1)

$$f'' + \frac{2}{\rho} f' + \frac{2}{\rho} f = \epsilon f \tag{7.14}$$

where

$$\epsilon = -\frac{2\hbar^2}{m k^2 e^4} E \tag{7.15}$$

In order to be normalizable, the acceptable solutions of Equation 7.14 will have to go to zero as ρ goes to infinity. A very simple function with this property is the negative exponential $e^{-\lambda\rho}$. Let us try a solution of the form

$$f(\rho) = c\, e^{-\lambda\rho}$$

Putting this into Equation 7.14 and then cancelling a common factor of $c\, e^{-\lambda\rho}$ gives

$$\lambda^2 - \frac{2(\lambda - 1)}{\rho} = \epsilon$$

which is satisfied for all ρ only if

$$\lambda = 1$$

and

$$\epsilon = \lambda^2 = 1$$

Our solution is therefore

$$f(\rho) = c\, e^{-\rho}$$

Using the definition of ϵ in terms of E (Equation 7.15) we see that the energy eigenvalue associated with this solution of the Schroedinger equation is

$$E = -\frac{m k^2 e^4}{2\hbar^2} = -2.18 \times 10^{-18}\, J$$

which is just the experimental ground state energy. Written in terms of r, the ground state wave function, which we call u_1, is

$$u_1 = c\, e^{-r/a_o}$$

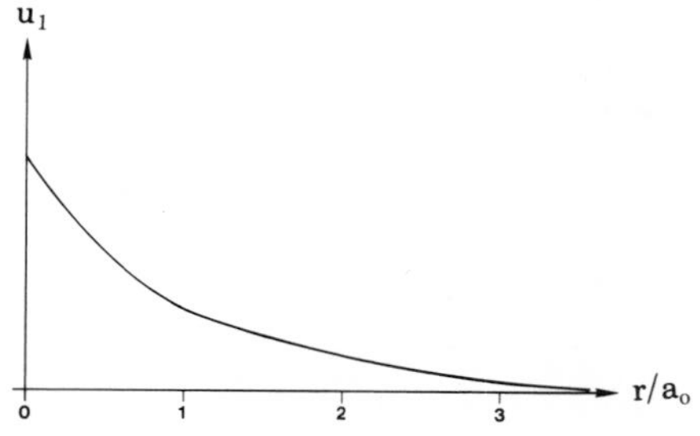

FIGURE 7.3

The hydrogen ground state wave function.

If we use the harmonic oscillator solutions as a guide we are led to try solutions of the form of a polynomial times an exponential. That is

$$f(\rho) = (a_o + a_1\rho + a_2\rho^2 + \cdots + a_N\rho^N)\, e^{-\lambda\rho}$$

The details of this procedure will be left to the problems. Here we shall simply state the results. For each $N \geqslant 0$ there is one solution of this form. The values of λ and ϵ for that solution are

$$\lambda = \frac{1}{N+1} \quad \text{and} \quad \epsilon = \frac{1}{(N+1)^2}$$

Written in terms of r the first few normalized radial solutions are given in Table 7.1 along with the corresponding energy eigenvalues. It is clear that the energy spectrum we are generating in this way agrees with the Bohr formula for the energy levels of hydrogen. That is

$$E_n = -\frac{m k^2 e^4}{2\hbar^2 n^2} \qquad (n = N+1 = 1, 2, 3, \cdots) \tag{7.16}$$

TABLE 7.1

Radial Solution	Associated Energy*
$u_1 = (\pi a_o^3)^{-1/2} e^{-r/a_o}$	$E_1 = -A$
$u_2 = (8\pi a_o^3)^{-1/2}(1 - \frac{r}{2a_o})\, e^{-r/2a_o}$	$E_2 = -A/4$
$u_3 = (27\pi a_o^3)^{-1/2}(1 - \frac{2r}{3a_o} + \frac{2r^2}{27a_o^2})\, e^{-r/3a_o}$	$E_3 = -A/9$

$^*A = 2.18 \times 10^{-18}\,J = 13.6\,eV$

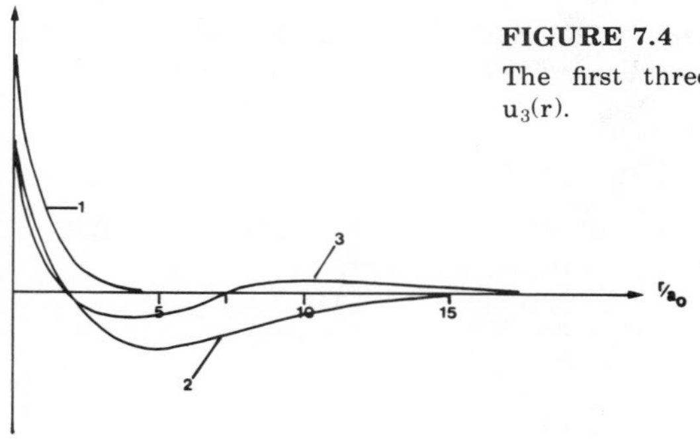

FIGURE 7.4
The first three radial solutions, $u_1(r)$, $u_2(r)$, and $u_3(r)$.

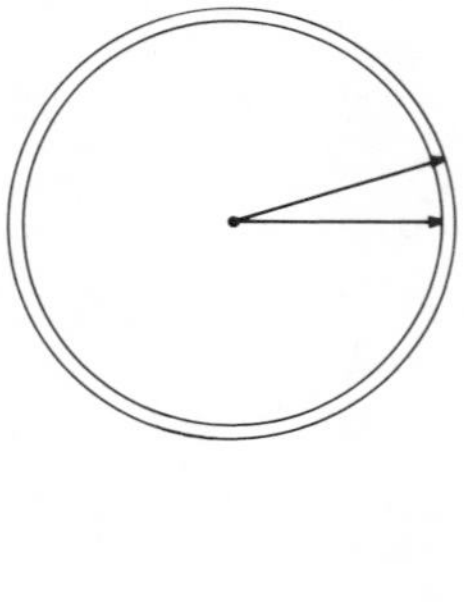

FIGURE 7.5

QUESTION: In the ground state, $u_1(r)$, what is the probability that the electron will be found in the distance range from r to r+dr from the proton?

ANSWER: What we are asking is, what is the probability that the electron will be found within the thin spherical shell shown in Figure 7.5. The probability density is constant within the shell and equal to $u_1^2(r)$. The total volume of the shell is $dV = 4\pi r^2 dr$. Thus

$$\begin{matrix}\text{Probability of finding} \\ \text{electron in range r to r+dr}\end{matrix} = 4\pi r^2 u_1^2(r)\,dr = \frac{4}{a_0^3} r^2 e^{-2r/a_0} dr \equiv P_1(r)\,dr$$

where $P_1(r)$ is the *radial probability function* for the state $u_1(r)$.

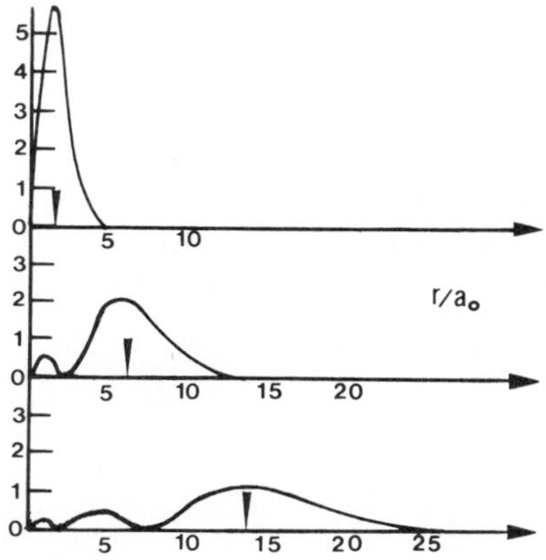

FIGURE 7.6

The radial probability function $P_n = 4\pi r^2 u_n^2$, for n = 1, 2, and 3. $P_n dr$ is the probability that an electron in state $u_n(r)$ will be found within the spherical shell shown in Figure 7.5. The most probable values of r for the three states are indicated by arrows.

In Figure 7.6 we have plotted the radial probability functions, $P_n(r) = 4\pi r^2 u_n^2(r)$, for the first three radial solutions from Table 7.1. One can see that, as the quantum number n increases, the most probable value of r (the value of r at which $P_n(r)$ is a maximum) also increases in rough agreement with the Bohr theory. However, it is also clear from the diffuseness of the probability function that there is no such thing as an electronic orbit defined for the quantum state. It is only for orbits much larger than a_0 that one can find wave packet solutions of the time-dependent Schroedinger equation that move on a classical orbit.

7.5. NONRADIAL HYDROGENIC WAVE FUNCTIONS

Now that we have generated all the energy levels in the hydrogen spectrum and can compute a corresponding wave function for each energy level, it might seem that we are about finished with the quantum theory of hydrogen, having obtained all the possible quantum states. This is very far from the truth. We have obtained only the *radial* solutions. That is, the wave functions that have the form $u(x,y,z) = F(r)$. Actually all the hydrogen energy levels except the ground state energy are degenerate. For the energy level E_n there are n^2 different eigenfunctions that give this eigenvalue. Only one of those n^2 functions

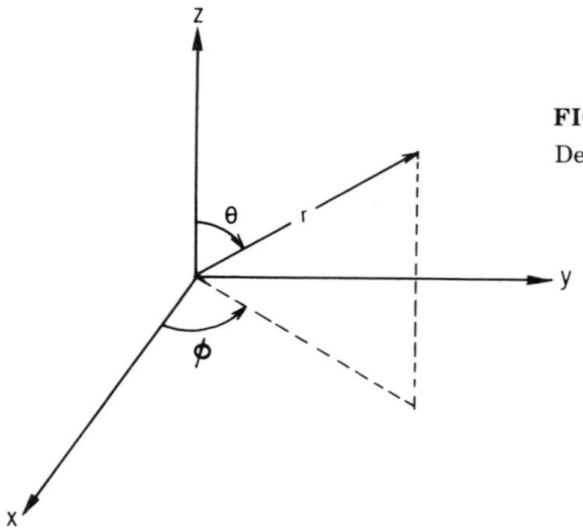

FIGURE 7.7

Definition of the spherical coordinates r, θ, and ϕ.

is of the radial type. The others are more complicated functions. The mathematical manipulations needed to derive all of the hydrogenic wave functions are very complex and will therefore only be sketched in this section. One first introduces the set of three-dimensional *spherical coordinates* shown in Figure 7.7. The relationship between the set of rectangular coordinates (x,y,z) and the set of spherical coordinates (r, θ, ϕ) is

$$x = r \sin\theta \cos\phi$$

$$y = r \sin\theta \sin\phi \tag{7.17}$$

$$z = r \cos\theta$$

One then looks for solutions of the Schroedinger equation (Equation 7.9) that have the form

$$u(x,y,z) = F(r)\, G(\theta)\, H(\phi)$$

Notice that the radial solutions, u = F(r), are a special case of these more general solutions in which G and H are set equal to one.

Using the Schroedinger equation and the coordinate transformation equations (Equation 7.17) it is possible to derive differential equations for the separate functions F, G, and H. With a great deal of work one can then show that there is one solution, which we shall call u_{nlm}, for each set of the three integer quantum numbers n, l, and m, that satisfies the following conditions.

1. n can by any positive integer
2. l can be any nonnegative integer less than n
3. m can be any integer between $-l$ and $+l$.

This can be written more compactly as

$$n = 1, 2, \cdots \tag{7.18.a}$$

$$l = 0, 1, \cdots, n-1 \tag{7.18.b}$$

$$m = 0, \pm 1, \pm 2, \cdots, \pm l \tag{7.18.c}$$

These quantum numbers, whose physical meaning we shall explain shortly, have the following names.

n is called the *principal* quantum number,

l is called the *angular momentum* quantum number, and

m is called the *magnetic* quantum number.

The principal quantum number, n, is just the same as the quantum number we have used to index the radial solutions. In fact, in this more general scheme the radial solutions u_n are those solutions that have the quantum numbers $(n,l,m) = (n,0,0)$. That is, u_n in the old notation is equal to u_{noo} in the new notation. The quantum number n determines the energy of the atom. An atom in state u_{nlm} has an energy E_n given by Equation 7.16

According to what we have just said the physical differences between two states of the same principal quantum number but different values of l and m cannot concern the energies of the two states. In fact, the quantum numbers l and m say nothing about the energy of the state but rather they give information about the angular momentum of a hydrogen atom in quantum state u_{nlm}. In the state u_{nlm} a hydrogen atom has a total angular momentum, L, which is given by the formula

$$L^2 = l(l+1)\hbar^2 \tag{7.19}$$

We can now understand where the restriction $0 \leqslant l \leqslant n-1$ comes from. In Section 7.1 we mentioned that in any classical orbit of energy E the angular momentum must lie in the range (See Equation 7.7)

$$0 \leqslant L^2 \leqslant \frac{m k^2 e^4}{2|E|} \tag{7.20}$$

If we put $|E| = \dfrac{m k^2 e^4}{2\hbar^2 n^2}$ and $L^2 = l(l+1)\hbar^2$ this inequality becomes

$$0 \leqslant l(l+1) \leqslant n^2$$

which is completely equivalent to Equation 7.18.b. This restriction simply states the fact that in quantum theory, just as in classical theory, large angular momentum values are inconsistent with low energies. The magnetic quantum number, m, also gives us information about the angular momentum of an atom in state u_{nlm}. (The reason for the peculiar name will become clear later when we discuss the effects of placing the atom in a magnetic field.) Recall that the wave functions we are discussing are all assumed to be of the form $F(r)\,G(\theta)\,H(\phi)$ where the angles θ and ϕ must naturally be defined with respect to some particular set of x, y, and z axes. With reference to those same axes, the z component of the angular momentum of a hydrogen atom in the quantum state u_{nlm} is

$$L_z = m\hbar \tag{7.21}$$

The reason for the restriction $-l \leqslant m \leqslant l$ on the allowed values of the quantum number m is now obvious. It simply states that $|L_z|$ cannot be larger than the magnitude of the angular momentum vector.

We have already seen that even when we know the exact quantum state of a particle in a one-dimensional box or in a harmonic oscillator potential we cannot predict the precise value of the particle's position or momentum. We only have a probability distribution for those observables. The maximum information you can ever obtain about a quantum mechanical system is the system's exact wave function, but that wave function never supplies as much precise information about the system as it is possible to obtain about the equivalent classical system. In these hydrogen atom quantum states we have an excellent example of this general property. For a classical Kepler orbit (the negative energy orbits of a particle in a Coulomb force) all three components of the angular momentum vector are known precisely. This is not true for the quantum mechanical hydrogen atom. Given the exact quantum state of the system, we can predict the result of measuring the magnitude of **L**, namely $\sqrt{l(l+1)}\hbar$, and we can predict with certainty the result of measuring L_z (that is $m\hbar$). We cannot predict with certainty the result of a measurement of L_x or L_y. In fact, if we were to measure one of these components, let us say L_x, we would have a certain probability of getting any of the values

$$L_x = -l\hbar, \quad \cdots, +l\hbar$$

Using the rules presented in Chapter 6, it is possible to calculate the probability of getting any one of the above values for L_x, but the calculation is difficult and will not be done here.

7.6. THE DETAILED WAVE FUNCTIONS

The hydrogenic wave functions $u_{nlm}(r,\theta,\phi)$ are all of the form $F(r)\,G(\theta)\,H(\phi)$. The function $F(r)$ depends on the two quantum numbers n and l and is traditionally written in the form $F(r) = a_0^{-3/2}R_{nl}(\rho)$, where $\rho = r/a_0$ is the proton-electron distance measured in Bohr radii. The function $G(\theta)$ depends on the two quantum numbers l and m. It has the form of a polynomial in $\sin\theta$ and $\cos\theta$. The function $H(\phi)$ depends only on the quantum number m and is simply $(2\pi)^{-1/2}e^{im\phi}$. The combination $G(\theta)\,H(\phi)$ is usually written together as $Y_{lm}(\theta,\phi)$ and called a *spherical harmonic*. Thus, in traditional notation, the wave function u_{nlm} is written as

$$u_{nlm}(r,\theta,\phi) = a_0^{-3/2}R_{nl}(\rho)Y_{lm}(\theta,\phi)$$

All the functions, R_{nl} and Y_{lm}, for $n \leqslant 3$ are given in Table 7.2. The volume element in spherical coordinates is

$$dV = r^2\sin\theta\,dr\,d\theta\,d\phi$$

The functions R_{nl} and Y_{lm} are all normalized so that

$$a_0^{-3}\int_0^\infty R_{nl}^2 r^2\,dr = 1 \qquad (7.23)$$

and

$$\int_0^{2\pi} d\phi \int_0^\pi Y_{lm}^* Y_{lm}\sin\theta\,d\theta = 1 \qquad (7.24)$$

which guarantees that the product u_{nlm} is a normalized eigenfunction.

$$\int_0^\infty r^2dr \int_0^{2\pi} d\phi \int_0^\pi u_{nlm}^* u_{nlm}\sin\theta\,d\theta = 1 \qquad (7.25)$$

TABLE 7.2

The functions $R_{nl}(\rho)$ and $Y_{lm}(\theta,\phi)$.
Note that $\rho = r/a_o$ and $u_{nlm} = a_o^{-3/2} R_{nl}(\rho) Y_{lm}(\theta,\phi)$

$R_{nl}(\rho)$	$Y_{lm}(\theta,\phi)$
$R_{10} = 2e^{-\rho}$	$Y_{00} = \dfrac{1}{\sqrt{4\pi}}$
$R_{20} = \dfrac{1}{\sqrt{2}}(1-\rho/2)\,e^{-\rho/2}$	$Y_{10} = \sqrt{3/4\pi}\cos\theta$
$R_{21} = \dfrac{1}{2\sqrt{6}}\,\rho\, e^{-\rho/2}$	$Y_{1\pm1} = \sqrt{3/8\pi}\sin\theta\, e^{\pm i\phi}$
$R_{30} = \dfrac{2}{3\sqrt{3}}(1-\frac{2}{3}\rho+\frac{2}{27}\rho^2)\,e^{-\rho/3}$	$Y_{20} = \sqrt{5/4\pi}(\frac{3}{2}\cos^2\theta-\frac{1}{2})$
$R_{31} = \dfrac{8}{27\sqrt{6}}(\rho-\frac{1}{6}\rho^2)\,e^{-\rho/3}$	$Y_{2\pm1} = \sqrt{15/8\pi}\sin\theta\cos\theta\, e^{\pm i\phi}$
$R_{32} = \dfrac{4}{81\sqrt{30}}\,\rho^2 e^{-\rho/3}$	$Y_{2\pm2} = \frac{1}{4}\sqrt{15/2\pi}\sin^2\theta\, e^{\pm 2i\pi}$

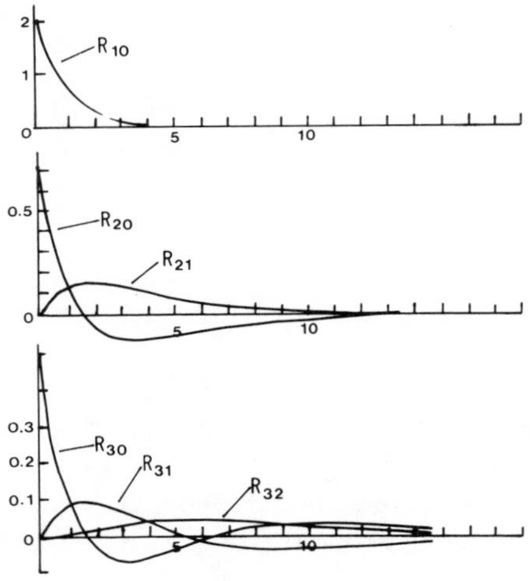

FIGURE 7.8

Graphs of the functions R_{nl} listed in Table 7.2.

QUESTION: For a hydrogen atom in the state u_{210}, what is the average distance of the electron from the nucleus?

ANSWER: The probability that an electron in the state u_{nlm} will be found in the volume element dV centered on location (r,θ,ϕ) is $P(r,\theta,\phi)dV$, where

$$P(r,\theta,\phi) = u^*_{nlm}(r,\theta,\phi)u_{nlm}(r,\theta,\phi)$$

The average value of r is thus

$$\bar{r} = \int P(r,\theta,\phi)\, r\, dV = a_o^{-3} \int_o^\infty R_{nl}^2 r^3 dr \int_o^{2\pi} d\phi \int_o^\pi Y_{lm}^* Y_{lm} \sin\theta\, d\theta$$

$$= a_o^{-3} \int_o^\infty R_{nl}^2 r^3 dr$$

For the state u_{210} this gives

$$\bar{r} = \frac{a_o^{-3}}{24} \int_o^\infty e^{\rho} \rho^2 r^3 dr$$

Using the fact that $a_o^{-4} r^3 dr = \rho^3 d\rho$, we get

$$\bar{r} = \frac{a_o}{24} \int_o^\infty e^{-\rho} \rho^5 d\rho = 5 a_o$$

which is slightly larger than the rough estimate we mentioned previously of $r_n = n^2 a_o$.

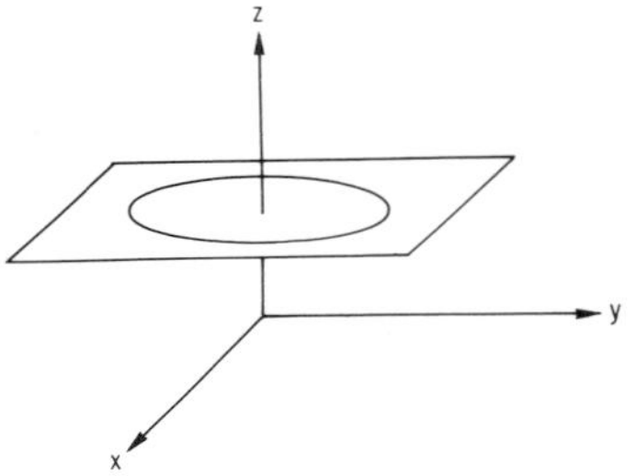

FIGURE 7.9

The angle θ is called the *polar angle*, while the angle ϕ is referred to as the *azimuthal angle.* The dependence of the function u_{nlm} on the azimuthal angle is very simple. The function contains a factor of $e^{im\phi}$. That such a factor is appropriate for a system with $L_z = m\hbar$ can be seen from the following suggestive, but not rigorous argument. Let us draw a circle of arbitrary radius, R, with its center on the z-axis and its plane perpendicular to that axis (Figure 7.9). The values of r and θ are the same at every point on that circle. Therefore, on that circle, the function u_{nlm} has the form of a constant times $e^{im\phi}$.

$$u_{nlm} = c\, e^{im\phi}$$

But this is exactly the eigenfunction we obtained before (See Equation 5.33) for a particle moving in a circular loop with angular momentum $m\hbar$. The direction of the particle motion around the circle is right-handed if m is positive and left-handed if m is negative. This visualization of the meaning of the wave function in terms of motion of a particle on a circle about the z-axis is not based on any exact principles of quantum theory. However, the following picture is based on precise quantum mechanical principles. Given any solution of the Schroedinger equation, u(x,y,z), it is possible to compute a vector field **f**(x,y,z) which has the interpretation that **f**(x,y,z) represents the particle flux (or the local current of particles) at location (x,y,z). Without actually doing the calculation we shall simply state the result that when the particle flux is calculated for the state u_{nlm} the resulting particle current has the form illustrated in Figure 7.10. It clearly represents a state in which there is a persistent right-handed flow of electrons around the z-axis if $m > 0$.

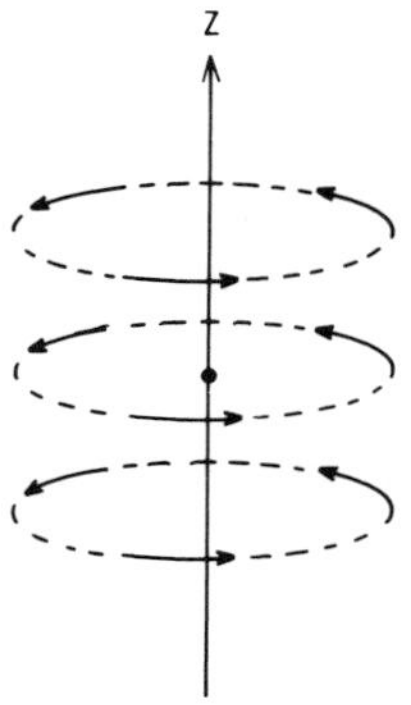

FIGURE 7.10

A wave function u_{nlm} has a right-handed electron flux around the z-axis associated with it if $m > 0$. For $m < 0$ the electron flux is left-handed.

7.7. THE ZEEMAN EFFECT

According to the physical interpretation we have given to the wave function, the state u_{nlm} has associated with it an electron flux or current around the z-axis if m is not zero. If m is positive, there is a right-handed flux about the z-axis. If m is negative, the flux about the z-axis is left-handed. Since the electron has a negative electric charge, a persistent right-handed electron current about the z-axis would constitute a left-handed electric current about the axis. This is just the current pattern one would need to produce a magnetic dipole. Essentially the electron flux associated with the quantum state acts like a small current loop around the z-axis. The current is in a direction opposite to the electron flux due to the fact that the electron charge is negative. Thus a hydrogen atom in a state with a nonzero m value has a magnetic moment that points in a direction opposite to its angular momentum. The electron circulation is responsible for both the angular momentum and the magnetic moment of the quantum state. It can be shown, in both the classical and the quantum theory, that the ratio of the magnetic moment of the magnetic dipole to the angular momentum is exactly one half the ratio of the charge to the mass of the circulating particles. This is true of any system of orbiting particles as long as nonrelativistic mechanics can be used. If the circulation velocities are close to the velocity of light, this simple result is not valid. In order to get some idea of the origin of the relationship between angular momentum and magnetic dipole moment we shall consider a simple classical system. The system is a collection of N similar charges, each of mass m and charge q, moving in a circle at velocity v around the origin in the x-y plane (See Figure 7.11). The angular momentum of the system is N times the angular momentum of one of the particles.

$$L = N r m v \tag{7.26}$$

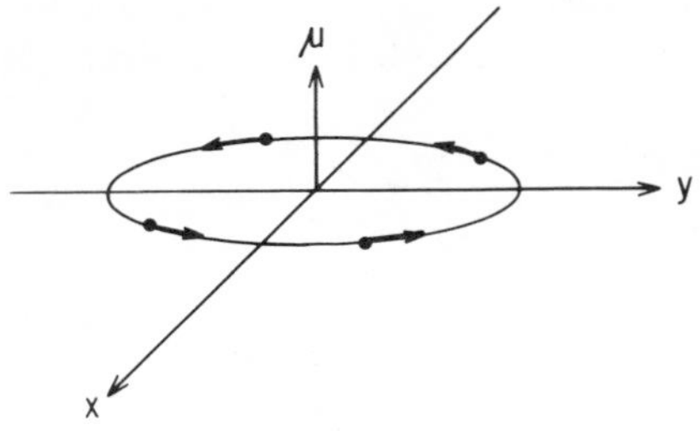

FIGURE 7.11

The magnetic moment vector μ for a system of positive charges moving on a circle.

The collection of moving particles constitutes a current loop. The area of the loop is

$$A = \pi r^2$$

The current in the loop is equal to the average rate at which charge passes any point in the loop. If T is the time it takes for one charge to complete a circular orbit, then during a time T an amount of charge Nq passes any point in the loop. The time T is given by

$$T = \frac{2\pi r}{v}$$

The current in the loop is therefore

$$I = \frac{Nq}{T} = \frac{Nqv}{2\pi r}$$

The magnetic dipole moment of a loop current is equal to the product of the current and the area of the loop. Thus, if we call the magnetic dipole moment μ, we get

$$\mu = IA = \frac{Nqv}{2\pi r}(\pi r^2)$$

or

$$\mu = \tfrac{1}{2} Nqvr$$

If we compare this formula for μ with Equation 7.26 for L, we see that

$$\mu = \frac{q}{2m} L \tag{7.27}$$

It is true, although not obvious, that the same relationship holds between any component of the angular momentum and the corresponding component of the magnetic dipole moment for any quantum state of a charged particle. The above relationship between magnetic moment and angular momentum determines the way in which an atom reacts to being placed in a magnetic field. A magnetic dipole that is placed in a magnetic field **B** has an energy which depends upon its orientation with respect to the field (See Figure 7.12). In particular, the energy of the dipole is $E_{dipole} = -\vec{\mu}\cdot\mathbf{B}$. If the field is a uniform field pointing in the z direction, then the magnetic dipole energy is

$$E_{dipole} = -B\mu_z$$

where μ_z is the z-component of the magnetic dipole vector. Using Equation 7.27 (with $q = -e$) we can write the magnetic dipole energy of a hydrogen atom in a magnetic field in terms of the component of angular momentum in the direction of the field.

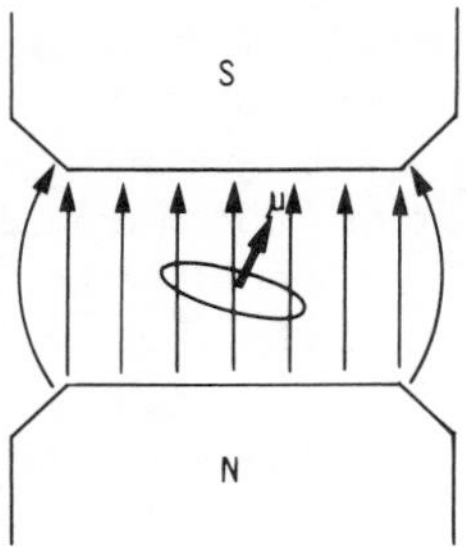

FIGURE 7.12
The energy of a magnetic dipole in a magnetic field depends on the angle between the dipole vector and the field.

$$E_{dipole} = \frac{e}{2m_e} B L_z$$

Because of the interaction of the atom with the magnetic field, we should expect that the energy value for quantum states will depend upon the quantum number m, because

$$L_z = m\hbar$$

The energy associated with the quantum state u_{nlm} would no longer depend only upon the quantum number n. It would be

$$E_{nlm} = E_n + \frac{e}{2m_e} B\hbar m \qquad (7.28)$$

The energy spectrum would therefore change in the manner shown in Figure 7.13. This change in the spectrum is called the *Normal Zeeman Effect.*

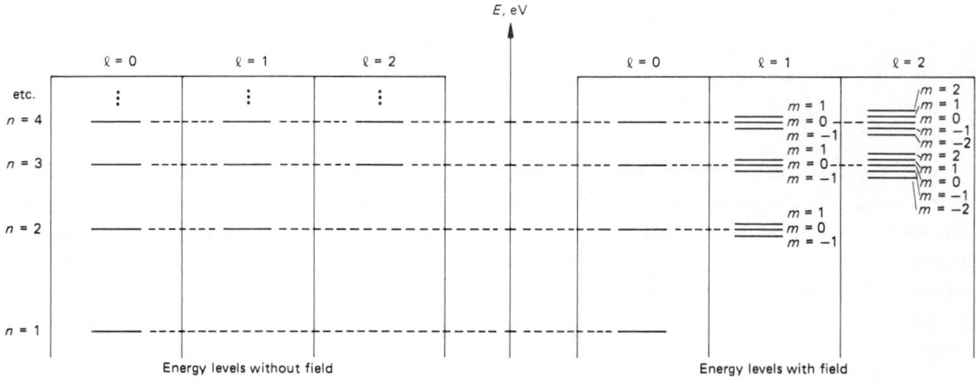

FIGURE 7.13 The predicted changes in the hydrogen energy levels when the atom is placed in a magnetic field. These predictions, called the Normal Zeeman Effect, are *not* in agreement with the observed shifts in hydrogen.

7.8. ELECTRON SPIN

If a sample of hydrogen is actually placed in a magnetic field and its energy spectrum is determined, what is found differs very much from the Normal Zeeman Effect. With no field the energy levels are given by Equation 7.16 as expected. As the field is turned on, the degenerate energy levels separate into a set of closely spaced levels called a *multiplet.* The separation is proportional to the magnitude of the magnetic field, B, as we would expect from Equation 7.28. The aspect of the spectrum that is very different from what we would expect is the degeneracy of each level. There are more levels in each multiplet than we would expect from the known combinations of m and l. Even the $l = 0$ states separate into two levels as if they have a magnetic moment which can have two different possible values. There is another source of angular momentum in the hydrogen atom besides the angular momentum due to the electron flux around the nucleus. That other source of angular momentum is an intrinsic *spin angular momentum* of the electron itself. One might picture the electron as spinning on its own axis so that the total angular momentum of the hydrogen atom is the vector sum of the spin angular momentum of the electron about an axis through its center plus the angular momentum of the electron due to its motion about the proton. This essentially classical picture of the electron as a little charged spinning ball should only be considered as an inadequate substitute for an honest theory of electron spin angular momentum. A proper theory does exist, but it is rather complicated. It predicts all properties of electrons and photons with astonishing accuracy. The full theory is called *quantum electrodynamics.* To the present no experimental contradiction of the predictions of quantum electrodynamics has ever been detected, even though some experiments have been able to test predictions with a precision of one part in 10^8. Quantum electrodynamics is a relativistic quantum theory of the electromagnetic and electron fields. It predicts that the particles associated with the electron field have an angular momentum whose magnitude s, is always

$$s = \sqrt{\tfrac{3}{4}}\,\hbar = \sqrt{\tfrac{1}{2}(\tfrac{1}{2}+1)}\,\hbar \tag{7.29}$$

Equation 7.29 should be compared with the equation, $L = \sqrt{l(l+1)}\,\hbar$, which gives the magnitude of the orbital angular momentum. The component of the electron spin angular momentum in any preassigned direction, which we shall call the z direction, can have either of the two values

$$s_z = \tfrac{1}{2}\hbar \quad \text{or} \quad s_z = -\tfrac{1}{2}\hbar$$

There is a magnetic dipole moment associated with the electron spin just as there is a magnetic moment associated with the electron flux around the nucleus. However, the relativistic quantum theory predicts that the ratio of the electron spin magnetic moment to the electron spin angular momentum is twice the value given by Equation 7.27. Thus

$$\mu_{z(\text{spin})} = -\frac{e}{m}s_z$$

When the electron spin magnetic moment is taken into account, the observed energy spectrum of a hydrogen atom in a magnetic field is explained completely. Because of the existence of the electron spin, the quantum numbers, n, l, and m are not adequate to completely specify the possible quantum states of an electron in a hydrogen atom. We must add one more quantum number, which we call σ, to define the z component of the spin angular momentum. The possible quantum states of the hydrogen atom are therefore labeled $u_{n\ell m\sigma}$ where σ can have the values of $+1$ or -1 and s_z is related to σ by

$$s_z = \sigma \hbar/2$$

Without a magnetic field present the energy value associated with the quantum state does not depend upon σ, and thus one does not detect the existence of the spin quantum number. When the hydrogen atom is placed in a magnetic field, the spin magnetic dipole moment interacts with the magnetic field and causes the otherwise degenerate energy levels to separate into the number of distinct levels observed.

7.9. THE EXCLUSION PRINCIPLE

We have so far restricted our discussion of atomic structure to the hydrogen atom because we must still introduce one more basic principle before we can present a quantum theory of many-electron atoms. That basic principle was first postulated by Wolfgang Pauli before the modern version of the quantum theory had been developed. It was later derived from the fundamental principles of the quantum theory. That derivation was also done by Pauli. This postulate, called the *Pauli exclusion principle* states that:

Two electrons will never be found in the same quantum state.

An equivalent way of stating the exclusion principle would be that:

No two electrons in a many-electron system can have exactly the same set of quantum numbers.

QUESTION: Do all types of fundamental particles satisfy the Pauli exclusion principle?

ANSWER: No. All known particles can be placed into either of two major groups. The particles in the first group, which includes electrons, protons, neutrons, and a number of other particles, are called *Fermi-Dirac particles* or fermions. They all have an intrinsic spin angular momentum which is some half-integer times $\hbar$ (i.e. $s = \frac{1}{2}\hbar, \frac{3}{2}\hbar$, etc.). In fact, all stable Fermi-Dirac particles have spin $\frac{1}{2}\hbar$ like the electron. (Unstable particles are particles that spontaneously decay into several lighter particles.) All Fermi-Dirac particles satisfy the Pauli exclusion principle.

Particles of the other major class are called *Bose-Einstein particles* or bosons. This group includes the photon, the π-meson, and others. Bose-Einstein particles all have integer spin ($s = 0, \hbar, 2\hbar$, etc.). For instance, the π-meson has zero spin while the photon has a spin angular momentum of $\hbar$. The particles in the Bose-Einstein class do not satisfy the Pauli exclusion principle. Therefore any number of photons or other Bose-Einstein particles may occupy the same single-particle quantum state. More will be said of these two groups when we discuss quantum statistical mechanics.

7.10. THE PERIODIC TABLE OF THE ELEMENTS

With a combination of the exclusion principle and an analysis of the quantum states of an electron in a spherically symmetric potential we can explain, at least qualitatively, the striking relationship between the chemical characteristics of the elements and their atomic numbers. The *atomic number* of an element is simply the number of electrons in a neutral atom of that element.

The modern picture of one atom of an element of atomic number Z is the following. Most of the mass of the atom is concentrated in a *nucleus* of extremely small volume and high density. The nucleus is composed of two types of particles, neutrons and protons. The neutrons have no electric charge but the protons each carry a positive electric charge of the same magnitude as the electronic charge. There are Z protons in the nucleus. Surrounding the nucleus is a distribution of Z electrons in Z different electronic quantum states. If the atom is in its lowest energy state (i.e., its ground state), then those electronic quantum states in which the Z electrons would be found are the Z states of lowest energy. The two lowest energy quantum states have quantum numbers $n = 1$ and $l = m = 0$. They differ from one another by the value of the z-component of spin angular momentum ($\sigma = +1$ and $\sigma = -1$). In the terminology commonly used by chemists and atomic physicists, both these states are called 1s states. The number one refers to the principle quantum number, n. The letter s indicates the angular momentum quantum number, l, in a scheme in which further values of the angular momentum are assigned letter symbols according to the code given in Table 7.3.

TABLE 7.3

The letter symbols used for the first six angular momentum values in the atomic spectroscopist's notation. Further states are given by letters in simple alphabetical order after h. The first three entries in this seemingly arbitrary sequence of letters described observed characteristics of the emission spectral lines associated with the states (s - sharp, p - principal, d - diffuse).

$l = 0$	s
$l = 1$	p
$l = 2$	d
$l = 3$	f
$l = 4$	g
$l = 5$	h

In the hydrogen atom (in its ground state) a single 1s state is occupied by the one electron. In helium both 1s states are occupied. The *configuration* of an atomic ground state is a listing of the occupied states. The configuration of helium is written as $1s^2$ where the superscript indicates that both 1s states are occupied. The ground state of the lithium atom, which has three electrons, has a configuration $1s^2 2s^1$. The configurations of the first eighteen elements are given in Table 7.4. From Table 7.4 we can immediately see that the noble gases, which are chemically quite unreactive, have what are called filled shells. All of the electronic states of one type such as the 1s or 2p are occupied by a maximum number of electrons while all further states are completely empty. Thus the noble gas helium occurs at the filling of the 1s shell. The noble gas neon occurs at the filling of the 2p shell, and the noble gas argon occurs at the filling of the 3p shell.

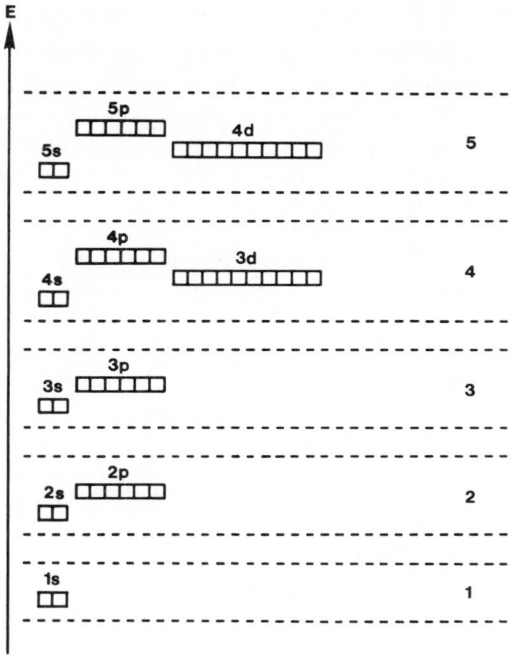

FIGURE 7.14 A diagram showing the order in which states of the first five major shells are filled. For an atom of atomic number Z the Z lowest energy states are occupied.

TABLE 7.4
The ground state configurations of the first eighteen elements.

Name	Symbol	At. No.	Configuration
Hydrogen	H	1	$1s^1$
Helium	He	2	$1s^2$
Lithium	Li	3	$1s^2 2s^1$
Beryllium	Be	4	$1s^2 2s^2$
Boron	B	5	$1s^2 2s^2 2p^1$
Carbon	C	6	$1s^2 2s^2 2p^2$
Nitrogen	N	7	$1s^2 2s^2 2p^3$
Oxygen	O	8	$1s^2 2s^2 2p^4$
Fluorine	F	9	$2s^2 2s^2 2p^5$
Neon	Ne	10	$1s^2 2s^2 2p^6$
Sodium	Na	11	$1s^2 2s^2 2p^6 3s^1$
Magnesium	Mg	12	$1s^2 2s^2 2p^6 3s^2$
Aluminum	Al	13	$1s^2 2s^2 2p^6 3s^2 3p^1$
Silicon	Si	14	$1s^2 2s^2 2p^6 3s^2 3p^2$
Phosphorus	P	15	$1s^2 2s^2 2p^6 3s^2 3p^3$
Sulfur	S	16	$1s^2 2s^2 2p^6 3s^2 3p^4$
Chlorine	Cl	17	$1s^2 2s^2 2p^6 3s^2 3p^5$
Argon	Ar	18	$1s^2 2s^2 2p^6 3s^2 3p^6$

7.11. THE SELF-CONSISTENT FIELD APPROXIMATION

From the above discussion of many-electron atoms the reader might easily get the erroneous impression that the ground state energy of an element of atomic number Z could be calculated using the following simple prescription. (1) Calculate the energy levels of a single electron in the field of a nucleus of charge Ze. (2) Add the energy values of the first Z quantum states in order to obtain the total ground state energy of the Z-electron atom. In fact, this procedure would give a total energy value that, except in the case of hydrogen ($Z = 1$), would be substantially lower than the true ground state energy of element Z. The source of the discrepancy would be that the just-mentioned procedure completely neglects the electrostatic interactions between the electrons. Electrons are negatively charged particles. Therefore they not only interact with the positively charged nucleus, but they also interact with one another. If we neglect relativistic effects, it is possible to write an exact Schroedinger equation for a many-electron atom, taking electron-electron interactions into account. No exact solution of the equation has been found for any many-electron atom. The problems involved in finding acceptably accurate approximate solutions of the equation, using high-speed digital computers, are very great. If we take as our criterion of acceptable accuracy for an approximate solution the property that it can be used to reliably predict the physical and chemical properties of the element, then acceptable approximate ground state wave functions are not yet known for many of the elements although the situation is steadily improving.

Instead of making any attempt at dealing with the exact Schroedinger equation of a many-electron atom, we shall base our discussion entirely on a method of treating the electron-electron interaction in an approximate way. The method is called the self-consistent field (or SCF) method. It is best to describe the SCF method by showing in some detail how it would be applied to a calculation of the ground state of a particular atom. We choose the neon atom ($Z = 10$) as our example.

(1) Our first step would be to make a guess of the electron density as a function of r for the ground state of the neon atom. Let us call our initial guess of the electron density function $\rho_0(r)$. Since, by assumption, we do not already know the ground state of the neon atom, our initial guess will not be the exact ground state electron density. The error in $\rho_0(r)$ will not effect the accuracy of our final result. However, a good initial guess will substantially reduce the amount of computing needed to obtain our final result. The second step in the procedure is as follows.

(2) Using Gauss' law, compute the electrostatic potential, V(r), that would be created by a nucleus, of charge Ze, located at the origin plus an electric charge density $-e\rho_0(r)$ surrounding the nucleus. An electron in that electric potential field would have a potential energy equal to $-e\,V(r)$.

(3) We now find the five lowest energy eigenfunctions of an electron in a spherically symmetric potential $-e\,V(r)$. That is we solve the equation

$$-\frac{\hbar^2}{2m}\nabla^2 u_n - e\,V(r)\,u_n = E_n u_n$$

for the normalized eigenstates $u_1, \cdots, u_5$.

(4) Having computed the first five eigenstates, we then assume that each of these eigenstates is occupied by two electrons of opposite spin. This would yield a neon ground state with an electron density given by

$$\rho(\mathbf{r}) = 2 \sum_{n=1}^{5} |u_n(\mathbf{r})|^2$$

If $\rho(\mathbf{r})$ differs significantly from the electron density, $\rho_0(\mathbf{r})$, which we used in step (2), then we must replace ρ_0 by ρ and go back to step (2). We repeat this cycle until we find that the input density and the output density agree. This means that the electrostatic potential that is being used in Schroedinger's equation for the neon ground state is the one that is actually produced by the distribution of electrons in neon.

7.12. LENGTHS OF THE PERIODS

The energy of an electron in a simple Coulomb potential depends only upon the value of the principal quantum number, n. If the energy eigenvalues of electrons in the actual potential fields of many-electron atoms also had the property of being independent of the quantum numbers, l, m, and σ, then the

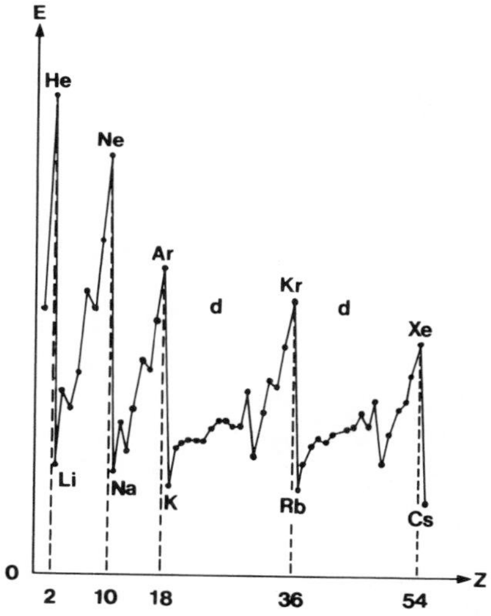

FIGURE 7.15 Ionization energies for atoms of the first five periods. The ionization energy is defined as the energy required to remove a single electron from the neutral atom of that element. The effects of atomic shell structure are apparent. If the outer electron in the atom is alone in a major shell, as it is for Li, Na, K, Rb, and Cs, it is easily removed. In contrast, if the last electron of the atom just fills a major shell then a large amount of energy must be supplied to remove it.

length of the nth period in the periodic table (i.e., the number of elements in the nth horizontal row) would be equal to $2n^2$. This is a consequence of the fact that there are $2n^2$ different combinations of l, m, and σ that can be combined with the principal quantum number n. One would therefore expect periods of lengths 2, 8, 18, 32, etc. Only the first two of these numbers match the actual periods. The reason for this lack of agreement is that the energy levels of an electron in a potential V(r) that is not a pure Coulomb potential depend on both quantum numbers, n and l. They are, however, still independent of m and σ. Thus the energy level must be written with two indices, E_{nl}, and has a degeneracy equal to $2l+2$. A noble gas occurs in the periodic table whenever there is a large gap between one degenerate energy level and the next. For exactly what values this will occur can only be ascertained by making detailed calculations of energy levels in realistic atomic potentials.

7.13. MOLECULAR STRUCTURE

There are two different pictures one can form of a molecule such as CO_2. One picture is patterned on the "ball and stick" models of molecules used in chemistry lectures. According to this picture CO_2 is composed of a ball called a carbon atom (usually black) and two balls called oxygen atoms that are stuck to the carbon atom by "covalent bonds", the covalent bonds being represented by little pegs that fit snugly into holes in the atom at its bonding sites. In the other picture CO_2 is composed of 25 particles, namely one carbon nucleus, two oxygen nuclei, and 22 electrons. The only forces allowed in this second picture are the attractive or repulsive electrostatic forces between the charged particles. There are no "covalent forces" between any of the particles. In most discussions of chemical reactions and molecular structure the first picture is the more practical one. However, in spite of the practical usefulness of the first picture it is the second picture that is the more truthful one. The basic aim of the quantum theory of molecules is to connect these two pictures in an exact and logical way. Starting from the fundamental picture of a molecule as a system of point particles* interacting with electrostatic forces it should be possible, by using the Schroedinger equation, to show that the nuclei and the electrons distribute themselves in such a way as to justify the use of the ball and stick picture as a reasonable approximation. Also, by looking at the energy eigenvalues of the Schroedinger equation it should be possible to predict the chemical bonding energies of the molecules. The configuration of the molecule, including all the distances between the atomic nuclei should come out of the calculation with no further assumptions. This is, of course, a very ambitious plan. It has been carried out successfully for almost all small molecules. The calculations are extremely complex and so we will not try to present them here. What we will do is:

a. We will describe the methods used in such a calculation for the simplest possible molecule, namely the hydrogen molecular ion.

b. We will discuss in a very qualitative way how the principles of quantum theory lead to the known tetrahedral bonding configurations in carbon compounds.

These two subjects should give the reader a small introduction to the very large field that has come to be called *quantum chemistry.*

---

*They can very well be considered point particles since the electron has no detectable radius and the size of a typical nucleus is .00001 times the size of a typical atom.

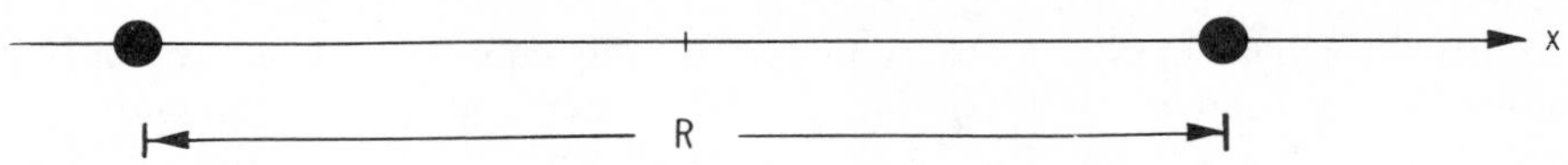

FIGURE 7.16 Loctions of the protons in H_2^+.

7.14. THE HYDROGEN MOLECULAR ION - H_2^+

By definition, a molecule is a bound system of electrons and nuclei containing more than one nucleus. Thus the simplest possible molecule is composed of two nuclei and one electron. The only molecule of this type is the hydrogen molecular ion, containing two hydrogen nuclei and an electron. We shall assume that the hydrogen nuclei are simply protons although a hydrogen nucleus could be a deuteron, made up of a proton plus a neutron, or a triton, composed of a proton plus two neutrons. At the end of this section we will discuss the question of why no other combinations of one electron plus two nuclei give bound molecules. The calculation we describe will begin by making the following important approximation.

The nuclei will be treated according to classical physics. That is, they will be assumed to have some definite fixed positions. They will not be described by a quantum mechanical probability distribution. We shall see shortly how their fixed positions will be determined. Treating the nuclei as classical particles is somewhat justified by the fact that they are very much heavier than the electron, and the quantum mechanical uncertainty in a particle's position is smaller for a heavier particle than for a lighter one. An even better justification is the fact that, once we have finished the description of this calculation, we shall see how one can improve it by treating the nuclear motion quantum mechanically. Thus, we picture two protons fixed at locations $(x, y, z) = (\pm R/2, 0, 0)$ (See Figure 7.16). The starting value of R, the distance between the protons we choose as some arbitrary but reasonable distance. (One or two Angstroms, not three meters.) We now want to set up the Schroedinger equation for the electron wave function. An electron at location (x, y, z) would experience a potential V(x, y, z) given by

$$V = -k e^2 \left(\frac{1}{r_1} + \frac{1}{r_2}\right)$$

where

$$r_1 = ((x - R/2)^2 + y^2 + z^2)^{1/2}$$

and

$$r_2 = ((x + R/2)^2 + y^2 + z^2)^{1/2}$$

The Schroedinger energy equation is therefore

$$-\frac{\hbar^2}{2m}\left(\frac{\partial^2 u}{\partial x^2} + \frac{\partial^2 u}{\partial y^2} + \frac{\partial^2 u}{\partial z^2}\right) - k e^2 \left(\frac{1}{r_1} + \frac{1}{r_2}\right) u = E u \qquad (7.30)$$

This equation has not been solved analytically. That is, no explicit function u(x,y,z) has ever been found that exactly satisfies the equation. However, with a digital computer it is a fairly simple task for an experienced quantum chemist to construct excellent approximate solutions to the equation. One can obtain an approximate ground state wave function $u_0(x,y,z)$ and an approximate ground state energy value E_0 such that Equation 7.30 is satisfied at all values of x, y, and z with a fractional error less than 10^{-8}.

The ground state wave function, so obtained has reasonable mathematical properties, such as:

1. It is symmetric with respect to reflection through the y-z plane. That is, $u_0(x,y,z) = u_0(-x,y,z)$.
2. It is symmetric with respect to rotation around the x axis. That is, u_0 has the same value for all points that have the same x coordinate and the same distance from the x axis.
3. It has its maximum values at the location of the protons and is generally large in the neighborhood of the two protons and near the line segment joining the two protons. That is to say, the electron probability density is large wherever the potential function V is small.

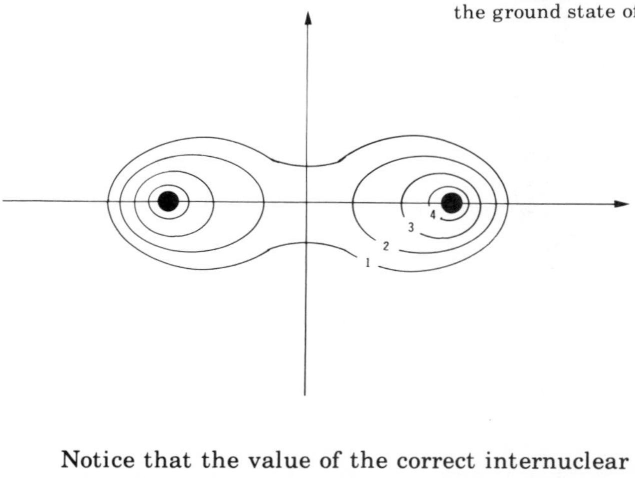

FIGURE 7.17
A contour plot of the electron probability density in the ground state of H_2^+.

Notice that the value of the correct internuclear distance R has still not been determined. It will be determined by the following rule. *The internuclear distance,* R, *is chosen so as to make the total energy of the system a minimum.*

The electronic ground state energy value, E_0, which is obtained by solving the Schroedinger equation (Equation 7.30) depends on the value of R we have chosen in setting up the equation. It represents the minimum possible energy of the electron if the nuclei are held fixed with a separation R. We will explicitly indicate its R dependence by writing it as $E_0(R)$. We must now add to $E_0(R)$ the energy of the protons in order to get the total energy of the system. Since we have assumed that the protons are stationary, they have no kinetic energy. But due to their mutual Coulomb repulsion they do have a potential energy equal to ke^2/R. (The potential energy due to the Coulomb attraction between the protons and the electron has already been taken into account in setting up the

Schroedinger equation and is therefore included in $E_o(R)$.) Thus the total energy of the system, with stationary protons separated a distance R, is

$$U(R) = E_o(R) + \frac{k e^2}{R} \tag{7.31}$$

The function U(R) is shown in Figure 7.18. From that figure one can pick out R_o, the minimizing value of the internuclear distance. R_o is the quantum mechanical prediction of the bond length in H_2^+. The quantum mechanical prediction of the bonding energy is given by $U(\infty) - U(R_o)$. This is the energy that would have to be supplied to separate the two protons from their equilibrium distance to infinity.

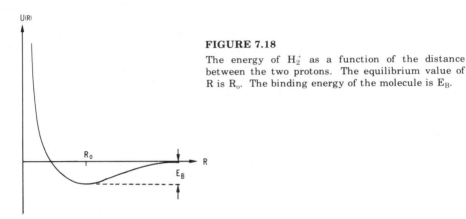

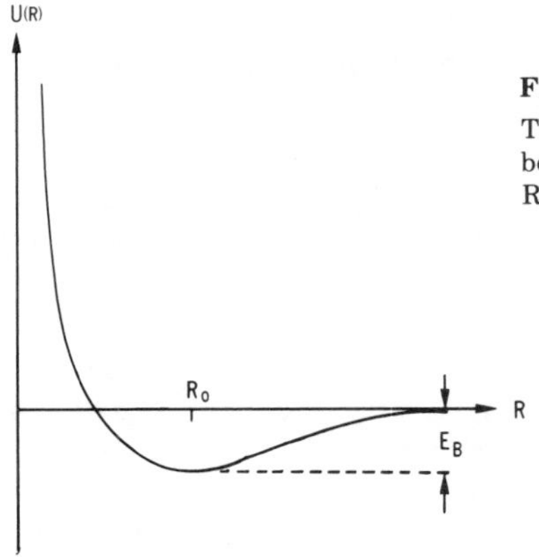

FIGURE 7.18

The energy of H_2^+ as a function of the distance between the two protons. The equilibrium value of R is R_o. The binding energy of the molecule is E_B.

QUESTION: Why are there no one-electron molecules except H_2^+?

ANSWER: One could carry out the calculation that has just been described on a system of, let us say, two helium nuclei plus an electron. This calculation of U(R) could be accomplished in much the same way as for H_2^+ but, due to the large contribution of the purely repulsive internuclear Coulomb force, the function $U(R) = E_o(R) + 4ke^2/R$ would have its minimum value at infinity. Thus the lowest energy state would be one in which the two nuclei were completely separated with the electron attached to one of them. The same thing happens for every combination of nuclei except two hydrogens.

7.15. MOLECULAR VIBRATION

In the last section we described how to calculate the nuclear configuration of minimum energy for H_2^+ using the approximation that the nuclei were stationary classical particles. We now want to improve on that analysis by removing that approximation. We will do so in two steps. First, continuing to treat the nuclei as classical particles, we will consider the problem of how to calculate their motion in case they are not initially stationary and at their equilibrium separation. This will immediately suggest how to write a Schroedinger equation for the nuclear wave function.

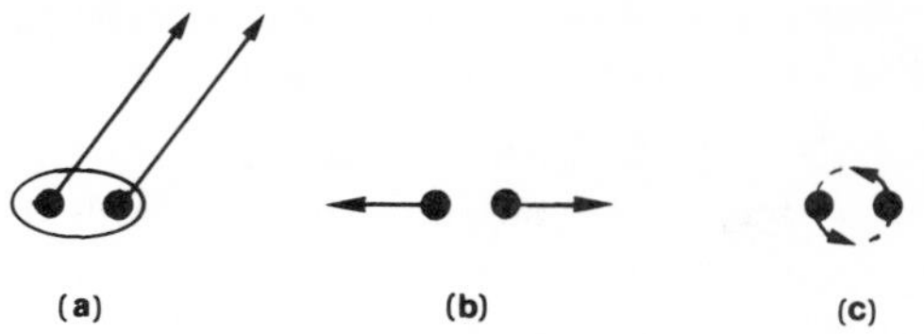

FIGURE 7.19
Three types of motion for a diatomic molecule.
(a) translation
(b) vibration
(c) rotation

One can easily picture three distinctly different types of nuclear motion (See Figure 7.19).

1. **Translation**. In this type of motion the relative position vector, **R**, remains constant while the molecule moves freely through space at constant velocity. This kind of motion is completely uninteresting and can be eliminated entirely by using a coordinate frame in which the center of mass of the molecule is at rest.

2 **Vibration**. In this type of motion the two nuclei oscillate in opposite directions about their equilibrium position while the center of mass remains at rest. The magnitude of the relative position vector, **R**, oscillates but its direction remains constant.

3. **Rotation**. In the third type of motion the magnitude of **R** remains constant but the vector rotates at constant angular velocity about its center. In order to produce the necessary centripetal force the constant magnitude of **R** would have to be somewhat larger than its equilibrium value just as two masses connected by a spring will cause the spring to stretch beyond its equilibrium length if the whole system is rotating about its center of mass.

It is obvious that, if the molecule were translating and vibrating simultaneously, those two motions would not affect one another. In fact, as we mentioned above, one could simply observe the molecule in its center of mass frame in order to eliminate the translation entirely so that the motion would become purely vibrational. It is also clear that if the molecule were simultaneously translating and rotating a similar thing would be true. What is not so obvious, but can be proven, is that if the molecule is simultaneously rotating and vibrating, then, as long as the amplitude of vibration is much less than R, those two motions also do not affect one another. That is, we could observe the molecule in a rotating frame in which its motion would look like a simple vibration. Thus we can analyze vibrational and rotational motions separately and assume that, in the general case, both things will be going on simultaneously. We will look at vibration first.

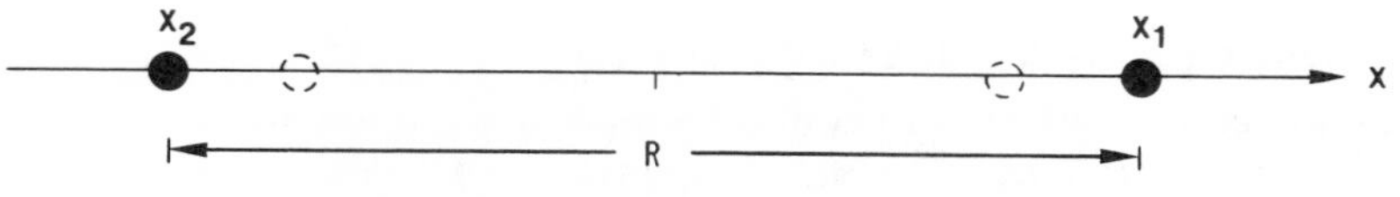

FIGURE 7.20 For pure vibrational motion in H_2^+ the displacements from equilibrium of the two protons are equal and opposite.

We defined the quantity U(R) as the total energy of the molecule when the nuclei were stationary with separation R. If we assume that the nuclei are moving, then we must add the kinetic energy of the nuclear motion to U(R) in order to obtain the total energy of the system. Referring to Figure 7.20 we see that the total energy E is given by

$$E = \frac{1}{2}M(\frac{dx_1}{dt})^2 + \frac{1}{2}M(\frac{dx_2}{dt})^2 + U(R)$$

But $x_1 = \frac{1}{2}R$ and $x_2 = -\frac{1}{2}R$ which gives

$$E = \frac{1}{2}\mu(\frac{dR}{dt})^2 + U(R) \qquad (7.32)$$

where the *reduced mass*, $\mu = \frac{1}{2}M$.

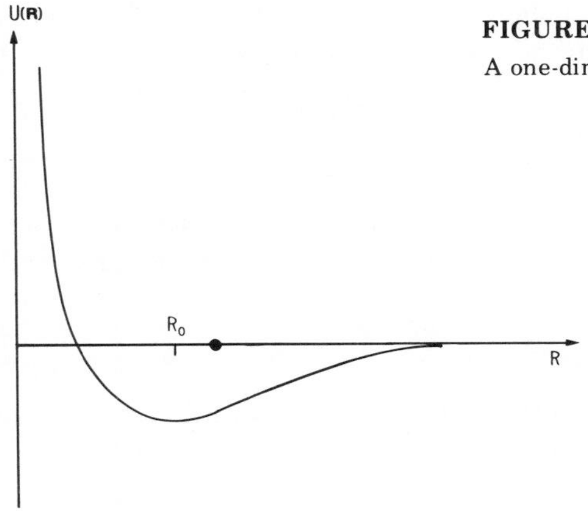

FIGURE 7.21
A one-dimensional particle in the potential U(R).

This is exactly the energy equation one would obtain for a one-dimensional particle of mass μ, with coordinate R, in a potential U(R). Thus the internuclear separation, R(t), behaves in the same way as the coordinate of the imaginary particle shown in Figure 7.21. If R is initially equal to the equilibrium value, R_o, and dR/dt is initially zero, then the particle will remain at rest at the location of the potential minimum. If it is slightly disturbed it will oscillate about the equilibrium position. For small amplitude oscillations one can approximate the function U(R) by a parabola that matches the value and curvature of U(R) at the point R_o. That is, one can use the approximation

$$U(R) \approx U_o + \frac{1}{2}k(R - R_o)^2 \qquad (7.33)$$

with $U_o = U(R_o)$ and $k = U''(R_o)$. With this approximation the motion becomes simple harmonic motion with an angular frequency, ω_o, given by

$$\omega_o = \sqrt{k/\mu} \qquad (7.34)$$

We have been using classical mechanics to describe the vibrational motion of the H_2^+ molecule. We found that the motion was indistinguishable from the motion of a one-dimensional particle of mass μ in a potential U(R). This correctly suggests that a quantum mechanical wave function v(R) for molecular vibrations can be obtained from the one-dimensional Schroedinger equation

$$-\frac{\hbar^2}{2\mu}v''(R) + U(R)\,v(R) = E\,v(R) \qquad (7.35)$$

The lowest quantum mechanical energy states have wave functions that are concentrated around the potential minimum. That is, their wave functions become essentially zero far from the potential minimum at R_o. For those states we can use the quadratic approximation to the potential given in Equation 7.33. The Schroedinger equation then becomes the equation of a quantum mechanical harmonic oscillator. We have already discussed the energy spectrum of such a system. When we take into account the fact that the minimum of the quadratic potential in Equation 7.33 is U_o while the minimum of the harmonic oscillator potential considered in Chapters 4 and 5 was zero we obtain the energy spectrum

$$E_n = U_o + (n + \tfrac{1}{2})\hbar\omega_o \qquad (n = 0, 1, 2, \cdots) \qquad (7.36)$$

where ω_o is the classical angular frequency of vibration, given in Equation 7.34. Thus we would expect to obtain an absorption and emission spectrum with angular frequencies ω_{mn} given by

$$\omega_{mn} = \frac{E_m - E_n}{\hbar} = (m - n)\omega_o \qquad (m > n)$$

We will postpone the comparison between this theoretical prediction and the observed spectrum until after we have discussed the effects of molecular rotation.

7.16. MOLECULAR ROTATION

The relationship between the kinetic energy of a rotating object and the angular momentum L of the object involves a quantity, I, called the moment of inertia of the object. It is

$$E = \frac{L^2}{2I} \qquad (7.37)$$

In calculating the moment of inertia of the H_2^+ molecule we will ignore the small contribution of the electron mass. In order to allow the later application of our results to a general diatomic molecule, we will calculate the moment of inertia of a system composed of two unequal masses M_1 and M_2, rotating about their center

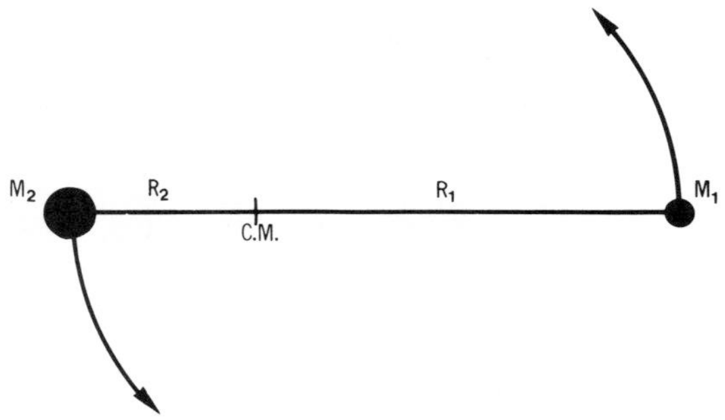

FIGURE 7.22 Rotation of a diatomic molecule about its center of mass. The center of mass is defined by the relation $M_1 R_1 = M_2 R_2$.

of mass. For such a system I is given by

$$I = M_1R_1^2 + M_2R_2^2 \tag{7.38}$$

where R_1 and R_2 are the distances of M_1 and M_2 from the center of mass. But, by the definition of the center of mass (see Figure 7.22),

$$R_1 = \frac{M_2}{M_1 + M_2} R$$

and

$$R_2 = \frac{M_1}{M_1 + M_2} R$$

Using these formulas for R_1 and R_2 in Equation 7.38 we get an equation for I in terms of the internuclear distance R and the reduced mass $\mu = \dfrac{M_1 M_2}{M_1 + M_2}$.

$$I = \mu R^2 \tag{7.39}$$

Notice that, for the case H_2^+, in which $M_1 = M_2 = M$ the reduced mass $\mu = \frac{1}{2} M$ and is therefore identical with the quantity μ used in the discussion of vibrational motion. According to classical physics the quantity L^2 can have any nonnegative value. The transition to quantum mechanics is accomplished simply by restricting the possible values of L^2 to those given in Equation 7.19. That is

$$L^2 = l(l+1)\hbar^2 \quad (l = 0, 1, 2, ...) \tag{7.40}$$

Substituting the expressions for I and L^2 given in Equations 7.39 and 7.40 into Equation 7.37 for the rotational energy gives the rotational energy spectrum of a diatomic molecule such as H_2^+.

$$E_l = \frac{l(l+1)\hbar^2}{2\mu R^2} \tag{7.41}$$

7.17. SELECTION RULES FOR MOLECULAR TRANSITIONS

The general calculational methods that have been described with reference to the spectrum of H_2^+ can be directly extended to arbitrary diatomic molecules. The two nuclei, which in the general case are no longer protons, are first treated as stationary classical particles with a separation R. The ground state energy of the system of electrons within the Coulomb fields of the two nuclei must then be calculated. This is usually done by means of the Self-Consistent Field Approximation. That calculation yields the total electronic energy as a function of R. The potential energy due to the Coulomb repulsion between the nuclei (namely $k Z_1 Z_2 e^2/R$) must be added to the electronic energy to obtain the function U(R) which can then be used in the calculation of the vibrational energy spectrum. The vibrational and rotational energy levels have the same form as in H_2^+. (Naturally one must use the appropriate values of k, μ, and R.)

Since the vibrational and rotational motions do not interact, one can obtain the total energy of a diatomic molecule in the nth vibrational state and the lth rotational state simply by adding the expressions for the vibrational and rotational energies. Thus the complete energy spectrum is indexed by the pair of indices, n and l.

$$E_{nl} = U_o + (n+\tfrac{1}{2})\hbar\omega_o + \frac{l(l+1)\hbar^2}{2\mu R^2} \tag{7.42}$$

One would expect to see terms in the emission and absorption spectrum of a diatomic molecule corresponding to every possible transition from one vibrational-rotational energy state to another. In fact one sees only terms corresponding to transitions $(n,l) \rightarrow (n',l')$ that satisfy the *angular momentum selection rule*

$$l - l' = \pm 1 \tag{7.43}$$

The physical basis of this selection rule is quite simple. As mentioned in Section 7.9 the photon is a particle with an intrinsic angular momentum of one unit. When a photon is emitted, the angular momentum of the molecule must therefore change by one unit in order for the total angular momentum to be conserved. Because angular momentum is a vector quantity, its conservation can be accomplished by having the molecule's angular momentum either increase or decrease

7.18. THE VIBRATIONAL-ROTATIONAL SPECTRUM OF DIATOMIC MOLECULES

The frequencies in the vibrational-rotational spectrum of a diatomic molecule can be separated into two groups depending upon whether the angular momentum increases or decreases by one unit in the emission process. In the first case, $l \rightarrow l+1$, $n \rightarrow n' = n - \Delta n$ and the emitted photon has an angular frequency ω given by

$$\hbar\omega = (n-n')\hbar\omega_o + \frac{\hbar^2}{2\mu R^2}[l(l+1) - (l+1)(l+2)]$$

This gives

$$\omega = \Delta n\,\omega_o - \frac{\hbar}{\mu R^2}(l+1) \tag{7.44}$$

where $l = 0, 1, 2, \cdots$ In the process $l \rightarrow l-1$, $n \rightarrow n' = n - \Delta n$ the angular frequency of the emitted photon is

$$\omega = \Delta n\,\omega_o + \frac{\hbar}{2\mu R^2}[l(l+1) - (l-1)l] = \Delta n\,\omega_o + \frac{\hbar}{\mu R^2}l \tag{7.45}$$

where $l = 1, 2, 3, \cdots$.

For typical diatomic molecules the vibrational angular frequency is about 10^{14} Hz. The wavelength of a photon of that frequency ($\lambda = 2\pi c/\omega_o$) is about $20\,\mu$m. This wavelength is in the infrared region (The longest wavelength of visible light is $.7\,\mu$m). According to Equations 7.44 and 7.45, the separation between emission lines that differ by one unit of l is equal to $\hbar/\mu R^2$. This quantity is called the *rotational constant* of the molecule. The rotational constant for typical diatomic molecules may be estimated by taking μ as 15 atomic mass units and R as $2\,\text{Å}$. This gives an angular frequency of 10^{11} radians/sec, which is much smaller than the typical vibrational angular frequencies. Therefore the vibrational-rotational spectrum has the structure of a series of closely spaced *rotational bands* as shown in Figure 7.23. The frequency of a line in the emission or absorption spectrum is mainly determined by the value of Δn, but for any given Δn there are a large number of evenly and closely spaced lines that depend on the value of l. Notice that there is no line of angular frequency equal to $\Delta n\,\omega_o$ exactly.

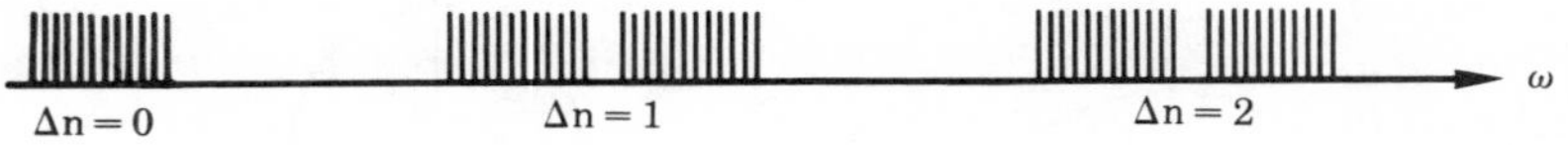

FIGURE 7.23 The rotational-vibrational spectrum of a diatomic molecule consists of a sequence of closely spaced rotational bands. In reality the ratio of the spacing within a band to the distance between neighboring bands is much more extreme than is shown here.

QUESTION: The distance between adjacent lines in the rotational bands of KCl is 7.6×10^9 Hz. What is the internuclear distance in the KCl molecule?

ANSWER: The masses of the K and Cl atoms are respectively

$$M_K = 39\,u = 6.5 \times 10^{-26} kg$$

and

$$M_{Cl} = 35.5\,u = 5.9 \times 10^{-26} kg$$

Therefore $\mu = \dfrac{M_K M_{Cl}}{M_K + M_{Cl}} = 3.1 \times 10^{-26}$ kg. From Equation 7.45 we can see that the frequency difference between two lines that differ by one unit of l is

$$\Delta \nu = \frac{\hbar}{2\pi \mu R^2}$$

Setting $\Delta \nu$ equal to 7.6×10^9 Hz and solving for R gives

$$R = 2.6\,\text{Å}$$

7.19. COVALENT BONDING IN CARBON COMPOUNDS

Our aim in this section is to obtain a qualitative or pictorial understanding of how the principles of quantum mechanics and some of the simple properties of atomic quantum states leads to the tetrahedral geometry of many carbon compounds. The carbon atom contains two electrons in the 1s state and four electrons in states of the second major shell. The 1s electrons are so tightly bound and so close to the carbon nucleus that they are very little affected by the presence of other nearby atoms. Thus we will simply ignore them except for their effect of keeping other electrons out of the 1s state. We will first consider the problem of constructing a methane molecule (CH_4) beginning with the C^{4+} ion and adding one proton plus two electrons at a time until we have added the full complement of four protons and eight more electrons.

Since the C^{4+} ion is spherically symmetric it obviously does not matter in what direction we bring in the first proton. We will bring it in along the z-axis. We want to estimate the nature of the electronic quantum state of the pair of electrons that will be added along with the proton. The proton has an electric charge of +e but the C^{4+} ion has a charge of +4e and is therefore much more

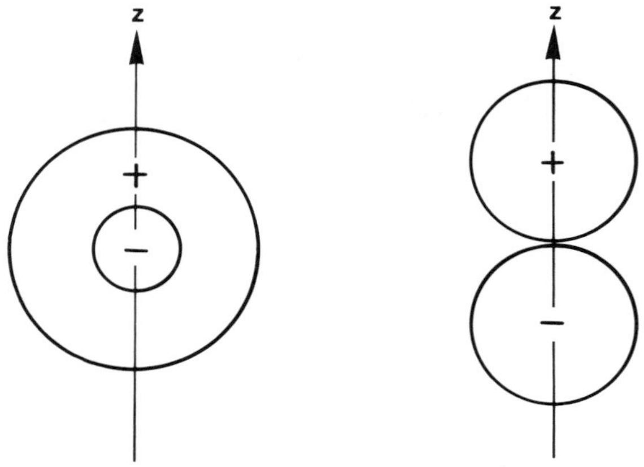

FIGURE 7.24 The states u_{200} and u_{210}. The wave function is positive in the regions marked + and negative in those marked −.

attractive to the electron pair than is the proton. The Coulomb field of the ion will dominate the field of the proton and therefore we can get a rough idea of the final electronic state by first considering the electronic states one could obtain with the ion electric field alone. For reasons that will shortly be made clear we will concentrate our attention on the following two second shell states; namely the states with (n,l,m) equal to $(2,0,0)$ and $(2,1,0)$. A pictorial representation of the states u_{200} and u_{210} is given in Figure 7.24. In the field of the C^{4+} ion alone these states have almost the same energy. However, if a proton were placed at the location on the z-axis where the electron density is a maximum, then there would be some advantage in having the pair of electrons that come with it occupy the u_{210} state since its electron density is somewhat more concentrated and in that state the electrons would be more likely to be found in the low potential energy region near the proton. While there would be some advantage in having the electrons occupy the u_{210} state rather than the u_{200}, it is clear that the advantage would not be very great because in the u_{210} state an electron is just as likely to be found on the negative z side of the atom, far from the proton as on the positive z side where its potential energy is lowest. We now want to show that, if we take a linear combination of the two states, u_{200} and u_{210}, we can find a quantum state that gives a much lower potential energy than either of the two states alone. Such a state is called a **hybrid** state. Mathematically it is of the form

$$v = a\,u_{200} + b\,u_{210}$$

where a and b are constants. Before we discuss the question of how to choose the two constants, let us see that the hybrid state is a solution of the Schroedinger equation for the Coulomb potential produced by the C^{4+} ion alone and that it has the same energy as either of the states u_{200} or u_{210}. (We are here using the approximation that those two states are degenerate. This would be exactly true if the ion produced a pure Coulomb field.) This is a consequence of the following theorem. We shall state the theorem for the case of a one-dimensional system but it obviously is also true in three dimensions.

Theorem: If the functions $u_1(x)$ and $u_2(x)$ are both solutions of the Schroedinger energy equation with the same energy, E, then the function $v(x) = a_1u_1(x)+a_2u_2(x)$ is also a solution of the Schroedinger equation with the energy E.

The proof of this theorem is left as an exercise (See Problem 7.15). Consider the hybrid state

$$v = \frac{1}{2}u_{200} + \frac{1}{2}u_{210}$$

As shown in Figure 7.25 the electron density associated with this state is concentrated along the positive z-axis.

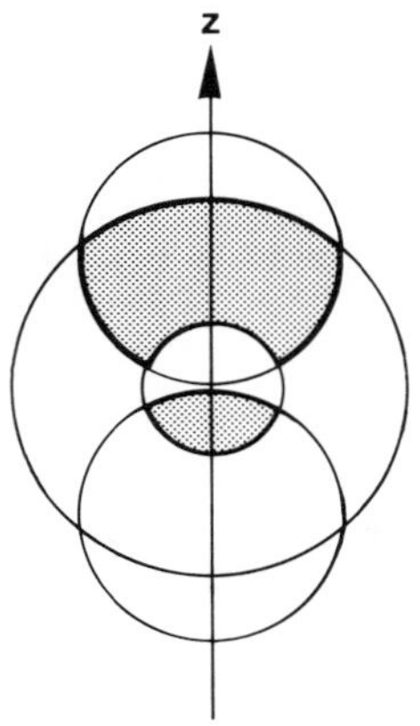

FIGURE 7.25 The two functions u_{200} and u_{210} have the same sign in the shaded regions. Therefore the function v is largest in those regions.

It would be better, from the point of view of minimizing the total energy, to have the two electrons occupy the state v and have the proton located at the location of maximum electron density rather than having the electrons in either of the states u_{200} or u_{210}. A hybrid state in which the electron density is concentrated on one side of the nucleus is called a hybrid bonding state. Starting from the four states in the second shell, namely $(n,l,m) = (2,0,0), (2,1,0), (2,1,-1)$, and $(2,1,1)$, we can, by taking linear combinations of the form

$$v = a\,u_{200} + b\,u_{210} + c\,u_{211} + d\,u_{21-1}$$

and choosing the constants appropriately, construct a total of four different hybrid bonding states. The states so constructed have electron density lobes that stick out from the nucleus in the directions of the vertices of a tetrahedron.

These hybrid bonding states are not really solutions of the Schroedinger equation for the methane molecule. They are solutions of the Schroedinger equation with the Coulomb potential of the C^{4+} ion alone. When the potential created by the four protons and the potential created by electrons themselves are added to the Schroedinger equation, the true bonding states differ significantly from these simple hybrid states, particularly close to the protons where the true electron density is much more concentrated at the location of the proton. However, the basic tetrahedral geometry of the simple hybrid bonding states is maintained for the exact bonding states.

FIGURE 7.26 The density lobes of the four hybrid states point toward the vertices of a tetrahedron.

QUESTION: In the last section we constructed the first hybrid bonding state so that it pointed in some arbitrary direction that we designated the "z-axis". Since we can obviously choose the z-axis in an infinity of different directions and then, once we have chosen it, construct a hybrid bonding state in that direction, how can we say that there are only four different hybrid bonding states and that they point toward the vertices of a tetrahedron?

ANSWER: It is true that we could choose the initial hybrid state so that it pointed in any desired direction. However, once we chose the direction of the first hybrid state and occupied the state with two electrons, then, according to the exclusion principle, the next pair of electrons would have to go into a *different* state. We are now faced with the problem of what constitutes a different state in the context of the exclusion principle. If, for a spherically symmetric potential, such as that of C^{4+}, we take any state of nonzero angular momentum and rotate the wave function a little bit, then we get a new wave function that is still a solution of the Schroedinger equation. If it were allowable to put a pair of electrons into the original state and another pair into the slightly rotated state and still another pair into the same state rotated a slight bit more, etc., it is clear that the exclusion principle would have virtually no effect. It would then be possible to occupy the second shell with an arbitrarily large number of electrons. Somehow just slightly rotating a state cannot make it sufficiently different to be counted as a separate state. There is a certain mathematical condition that two wave functions have to satisfy in order for them to qualify as independent states that can each be occupied by a pair of electrons. If the two states are $u_1(x,y,z)$ and $u_2(x,y,z)$, the required condition is that

$$\int u_1^* u_2 \, dx \, dy \, dz = 0 \tag{7.46}$$

Two wave functions that satisfy Equation 7.46 are said to be *orthogonal* to one another. The exclusion principle demands that, if the state u(x,y,z) is occupied by a pair of electrons, then the next electron must go into a state that is orthogonal to u. In our previous application of the exclusion principle to atomic structure all the states involved were mutually orthogonal. It is the orthogonality condition that prevents arbitrary angles beween the hybrid bonding states. We can choose the first state in any direction we like, but if we then want to construct a second hybrid bonding state that is orthogonal to the first, we find that the angle between the two states is 109.5° (the tetrahedral angle). If we then try to construct a third hybrid bonding state that is orthogonal to the first two, we find that it must point toward the third vertex of the tetrahedron determined by the orientation of the first two states and so forth. Of course, we have not proven any of the statements we have just made. The proofs are rather complicated and cannot reasonably be given in an introductory course.

7.20. OTHER TETRAHEDRAL COMPOUNDS

If we were to add one unit of charge to the carbon nucleus in methane we would obtain a nucleus with the atomic number of nitrogen. By removing one of the outer protons we could then get an electrically neutral nitrogen compound. It would be ammonia, NH_3. In order to remain concentrated near the three protons the electron pairs would have to remain in something close to hybrid bonding states. Because of the tetrahedral orientation of the hybrid states one would expect the ammonia molecule to have the geometrical arrangement of a decapitated methane molecule. This expectation is in fairly close agreement with the facts (See Figure 7.27). Once electron pairs have been put into each of three hybrid bonding states in order to concentrate their negative charge densities near the positive protons, the only state left for the last electron pair is the fourth hybrid state directed toward the remaining tetrahedral vertex. Thus the last electron pair wave function sticks out toward the fourth vertex even though there is no proton there to attract the electronic charge. It is forced into that state by the exclusion principle. The resulting electronic charge distribution gives the ammonia molecule a large electric dipole moment. That large dipole moment accounts for many of the physical and chemical properties of ammonia, such as

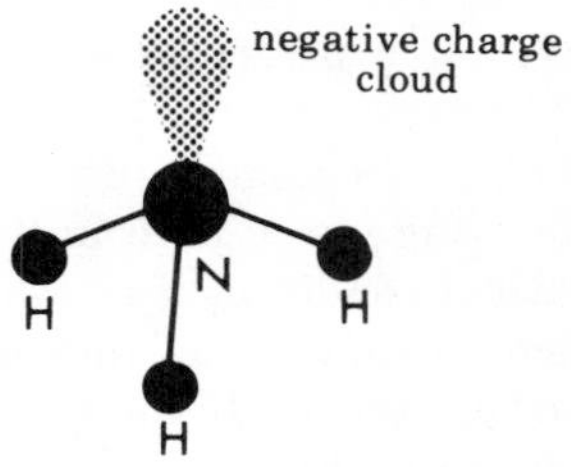

FIGURE 7.27

In the NH_3 molecule one electron pair forms a negative charge cloud pointing away from the tripod of hydrogen atoms. This gives the molecule a strong electric dipole moment.

its large solubility in water and other liquids of high dielectric constant. (The electrostatic energy of any fixed charge distribution is inversely proportional to the dielectric constant of the medium it is in. Therefore polar molecules have a lower energy in fluids of high dielectric constant.)

The element after carbon and nitrogen in the second row of the periodic table is oxygen. Therefore the next molecule in the sequence CH_4, NH_3, is H_2O. (Why not OH_2?) The angle between the two O-H bonds in the water molecule is also quite close to the expected tetrahedral angle. The tetrahedral geometry of bonding angles is maintained in almost all carbon compounds except those containing double or triple bonds involving the carbon atom. The subject of multiple bonds in carbon compounds will not be taken up here. We have done enough. The rest can be left to a course in organic chemistry.

SUMMARY

The hydrogen atom is a bound system composed of an electron and a proton. The *relative position vector*, **r**, is a vector from the proton to the electron.

The equation of motion for the relative position vector is the same as that for an electron of mass μ (the *reduced mass*) moving around an absolutely stationary proton. μ is given by

$$\mu^{-1} = m_e^{-1} + m_p^{-1}$$

The Schroedinger equation for an electron in a Coulomb field has positive and negative energy solutions.

The positive energy solutions describe the scattering of elections by protons and are therefore irrelevant to the structure of the hydrogen atom.

The negative energy solutions exist only for the energy values

$$E_n = -\frac{m\,k^2e^4}{2\hbar^2 n^2}$$

where $n = 1, 2, 3, \cdots$.

Classical mechanics gives a reasonably accurate description of states of the hydrogen atom in which $r >> a_0$, where a_0 is the *Bohr radius.*

$$a_0 = \frac{\hbar^2}{m\,k\,e^2}$$

A *radial solution* of the Schroedinger equation for hydrogen is a solution of the form $u(x,y,z) = f(\rho)$, where $\rho = r/a_0$.

The function $f(\rho)$ must satisfy the *radial Schroedinger equation*

$$f'' + \frac{2}{\rho} f' + \frac{2}{\rho} f = \epsilon f$$

where $\epsilon = -\dfrac{2\hbar^2}{m\,k^2e^4} E$

Each normalizable solution of the radial equation has the form of a polynomial times an exponential function.

$$f(\rho) = (a_0 + a_1\rho + \cdots + a_N\rho^N)e^{-\lambda\rho}$$

For each value of N there is one solution of this type. It has $\lambda = 1/(N+1)$ and $\epsilon = 1/(N+1)^2$.

The general solutions of the Schroedinger equation for hydrogen can be written in terms of the spherical coordinates r, θ, and ϕ as product functions.

$$u(x,y,z) = F(r)\,G(\theta)\,H(\phi)$$

There is one solution for each set of the three quantum numbers, n, l, and m that satisfies the following conditions

$$n = 1, 2, 3, \cdots$$

$$l = 0, 1, \ldots, n-1$$

$$m = 0, \pm 1, \pm 2, \ldots, \pm l$$

n is called the *principal* quantum number. It determines the energy of the quantum state

l is called the *angular momentum* quantum number. In the state u_{nlm} a hydrogen atom has a total angular momentum L given by

$$L^2 = l(l+1)\hbar^2$$

m is called the *magnetic* quantum number. The z-component of the angular momentum of a hydrogen atom in state u_{nlm} is

$$L_z = m\hbar$$

The wave function u_{nlm} is usually written as

$$u_{nlm}(r, \theta, \phi) = a_o^{-3/2} R_{nl}(\rho) Y_{lm}(\theta, \phi)$$

where the function $Y_{lm}(\theta,\phi)$ is called a *spherical harmonic.* The functions R_{nl} and Y_{lm} are normalized so that

$$a_o^{-3} \int_0^\infty R_{nl}^2 r^2 dr = 1$$

and

$$\int_0^{2\pi} d\phi \int_0^{\pi} Y_{lm}^* Y_{lm} \sin\theta \, d\theta = 1$$

If electron spin is ignored, the hydrogen atom would be expected to have a magnetic moment μ, related to its angular momentum by

$$\mu = \frac{e}{2m} L$$

This would lead to a splitting of energy levels for a hydrogen atom placed in a magnetic field. The expected change in the energy spectrum is given by the *Normal Zeeman Effect* formula

$$E_{nlm} = E_n + \frac{e}{2m} B\hbar m$$

What is actually found is a more complicated splitting due to the fact that the electron has a *spin angular momentum* of magnitude

$$s = \sqrt{\tfrac{1}{2}(\tfrac{1}{2}+1)}\,\hbar$$

The z component of the spin angular momentum has only two possible values; $\pm \hbar/2$

The *Pauli Exclusion Principle* states that:

No two electrons in a many-electron system can have exactly the same set of quantum numbers.

In a neutral atom of atomic number Z:

1. Most of the mass is concentrated in a *nucleus* of extremely small volume and high density. The nucleus is composed of a number of neutral *neutrons* and Z positive *protons.*
2. Surrounding the nucleus is a distribution of Z electrons in Z different quantum states. The *configuration* of an atomic ground state is a listing of the occupied states.

H_2^+ is a system composed of two hydrogen nuclei and one electron. The ground state of H_2^+ is calculated by the following steps:

1. For a fixed distance R between the nuclei, solve the Schroedinger equation for the electron in the combined Coulomb fields of the two nuclei. This gives the electronic energy function $E_o(R)$.
2. Add the nucleus-nucleus interaction potential to $E_o(R)$ to obtain the total energy function $U(R) = E_o(R) + k\,e^2/R$.
3. The equilibrium distance between the nuclei in H_2^+ is given by the value of R that minimizes U(R).

The motion of a diatomic molecule can be considered as composed of three separate types of motions, occurring simultaneously. They are:

Translation. Uniform motion of the center-of-mass with the relative position vector **R** remaining constant.

Vibration. Oscillation of the two nuclei in opposite directions with fixed center-of-mass and fixed direction of **R**. (Only the magnitude of **R** varies.)

Rotation. The magnitude of **R** remains constant but the vector rotates at constant angular velocity about the fixed center-of-mass.

For small amplitude vibrations the quantized vibrational motion leads to a harmonic oscillator vibrational energy spectrum with energy levels

$$E_n = U_o + (n + \tfrac{1}{2})\hbar\omega_o$$

where U_o is the minimum of U(R), $\omega_o = \sqrt{k/\mu}$, μ is the reduced mass, and $k = U''(R_o)$.

The rotational motion alone (ignoring the vibrational motion) gives a rotational energy spectrum with levels

$$E_l = l(l+1)\frac{\hbar^2}{2\mu R^2}$$

The combination of vibration and rotation occurring simultaneously gives the *vibrational-rotational energy spectrum* which consists of a series of closely spaced *rotational bands.*

$$E_{nl} = U_o + (n + \tfrac{1}{2})\hbar\omega_o + l(l+1)\frac{\hbar^2}{2\mu R^2}$$

The allowed transitions must satisfy the *angular momentum selection rule.* $l \rightarrow l \pm 1$ In the case $l \rightarrow l+1$ *and* $n \rightarrow n - \Delta n$ the emitted photon has an angular frequency

$$\omega_{ph} = \Delta n\,\omega_o - (l+1)\frac{\hbar}{\mu R^2} \qquad (l = 0, 1, 2, \cdots)$$

In the case $l \rightarrow l-1$ and $n \rightarrow n - \Delta n$ the emitted photon has an angular frequency

$$\omega_{ph} = \Delta n\,\omega_o + l\frac{\hbar}{\mu R^2} \qquad (l = 1, 2, 3, \cdots)$$

A *hybrid state* is a combination of a 2s and a 2p state that has the electron density concentrated on one side of the nucleus.

From the single 2s and the three 2p states one can construct four orthogonal hybrid states whose electron densities point to the vertices of a tetrahedron.

PROBLEMS

7.1* Carry out the algebraic steps omitted in the derivation of Equation 7.13 and 7.14.

7.2* Derive the value of the normalization constant, c, for the ground state wave function of hydrogen $u_1 = c\,e^{-r/a_o}$.

7.3** (A) By trying a solution of the form $f = (a_o + a_1\rho)e^{-\lambda\rho}$ in Equation 7.14 obtain algebraic equations for the ratio a_1/a_o and the constants λ and ϵ.

(B) Using the normalization condition $4\pi \int_o^\infty f^2 r^2 dr = 1$ and the ratio derived in (A) determine a_o and a_1 completely.

7.4*** (A) Calculate the first and second derivatives of the function $f = (a_o + a_1\rho + a_2\rho^2 + \cdots + a_N\rho^N)e^{-\lambda\rho}$.

(B) Substitute the results you obtained in (A) into Equation 7.14 and cancel a common factor of $e^{-\lambda\rho}$ to obtain a polynomial equation.

(C) In order for a polynomial to be zero the coefficient of each power must be zero. Setting the coefficients of ρ^N and ρ^{N-1} equal to zero obtain the values of ϵ and λ.

(D) By setting the coefficient of ρ^{K-1} equal to zero (where $K < N$) obtain an equation for the ratio of successive terms a_{K+1}/a_K.

7.5* For a hydrogen atom in its ground state what is the probability of finding the electron at any distance larger than two Bohr radii from the proton?

7.6** Given any function g(r) the average value of g is defined as

$$\bar{g} = \int_o^\infty g(r)\,P(r)\,dr$$

where P(r) is the radial probability function of the atom. For the first two states of hydrogen show that the average value of the potential energy is equal to twice the total energy. This relation is actually true for all states of hydrogen.

7.7* Invert Equation 7.17 to obtain r, θ, and ϕ as functions of x, y, and z.

7.8* The inequalities $r_o \leqslant r \leqslant r_o + dr$, $\theta_o \leqslant \theta \leqslant \theta_o + d\theta$, and $\phi_o \leqslant \phi \leqslant \phi_o + d\phi$ define a small region in space. What is the volume of that region?

7.9* For a given value of n prove that there are $2n^2$ allowed combinations of l, m, and σ.

7.10** For a hydrogen atom in state (n,l,m) = (3,2,1) what is the probability of finding the electron within the cone defined by the inequalities $0 \leqslant \theta \leqslant \pi/4$?

7.11* According to the normal Zeeman effect the four degenerate hydrogenic states with n = 2 should split into a triplet of energy levels when the hydrogen atom is placed in a magnetic field.

(A) For a magnetic field of 1T calculate the separation between levels within the triplet.

(B) Calculate the ratio of the separation between levels in the triplet to the separation between the unshifted n = 2 and n = 1 levels.

7.12 Suppose the electron had an intrinsic spin of $\frac{3}{2}\hbar$ so that s_z could have the values $-\frac{3}{2}\hbar$, $-\frac{1}{2}\hbar$, $\frac{1}{2}\hbar$, and $\frac{3}{2}\hbar$. How many electrons could then be put into the major shell with n = 2?

7.13* The function $E_o(R)$ for the ion H_2^+ is defined in Section 7.14. Give arguments to show that:

(A) $E_o(0) = -8.72 \times 10^{-18}$ J and

(B) $E_o(\infty) = -2.18 \times 10^{-18}$ J.

7.14** Suppose the function $E_o(R)$, defined in Section 7.14, could be approximated by $E_o = -A(1 + \frac{3}{1 + R/a_o})$, where $A = 2.18 \times 10^{-18}$ J and a_o = Bohr radius.

(A) What would be the equilibrium distance between nuclei in the hydrogen molecular ion?

(B) What would be the ground state energy of H_2^+?

(C) What would be the vibrational frequency of H_2^+?

(D) What would be the moment of inertia of the H_2^+ molecule?

7.15* Prove the theorem in Section 7.19

CHAPTER EIGHT

NUCLEAR PHYSICS

8.1. THE NUCLEAR FORCE

In the chapter on Atomic Structure the nucleus of the atom was taken to be nothing more than a positively charged massive point object. From the point of view of the atomic electrons surrounding it, this is quite a good description of the atomic nucleus. The size of the nucleus is extremely small on an atomic scale. If we expanded a carbon atom until it was the size of a typical room, its nucleus would be the size of a grain of finely ground salt. In spite of its very small size, it accounts for 99.995% of the mass of the whole atom. Therefore the mass density within a nucleus is much greater than the mass density of any ordinary substance. The repulsive electrostatic forces between the positively charged protons in this tiny object are billions of times stronger than the electrostatic forces involved in atomic structure. Surely there must be a powerful new attractive force to overcome the disruptive force of like charges brought so close to one another. There is. It is a totally different kind of force, called the *strong force*, which attracts protons to protons in spite of their electrostatic repulsion. It also attracts protons to neutrons, and neutrons to neutrons. The nuclear force is not only adequate to overcome the electrostatic repulsive forces. At typical nuclear distances it is so strong that in the lighter nuclei such as carbon it is an excellent approximation to completely neglect the comparatively weak electrostatic force in analyzing the internal structure of the nucleus.

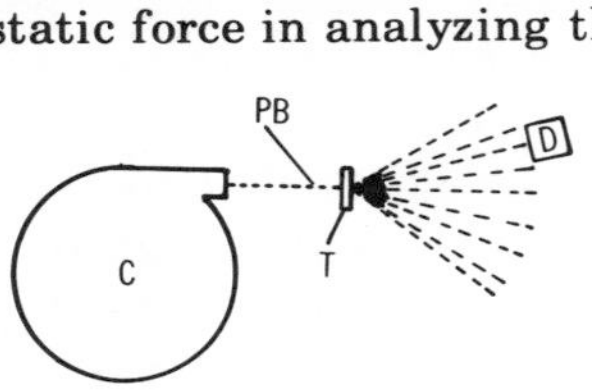

FIGURE 8.1

A proton-proton scattering experiment. The proton beam (PB) from the cyclotron (C) strikes a target (T), containing protons and is scattered into the detector (D).

The most direct way to study the forces between nuclear particles is by means of two-nucleon scattering experiments. A nucleon is a single term used to indicate either a proton or a neutron. In two-nucleon scattering a substance rich in hydrogen nuclei (i.e. protons) such as paraffin is used as a target. The target is placed directly in a beam of high velocity protons or neutrons. The flux of particles of each type leaving the location of the target is measured carefully with an appropriate type of particle detector (Figure 8.1). One then compares the observed distribution of particle flux with the flux that is predicted by Schroedinger's equation for various conjectured nucleon-nucleon forces. Actually it is the potential function associated with the nucleon-nucleon force that is directly used in the calculation. When the observed results of all two-nucleon scattering experiments can be theoretically reproduced by using the Schroedinger equation with a single potential function, one then says that one has an acceptable nucleon-nucleon potential. Of course, as scattering experiments are improved, the nucleon-nucleon potential may have to be refined. In broad outline the results of these investigations are as follows.

1. The nuclear force, in contrast to the electromagnetic force, is the same between two protons, two neutrons, or a proton and a neutron. It is thus said to be *charge independent.*
2. The nuclear force is very strong. Because of this, typical binding energies of nucleons in a nucleus are in millions of electron volts (MeV).
3. The nuclear force has a range of only about 10^{-15} m. The distance 10^{-15} m is called one *fermi* (1 fm). At a distance of a few fermis the nucleon-nucleon force is completely negligible.
4. The nuclear force is strongly repulsive at very short distances (less than about .5 fm) but attractive at larger distances (see Figure 8.2).

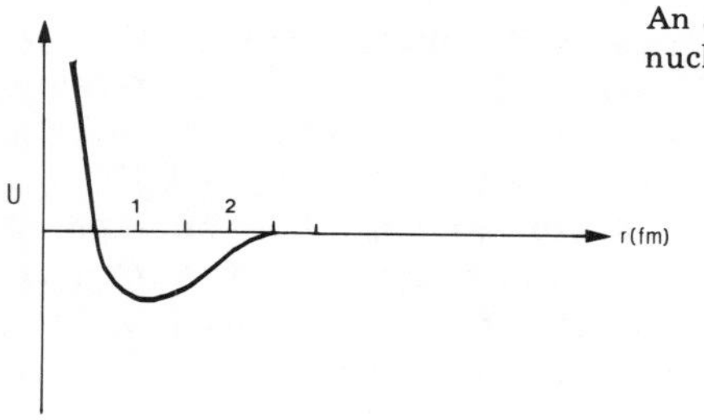

FIGURE 8.2

An approximate graph of the potential energy of two nucleons separated by a distance r.

5. The nuclear force is really much more complicated than Fig. 8.2 indicates. The proton and neutron are both particles with a spin angular momentum of $\hbar/2$. The force between two nucleons depends not only on their separation r but also upon the relative orientation of their spin axes and their relative velocities. We will not bother to describe it in complete detail.

QUESTION: The two-nucleon scattering experiments that have been described use proton targets. They would give information about the proton-proton force or the neutron-proton force. How can one determine the neutron-neutron force?

ANSWER: To measure the neutron-neutron force we use a target containing deuterium. Deuterium is a form of hydrogen in which the atomic nuclei are composed of a proton plus a neutron, bound together by the attractive neutron-proton force. If we scatter neutrons off deuterium then the scattering will result from two effects, namely scattering from the protons in the deuterium and scattering from the neutrons in the deuterium. Assuming that we have already carried out neutron-proton experiments in order to determine the neutron-proton force we can compute the effects of the scattering by the protons in the deuterium and eliminate them from the deuterium scattering data. This would give us information about the scattering of neutrons in the beam by neutrons in the deuterium. From that data it is possible to determine the neutron-neutron force.

8.2. NUCLEAR FLUID

We have seen that the sizes of atoms fluctuate greatly as the number of electrons is increased. There is only a slight trend for the size of an atom to increase as Z increases. If we compare atoms of similar type such as helium and krypton which are both noble gases, we see that krypton has less than twice the radius of helium, even though it contains eighteen times as many electrons. The electron density in krypton is therefore appreciably larger than the electron density in helium. In general, if we compare atoms with the same outer shell structure, the average electron density rises considerably as we go to atoms of higher and higher Z. This is quite understandable. The larger attractive electrical charge of the higher Z nucleus tends to pull in all the surrounding quantum states. This effect very much reduces the rate of growth of atoms in comparison to what one would expect if a fixed set of quantum states were filled with more and more electrons as Z increased.

Due to the fact that the nuclear force cuts off sharply at a few fermis and has a strongly repulsive core, there is no corresponding effect of continuously increasing density within the nucleus as the number of nucleons is increased. Because of the repulsive core, the nucleons in a larger nucleus must maintain a certain distance from one another. Because of the short range of the nuclear force, a particular nucleon only experiences the force of the other nucleons within its immediate neighborhood. Thus, if we consider nucleons in the interior of a nucleus, the potential energy due to nuclear forces experienced by a given nucleon is independent of the total number of nucleons in the nucleus (see Figure 8.3).

FIGURE 8.3
The force on any nucleon in the interior of the nucleus is due to its immediate neighbors and does not change when more nucleons are added to the system.

The basic structure of the nucleus bears a strong qualitative resemblance to the structure of a liquid drop. The similarity is only qualitative. In a nucleus the size of the repulsive core is much smaller in comparison to the average interparticle distance than it is in a liquid. Also the number of particles in a nucleus is vastly smaller than in a liquid, and the exclusion principle plays an important role in the structure of nuclei but not of liquids. The particle density of the nuclear fluid is about 1.5×10^{44} particles/m^3. Since the mass of each nucleon is about 1.67×10^{-27} kg, this gives a nuclear fluid mass density of

$$\rho_{\rm nuc} = (1.5\times 10^{44}\,\frac{\rm nucleons}{\rm m^3})(1.67\times 10^{-27}\,\frac{\rm kg}{\rm nucleon}) = 2.5\times 10^{17}\,\frac{\rm kg}{\rm m^3}$$

A cubic centimeter of nuclear fluid would weigh a quarter of a billion metric tons!

QUESTION: What is the diameter of a uranium-238 nucleus according to this liquid drop model?

ANSWER: The nucleus contains 238 nucleons. Setting

$$\tfrac{4}{3}\pi R^3 (1.5 \times 10^{44} \frac{\text{nucleons}}{\text{m}^3}) = 238$$

we get

$$R = 7.24 \times 10^{-15}\,\text{m}$$

The diameter is therefore 14.5 fm.

QUESTION: What is a neutron star?

ANSWER: It is a sphere of pure nuclear fluid with a typical diameter of twenty kilometers. This gives it a total mass of a few times the mass of the sun. Neutron stars are created when the compressive forces due to gravitational attraction within a star are so great that they overcome the forces that usually hold atoms apart from one another. The electronic structures surrounding the initially separate atomic nuclei collapse under the gravitational pressure and the nuclei finally fuse, forming a giant nucleus of kilometer dimensions.

TABLE 8.1
MASSES OF SELECTED NUCLEI

Isotope	N	Z	Mass of Nucleus (without electrons) In units* of u	In units of 10^{-27}kg
^{1}H	0	1	1.00728	1.67265
^{2}H	1	1	2.01356	3.34364
^{3}H	2	1	3.01551	5.00743
^{4}He	2	2	4.00152	6.64476
^{12}C	6	6	11.9967	19.9213
^{16}O	8	8	15.9906	26.5533
^{30}Si	16	14	29.9662	49.7606
^{38}Ar	20	18	37.9529	63.0231
^{38}Ca	18	20	37.9654	63.0439
^{48}Ca	28	20	47.9417	79.6100
^{54}Cr	30	24	53.9258	89.5471
^{56}Fe	30	26	55.9207	92.8507
^{60}Ni	32	28	59.9156	99.4934
^{108}Cd	60	48	107.878	179.134
^{124}Sn	74	50	123.878	205.707
^{232}Th	142	90	231.989	385.232
^{232}U	140	92	231.987	385.229
^{235}U	143	92	234.994	390.222
^{236}U	144	92	235.996	391.885
^{238}U	146	92	238.001	395.215

*Nuclear masses are often expressed in terms of *atomic mass units.* One atomic mass unit (1 u) is defined as exactly 1/12 of the mass of a neutral ^{12}C

The liquid drop analogy can be carried a little further in the following way. It is clear from Figure 8.3 that a nucleon near the surface of a nucleus would very likely feel a different potential than a nucleon well inside the nucleus. This is exactly the effect that, for a fluid, can be shown to lead to the phenomenon of surface tension. The nucleus also exhibits surface tension, and many dynamical effects which would be very difficult to analyze by considering the mechanics (or quantum mechanics) of individual particles in the nucleus can be treated more simply and with acceptable accuracy by treating the nucleus as a liquid drop of electrically charged fluid.

8.3. BINDING ENERGIES OF NUCLEI

An excellent example of the way in which the simple liquid drop model of the nucleus can be used to understand gross features of nuclei is the use of the model to explain the main effects determining nuclear binding energies. A nucleus is identified by two integers:

N = the number of neutrons in the nucleus
Z = the number of protons in the nucleus.

In terms of N and Z the total number of nucleons, A, which is also called the *mass number* of the nucleus, is given by

$$A = N + Z$$

The value of Z determines the chemical characteristics of an atom. It therefore tells us to what element a given nucleus belongs. For a given Z, the possible N values that can be coupled with it to form reasonably stable nuclei determine the different *isotopes* of the element of atomic number Z. For instance hydrogen can have N equal to 0, 1, or 2. Those three different isotopes are denoted ^1H, ^2H, and ^3H, where the number given as a left superscript is the mass number of the nucleus.

The masses of separated protons and neutrons are

$$m_p = 1.6727 \times 10^{-27}\,kg \quad \text{and} \quad m_n = 1.6748 \times 10^{-27}\,kg$$

However, the mass of a ^{12}C nucleus, which is composed of six protons and six neutrons, is (See Table 8.1.)

$$M(^{12}C) = 19.921 \times 10^{-27}\,kg$$

which is not equal to the combined masses of six protons plus six neutrons.

$$6\,m_p + 6\,m_n = 20.085 \times 10^{-27}\,kg$$

The mass of the combined system of twelve particles is less than the sum of the separated masses by an amount

$$\Delta m = 1.7 \times 10^{-28}\,kg$$

This is exactly the mass equivalent of the energy that would have to be supplied in order to separate the twelve particles in the nucleus against the action of the nuclear force (See Equation 3.12). This same energy would be released as heat and radiant energy if the separated nucleons were allowed to combine to form a nucleus of ^{12}C. This energy of combination is called the total *binding energy* of ^{12}C.

$$B(^{12}C) = \Delta m\, c^2 = 1.54 \times 10^{-11}\,J$$

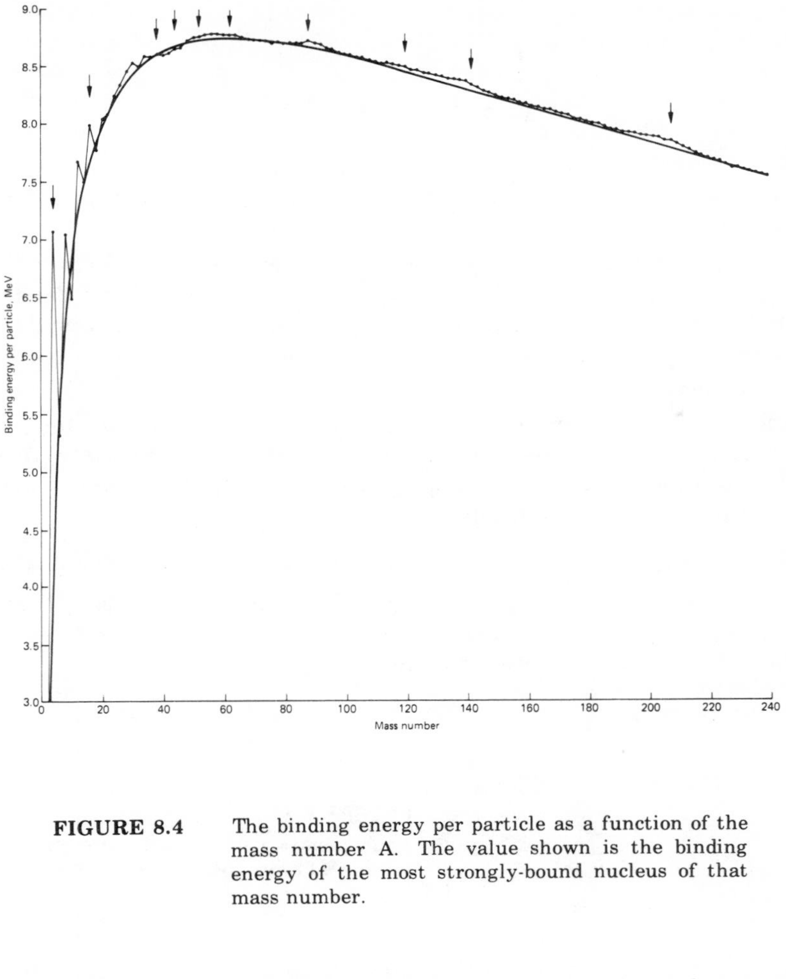

FIGURE 8.4 The binding energy per particle as a function of the mass number A. The value shown is the binding energy of the most strongly-bound nucleus of that mass number.

All known nuclei have masses that are smaller than the sum of the masses of the particles that compose them. For the nucleus of neutron number N and proton number Z, the binding energy is given in terms of the mass of the nucleus, M(N,Z), by

$$B(N,Z) = (N\,m_n + Z\,m_p - M(N,Z))c^2 \tag{8.1}$$

There is a formula, called the *Empirical Mass Formula* which predicts the binding energies of most nuclei remarkably well. We will first present the formula and then explain the origin of each term in the formula by means of the liquid drop model.

$$B(N,Z) = A\,b_{\text{volume}} - 4\pi\,R^2 b_{\text{surface}} - \tfrac{3}{5}\,\frac{k\,Z^2 e^2}{R} - \frac{(N-Z)^2}{A}\,b_{\text{symmetry}} \tag{8.2}$$

where $A = N+Z$ and R is the radius of a sphere containing A particles at the density of nuclear fluid. That is $A = \frac{4}{3}\,\pi\,R^3 \times (1.5\times 10^{44}\,\text{nucleons/m}^3)$ or

$$R = (1.2\times 10^{-15}\,\text{m})\times A^{1/3} = 1.2\,A^{1/3}\,\text{fm} \tag{8.3}$$

The constants b_{volume}, $b_{surface}$, and $b_{symmetry}$ must be determined experimentally. Actually they can be calculated from the Schroedinger equation using the nuclear potential but the calculation is extremely complex. The numerical values of the constants are

$$\begin{aligned} b_{volume} &= 2.49 \times 10^{-12}\,\mathrm{J} \\ b_{surface} &= 1.43 \times 10^{17}\,\mathrm{J/m^2} \\ b_{symmetry} &= 3.73 \times 10^{-12}\,\mathrm{J} \end{aligned} \tag{8.4}$$

Before we make any use of the formula, let us consider the physical meaning of each term in it.

b_{volume} is the binding energy per particle one would obtain for a very large piece of nuclear fluid with equal numbers of neutrons and protons if one could switch off the electrostatic interaction. All the other terms in the formula should be viewed as corrections to the term $A\,b_{volume}$ which tend to reduce the binding energy of the nucleus.

$-4\pi R^2 b_{surface}$ is the correction to take account of the fact that the particles near the surface of the nucleus do not form as many potential energy bonds as those in the interior. The constant $b_{surface}$ has exactly the same interpretation as the surface tension of a classical fluid.

$-\frac{3}{5}\frac{kZ^2e^2}{R}$ is a formula for the electrostatic energy of an amount of charge Ze distributed uniformly throughout a spherical volume of radius R. The fact that it is negative indicates that the repulsive electrostatic forces act to reduce the cohesion of a nucleus. They are disruptive forces.

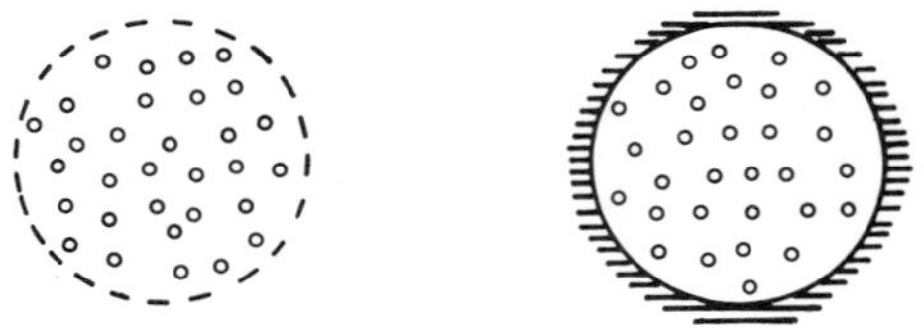

FIGURE 8.5
In our analysis the spherical surface of the nucleus is replaced by a spherical container.

$-\frac{(N-Z)^2}{A} b_{symmetry}$ is what is known as the symmetry energy. It can be understood from the following quantum theoretical argument. We know that the nucleus has a fairly definite surface. That is, the nucleons are very unlikely to be found at appreciable distances outside the surface. Let us therefore replace the spherical nucleus by a spherical volume of the same radius with hard walls. The interactions among the nucleons we will approximate by a constant potential throughout the volume which is equal to the average nuclear potential that a nucleon in the actual nucleus experiences. This is the same for protons and neutrons because of the charge independence of the nuclear potential. We ignore the electrostatic potential because its effects have already been taken into account in the electrostatic energy term. To calculate the energy of A nucleons in our simplified model of the nucleus we would first have to solve the Schroedinger energy equation (which has the same form for proton or neutron wave functions as for electron wave functions). From that equation we would obtain a sequence of quantum states u_1, u_2, $\cdots$. and corresponding energies E_1, E_2, $\cdots$. To get the lowest energy for a system of A nucleons in this spherical

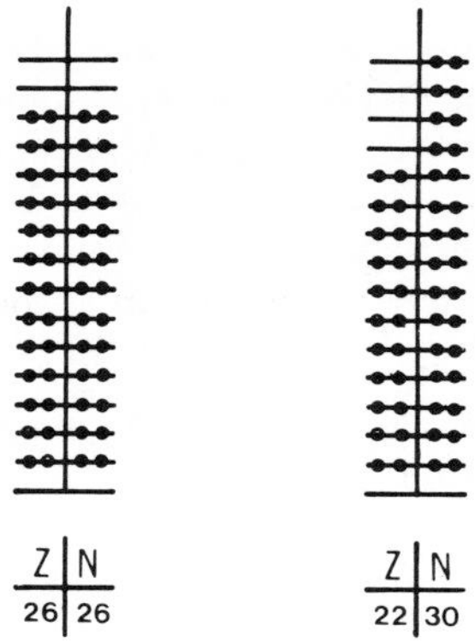

FIGURE 8.6

A. Occupation of the levels for the nucleus $(N,Z) = (26,26)$.
B. Corresponding diagram for the nucleus $(30,22)$. The energy of the nucleus $(30,22)$ is greater than the energy of the nucleus $(26,26)$ by the amount $\epsilon\,(N-Z)^2/8$, where $N-Z = 8$ in this case. It can be shown that ϵ is proportional to A^{1}. Thus the energy difference is proportional to $(N-Z)^2/A$.

box we fill up A of the quantum states in such a way as to obtain the lowest total energy. Both protons and neutrons must satisfy the Pauli exclusion principle. The lowest energy is obtained when we put two neutrons (with antiparallel spins) and two protons (also with opposite spins) in the lowest $A/4$ quantum states. (If A is not evenly divisible by four, then the last quantum state will not be completely filled). This would give us a nucleus with equal numbers of protons and neutrons. For that state $N-Z = 0$ and thus the expression for the symmetry energy would give zero. If we now want to modify the state so that we get a nucleus with the same number of nucleons A but unequal numbers of protons and neutrons (let us say $N > Z$), then we must remove a number of protons and replace them by neutrons. But when we do so we cannot put the neutrons into the same quantum states that the protons were in because those quantum states already are occupied by two neutrons each. We must therefore put the new neutrons into higher energy states than were occupied by the protons we removed. Thus the energy (not including the electrostatic energy) is greater for a nucleus with unequal numbers of protons and neutrons than for a nucleus of the same size with equal numbers of protons and neutrons. The symmetry energy term describes that increase in energy. (An increase in the total energy is a decrease in the binding energy.)

QUESTION: How do the surface, electrostatic and symmetry energy terms compare with the volume term for small and large nuclei?

ANSWER: As examples of small and large nuclei we will choose boron 10 and lead 208. All the desired quantities are calculated for the two cases in the table below.

	^{10}B	^{208}Pb
N	5	126
Z	5	82
R	2.6 fm	5.9 fm
$E_v = A\,b_{vol}$	2.5×10^{-11}J	5.2×10^{-10}J
$4\pi R^2 b_{sur}/E_v$	.5	.12
$\frac{3}{5}kZ^2e^2/RE_v$	.05	.3
$(N-Z)^2 b_{sym}/AE_v$	0	.07

For the small nucleus ^{10}B the surface effect substantially reduces the binding energy. For that nucleus the electrostatic energy is a small correction to the binding energy. For the large nucleus ^{208}Pb the relative importance of the surface and Coulomb energies is reversed. For both nuclei the symmetry energy is small. It does however play a significant role in determining the ratio of neutrons to protons in stable nuclei as we will see later.

Using Equation 8.3 to eliminate R in favor of A we can write the empirical mass formula in terms of the integer variables A, N, and Z.

$$B = [2.5\,A - 2.6\,A^{2/3} - .115\,Z^2/A^{1/3} - 3.7\,(N-Z)^2/A] \times 10^{-12}\,J \qquad (8.5)$$

NUCLEAR DECAY MODES

There are a number of ways in which a nucleus may spontaneously transform itself to reach a lower energy state. The three most important processes of spontaneous transformation are known as α-decay, β-decay, and γ-decay.

γ-decay is photo-emission. A nucleus that is in a state other than its ground state may make a transition to a lower energy state, emitting a photon which carries off the difference in energy between the two states. Since typical nuclear energies are in the Mev range, the photons emitted are in the energy range referred to as γ rays (see Figure 8.7).

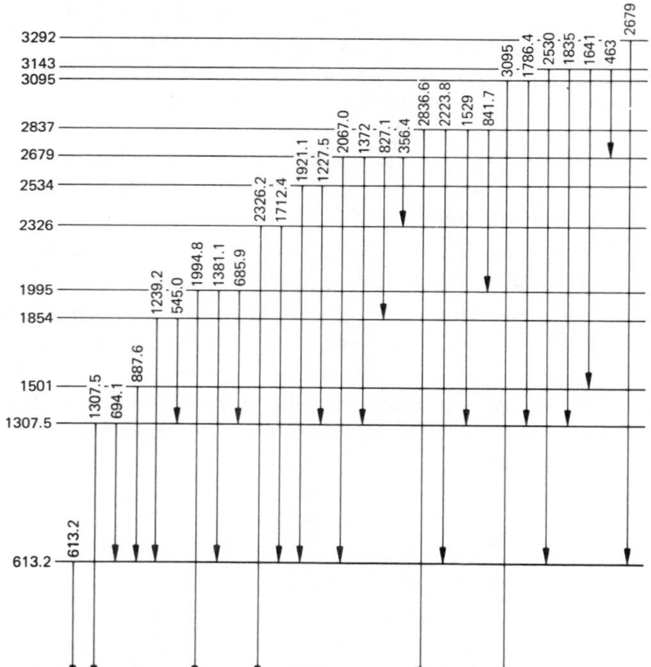

FIGURE 8.7 Some of the energy levels of the ^{78}Se nucleus, with the energies of observed emitted γ rays shown. All energies are in keV (kilo-electron-volts).

α-decay is a process in which a nucleus emits an α-particle. An α-particle is just a helium nucleus. It is a bound system composed of two protons and two neutrons. In an α-decay both the proton number and the neutron number of the nucleus are reduced by two. Thus the nucleus (N,Z) is transformed by an α-decay into the nucleus (N − 2,Z − 2). Because of the electrostatic energy term in the binding energy, all the large nuclei (Z > 82) are unstable with respect to α-decay. Left to themselves they will eventually transform by emitting an α-particle.

β-decay is a somewhat more exotic process. In the other two decay processes the particle ejected is either a massless uncharged particle (a photon) or a pre-existing part of the nucleus. In β-decay a charged massive particle is created or destroyed at the instant of decay. That particle is either an electron or a positron. The positron is a particle that is identical to an electron in every respect except its charge. It has a charge of +e rather than a charge of −e. Usually the number of electrons in a system cannot change because the total charge must remain constant. In β-decay the requirement that electric charge be conserved is satisfied by having a neutron change into a proton at the instant at which the electron is created. If the particle ejected is a positron, then a proton must be transformed to a neutron at the instant of decay. A β-decay in which the ejected particle is an electron is called a β^--decay. One in which the ejected particle is a positron is called a β^+-decay. One of the simplest and most important examples of β-decay is the β^--decay of free neutrons. An individual neutron has a mass that is larger than the combined masses of a proton and an electron.

$$m_n = m_p + m_e + 1.4 \times 10^{-30}\,\mathrm{kg}$$

Thus if the neutron underwent β^--decay, the resulting proton and electron would be expected to have a total kinetic energy of

$$(1.4 \times 10^{-30}\,\mathrm{kg})\,c^2 = 1.26 \times 10^{-13}\,\mathrm{J}$$

In actuality when β^--decay of neutrons is observed, the resulting particles have a wide range of kinetic energies but always less than the expected amount. Even more surprising is the fact that a neutron which is at rest may give rise to an electron and proton whose total momentum are not zero. Thus the decay process seems to violate both energy and momentum conservation (See Figure 8.8). The missing energy and momentum are always related by $E = cp$. The paradox has been resolved by the discovery that there is another particle involved in the β^--decay. The other particle is a zero mass particle of spin $\hbar/2$ called an antineutrino (written $\bar{\nu}$). The β^--decay reaction is actually of the form

$$n \rightarrow p + e^- + \bar{\nu}$$

The antineutrino has no electric charge. It has the peculiar property the its spin angular momentum is always parallel to its linear momentum vector. A particle whose spin vector is parallel to its momentum vector is said to have positive *helicity*. One whose spin vector is antiparallel to its momentum vector has

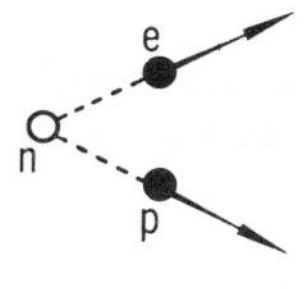

FIGURE 8.8

The β^--decay of a neutron does not seem to conserve momentum.

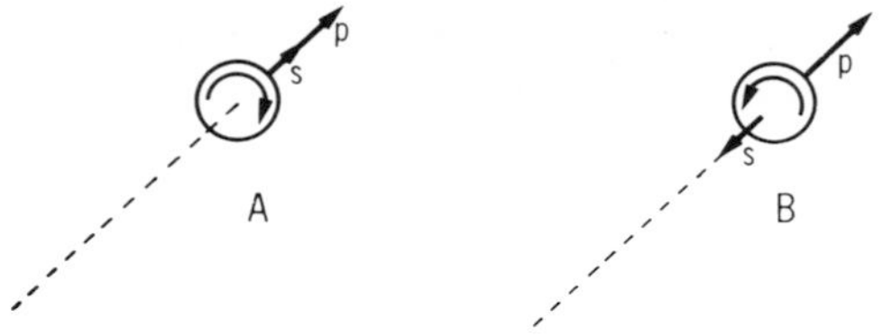

FIGURE 8.9 The antineutrino (A) and the neutrino (B) have opposite values of helicity.

negative helicity. The antineutrino is the antiparticle to another particle, the neutrino, which has exactly the same properties as the antineutrino but has its spin angular momentum vector always antiparallel to its momentum vector (i.e., it has negative helicity). (See Figure 8.9). The neutrino is written as ν in reaction equations. A neutrino is emitted along with the positron in β^+-decay*.

There is still another mode of β-decay called *electron capture.* In electron capture one of the orbital electrons is annihilated and, at the same instant, a proton is transformed into a neutron along with the emission of a neutrino.

$$p + e^- \rightarrow n + \nu$$

Since the electron involved in the electron capture reaction is almost always one of the electrons in the innermost shell, that is, the K shell, the process is sometimes called K-capture. One can remember which processes involve neutrinos and which antineutrinos by the following rule. We assign an integer parameter, called the *lepton number* to each type of particle that may be involved in β-decay. The lepton numbers of the neutron and proton are zero. The lepton numbers of the electron and neutrino are one and the lepton numbers of their *antiparticles*, the positron and the antineutrino, are minus one. The rule is then that β-decay processes conserve both charge and total lepton number.

In β^--decay the nucleus (N,Z) is changed to the nucleus(N−1, Z+1). In β^+-decay and electron capture the nucleus (N,Z) is changed to(N+1, Z−1).

8.4. THE LAW OF RADIOACTIVE DECAY

One characteristic is common to all modes of nuclear decay. It is that:

The probability that any given unstable nucleus will spontaneously decay during the short time interval dt is proportional to the time interval and independent of the past history of that nucleus and the state of any surrounding nuclei.

If at time t = 0 we begin with a sample containing N_0 unstable nuclei of one type they will undergo decay in a way which allows us to predict the overall rate of disintegrations quite accurately although we cannot predict the exact time when any particular nucleus will disintegrate. Let N(t) be the number of unstable nuclei still left at time t. During the interval from t to t + dt the

---

*Actually there are several different types of massless neutrinos and antineutrinos. The one we are describing here is the *electron neutrino.* Later, in the chapter on High Energy Physics we will also introduce the muon neutrino and the tau neutrino, each with its own antiparticle.

probability that any one of them will decay is equal to $\lambda\,dt$, where λ is some constant that depends upon the type of nucleus under consideration. The number of nuclei that will decay during the time interval dt is therefore $N(t)\,\lambda\,dt$. Thus the change in $N(t)$ during that time interval will be $dN = -N(t)\,\lambda\,dt$. This gives the basic equation of radioactive decay.

$$\frac{dN}{dt} = -\lambda\,N(t)$$

The solution of the equation which satisfies the requirement that $N(0)$ be equal to the given number N_0 is

$$N(t) = N_o\,e^{-\lambda t} \tag{8.6}$$

The constant λ is called the *decay constant* for that particular decay process. It has units of s^{-1}. There is another constant that is often used to describe the rate of a radioactive decay process. It is called the *half-life* of the nucleus. The half-life is defined as the time required for exactly half of the original nuclei to decay. Thus the half-life T is given by the equation $N(T) = \frac{1}{2}N_0$. Using Equation 8.6 we can relate the half-life to the decay constant.

$$\tfrac{1}{2}N_o = N_o\,e^{-\lambda T}$$

which gives

$$T = \frac{\ln 2}{\lambda}$$

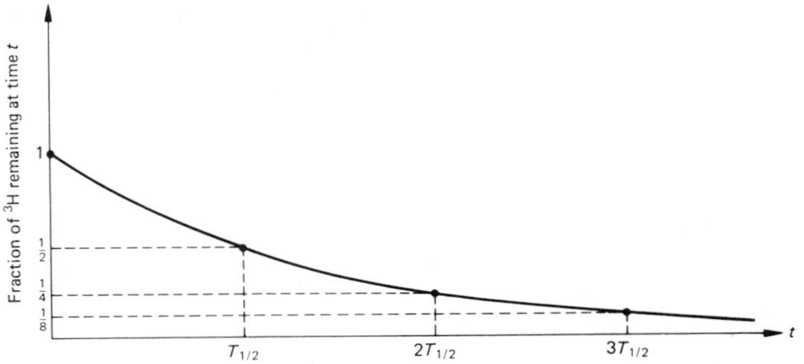

FIGURE 8.10 The β^--decay of ^3H to produce ^3He. In this case $T_{1/2} = 12.33\,yr$.

QUESTION: How are isotopes used as *tracers* in biology and medicine?

ANSWER: If we want to trace how a particular chemical, say strontium, is incorporated into the tissues of a biological organism we can feed (or inject into) that organism a known amount of some unstable isotope of that element. The unstable isotope is chemically indistinguishable from the more common stable isotope of the element. It will therefore be used by the organism in the same way as the stable isotope. However the nuclei of the unstable isotope will disintegrate at a known rate that is unaffected by their chemical environment within the organism. When unstable nuclei disintegrate they emit small amounts of various kinds of radiation. By detecting the emitted radiation one can track the movement of the chemical through the organism.

8.5. NUCLEAR LINE WIDTHS

The photons emitted in γ-decay have angular frequencies given by the Einstein relation

$$\hbar\omega = E_n - E_m$$

which seems to indicate that the γ emission spectrum should consist of perfectly sharp lines. Nothing real is perfect. Although the frequency distribution of the photons emitted consists of fairly sharp peaks they are not perfectly sharp. The photons associated with a particular transition $E_n \rightarrow E_m$ actually have a narrow distribution of energies. The width of that distribution is called the *line width* of the line. There are three principal sources of the finite line width.

Natural line width. If the transition $E_n \rightarrow E_m$ takes place spontaneously with a half-life of $T_{1/2}$ then, starting with the system in the upper level, the photon will be emitted within a time of order $T_{1/2}$. Thus the maximum spatial extent of the photon wave is about $c\,T_{1/2}$. This gives a finite (although usually quite large) bound on the uncertainty in position of the emitted photon; namely $\Delta x \approx c\,T_{1/2}$. The momentum of the photon must therefore have an uncertainty not less than $\hbar/c\,T_{1/2}$. Since $p = \hbar k = \hbar\omega/c$ we get that the minimum uncertainty in ω is

$$\Delta\omega \approx 1/T_{1/2} \quad \text{(Natural line width)}$$

This gives an absolute minimum to the unpredictability in the photon frequency for any transition with a finite half-life. Of course, all transitions must have a finite half-life in order ever to occur. This minimum uncertainty is called the *natural line width* of the transition.

Doppler broadening is the effect of the Doppler shifts caused by the random thermal motion of the atoms containing the emitting nuclei. If the radiating nucleus is moving toward the detector then its radiation will appear to the detector to be shifted upwards in frequency by an amount $\Delta\omega$, where $\Delta\omega/\omega = v/c$. If it is moving away from the detector then the frequency of its radiation will be downshifted by the same amount. Since any large collection of nuclei are always undergoing unpredictable thermal motions the distribution in frequency of the radiation they emit is broadened by this thermal Doppler effect. Typical thermal velocities can be estimated by setting $\frac{1}{2}mv^2$ equal to k_BT. Putting that value of v into the formula for $\Delta\omega/\omega$ we get an estimate of the thermal Doppler broadening of a spectral line at temperature T.

$$\frac{\Delta\omega}{\omega} \approx \frac{\sqrt{k_BT/m}}{c} \quad \text{(Doppler broadening)}$$

For a nucleus of mass number 60 and a temperature of 300 K this gives

$$\frac{\Delta\omega}{\omega} \approx 10^{-6}$$

Nuclear recoil. Since the emitted photon carries away momentum as well as energy the nucleus will recoil like a rifle firing a bullet. The recoiling nucleus will have kinetic energy and therefore the full energy released by the internal nuclear transition will not be available to the photon alone but will be shared between nucleus and photon. If we were dealing only with isolated nuclei, initially at rest, then the nuclear recoil effect would cause only a definite predictable downshift in the photon frequency. To calculate the magnitude of the frequency shift we set the recoil momentum of the nucleus equal to the momentum of the photon. $mv = \hbar k = \hbar\omega/c$, where m is the mass of the nucleus and ω is the angular frequency of the photon. The energy loss by the photon, $\hbar\Delta\omega$, is equal to the energy gain by the nucleus, $\frac{1}{2}mv^2$. This gives

$$\frac{\Delta\omega}{\omega} = \frac{\hbar\omega}{2mc^2} \quad \text{(Nuclear recoil)}$$

In reality, the nucleus is usually part of a solid or liquid sample. During the process of radiation some of the momentum and energy is transferred to the neighboring atoms in the sample and therefore the isolation assumed in our analysis does not exist and our final formula for the frequency shift can be taken only as an estimate of the effect. The emission line is broadened by the recoil effect and not merely downshifted.

QUESTION: Can the same analysis be applied to atomic and molecular emission lines?

ANSWER: Yes, if one additional line-broadening effect is taken into account. Nuclei are protected by the atomic shells around them from any significant interaction with other nuclei. Atoms or molecules are not. During their frequent collisions with one another the proximity of other atoms or molecules causes distortions in their wave functions and corresponding shifts in their energy levels. Atoms or molecules that radiate during such collisions will do so with altered emission frequencies. This effect is called *collision broadening.* It is very much enhanced by the fact that the very process of collision tends to stimulate the emission of a photon by an atom or molecule in an excited state.

8.6. THE MOSSBAUER EFFECT

It is not uncommon for the natural line widths of nuclear γ-decays to be very small. An important example is the decay, by photon emission, of a particular excited state of ^{57}Fe. The photon emitted has an energy of about 14.4 keV and the transition occurs with a half-life of about 7×10^{-8} s. Thus

$$\frac{\Delta\omega}{\omega} = \frac{1}{T_{1/2}\,\omega} = \frac{\hbar}{T_{1/2}(\hbar\omega)}$$

$$= \frac{(6.56\times10^{-16}\text{eV-s})^*}{(7\times10^{-8})(1.44\times10^{4}\text{eV})} = 6.5\times10^{-13}$$

$^*\hbar = 1.05\times10^{-34}$ J-s $= 6.56\times10^{-16}$ eV-s

In a *resonance absorption* experiment using ^{57}Fe one measures the absorption, in one sample of ^{57}Fe, of photons emitted by another sample of ^{57}Fe. If one could eliminate all other line broadening effects, then the emission and absorption lines would be so narrow that one could easily detect frequency shifts of one part in 10^{14} (and, with difficulty, even one part in 10^{15}). For example, one could detect the Doppler shift caused by a motion of one of the samples with a velocity of $10^{-14}c = 3\mu m/s$ or one could detect the redshift caused by having the emitter and absorber at different levels in the earth's gravitational field.

Our estimates of the thermal Doppler broadening and recoil effects seem to make it very clear that they would broaden the line by a factor of about 10^6 and thereby completely eliminate the extreme sensitivity suggested by the natural line width alone. During the late 1950's the German physicist, Rudolf Mossbauer realized that the arguments given above for estimating the recoil and Doppler broadening effects were not correct when applied to γ-decay by nuclei in solids at low temperatures. He carried out a more accurate quantum mechanical analysis and discovered that it should be possible to conduct resonance absorption experiments with no recoil or Doppler broadening effects at all. He then went on to support the correctness of his theoretical analysis by experimentally demonstrating the effect. It is now called the Mossbauer effect.

We can understand the basic mechanism of the Mossbauer effect without the fairly intricate mathematical analysis usually introduced to describe it by considering the simple system shown in Figure 8.11. A nucleus in an excited state is attached to a much heavier object by a Hooke's law force (i.e., a spring). The system has zero total momentum. We treat the internal vibration quantum mechanically and assume that, as a harmonic oscillator, it is in its ground state. The lighter particle clearly represents a radioactive nucleus, capable of γ-decay, that is imbedded in a solid. The heavy object represents the collection of all the other particles in the solid sample. In order to emphasize this relationship we will call the heavy particle *the crystal.* The spring is an approximate representation of the forces between a given radioactive atom and its neighbors in the crystal sample. We will assume that the nuclear excited state has a long half-life and therefore a well-defined energy. We will also assume something that superficially seems to directly contradict the last assumption; namely that the process of photon emission takes place essentially instantaneously. In particular, that it takes place in a time much shorter than a period of vibration of the oscillator. The apparent contradiction is based on an erroneous identification of the two times involved. To understand what is meant here, consider the following two facts. The average time period between occasions on which the author, who is of less than average agility, trips and falls flat on his face is about a month, but the time it takes to carry out the action is never more than a second. In a similar way, the average time that an excited nucleus lasts before it decays is

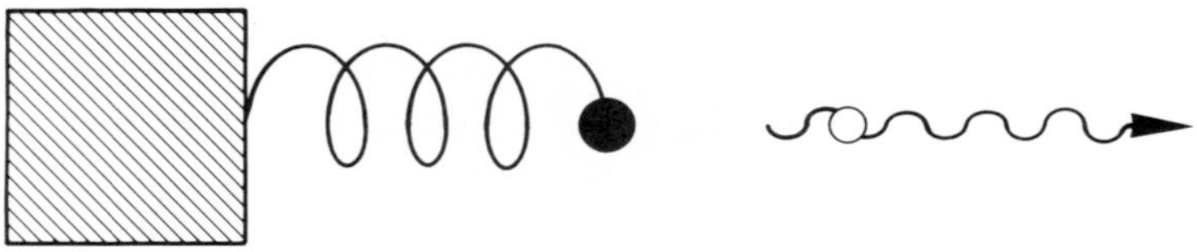

FIGURE 8.11 A radioactive nucleus attached to an object of large mass by a Hooke's law force.

about a half-life but the time taken for the process to occur is much less than a vibration period. Therefore, when the process occurs, a photon of momentum $p = \hbar k$ appears and the momentum of the nucleus undergoes an abrupt change from its initial value P_i to $P_f = P_i - p$. The kinetic energy of the nucleus changes abruptly by an amount

$$\Delta E = \frac{P_f^2 - P_i^2}{2m}$$

Since the nucleus moves a negligible amount during the photon emission process the potential energy does not change. Thus the formula above gives the change in the total energy of the system as a function of the photon momentum and the initial momentum of the nucleus. But the change in the energy of the oscillator must be equal to $n\hbar\omega_0$, where $n = 0$ if the oscillator stays in its ground state, $n = 1$ if it makes a transition to its first excited state, etc.. From Figure 8.12 one can see that there are only certain values of P_i that lead to allowed transitions. Since the harmonic oscillator ground state is not an eigenstate of the momentum of the nucleus there is a probability function for finding the nucleus with any given value of P_i (It looks very much like the momentum probability function for a particle in a box shown on page 162 and it is calculated in the same way). Using that momentum probability function one can calculate the relative probabilities of the different transitions. The details are not important, particularly since our model of a nucleus in a crystal lattice is an extremely crude one. What is significant is the simple fact that there is a finite probability that the system will emit the photon with no change in its internal quantum state. Let us carry the analysis a bit further in order to see what has happened to the recoil momentum of the photon. Although the initial state of the system was not an eigenstate of the momentum of the nucleus alone it was, by assumption, an eigenstate, with eigenvalue zero, of the combined momentum of nucleus plus crystal. This means that the momentum of the nucleus and the momentum of the crystal are highly correlated. There is zero probability of finding the nucleus with momentum P_i and simultaneously finding the crystal with any momentum other than $-P_i$. From Figure 8.12 it is clear that the $n = 0$ transition can only

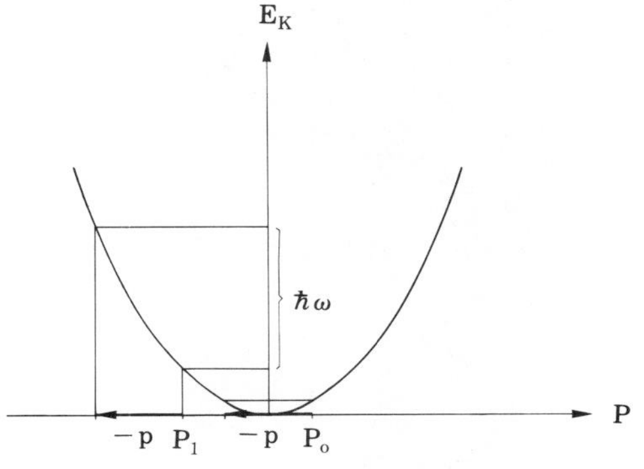

FIGURE 8.12 In order for the change in energy ΔE to be equal to $n\hbar\omega_0$ the initial momentum must have a certain value.

occur with $P_i = p/2$. After the transition the nucleus would have momentum $-p/2$ and the crystal, because of the above-mentioned correlation, would also have momentum $-p/2$. Thus the total momentum would be $-p$ which is just the opposite of the photon momentum. The recoil momentum is shared equally between the nucleus and the crystal, but notice that the momentum transferred by the photon was imparted entirely to the nucleus. This could be done with no transfer of energy because the energy loss, for a fixed amount of momentum transfer, depends upon the initial momentum of the nucleus and, for the value of P_i that leads to the $n = 0$ transition, the energy loss is zero. The emitted photon therefore carries the full energy available from the nuclear transition. In a transition of this sort there are no broadening effects at all. The linewidth for the $n = 0$ photons is the natural linewidth of the nuclear transition.

Now we must discuss what happens when one uses a more realistic model of a nucleus in a crystal. We will not say much about a realistic model of a crystal because that subject will be taken up at length in Chapter 10. There we will show that, in a crystal, there are a very large number of different modes of vibration with many different frequencies. Each of these modes of vibration acts like a harmonic oscillator with a discrete spectrum of energies. Thus there is a much larger number of possible internal quantum states to which the system can go after the photon is emitted. In some cases this large set of possible transitions, which do absorb energy and therefore give a shift in the photon frequency, overwhelm the $n = 0$ transition and thus the probability of unshifted photon emission becomes negligible. In other cases there is still an appreciable probability that the photon will be emitted without a change in the internal quantum state. Which case prevails depends upon the details of the crystal structure and we will not take up that question.

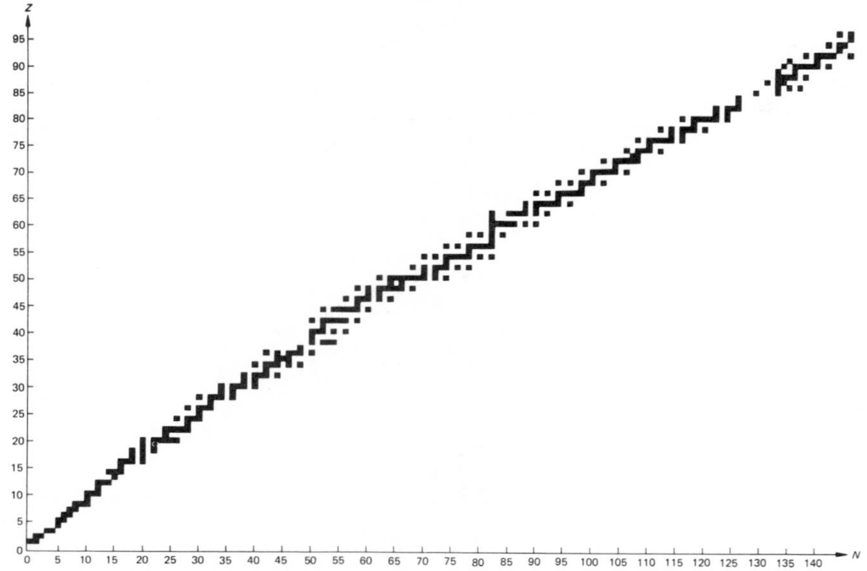

FIGURE 8.13 The values of N and Z that yield nuclei that are stable with respect to β-decay.

8.7. NUCLEAR STABILITY

By undergoing β^+- or β^--decay it is possible for a nucleus to adjust its electrical charge Z without affecting its mass number A. In the two types of β-decay the nuclear transformations are

$$\beta^- \ (N,Z) \rightarrow (N-1, Z+1) \tag{8.7a}$$

$$\beta^+ \ (N,Z) \rightarrow (N+1, Z-1) \tag{8.7b}$$

Most values of N and Z would give nuclei which were unstable with respect to β-decay. For given A there are typically very few values of Z which yield β stable nuclei. They are shown in Figure 8.13.

It is obvious that the β stable nuclei form a definite curve in the N−Z plane. We can understand the physical processes at work producing that curve, and even predict the detailed curve fairly well, by making use of the Empirical Mass Formula.

The total energy of the nucleus (N,Z) is the mass energy of the nucleons minus the binding energy (See Equation 8.1).

$$E(N,Z) = M(N,Z)c^2 = N m_n c^2 + Z m_p c^2 - B(N,Z) \tag{8.8}$$

The nucleus (N,Z) will be stable with respect to β-decay if either of the decays (Equation 8.7) yield a nucleus of higher energy. That is, if

$$E(N-x, Z+x) > E(N,Z) \tag{8.9}$$

for x equal to +1 or −1. Certainly the inequality 8.9 would be satisfied if $E(N-x, Z+x)$, considered as a function of x, had a minimum at $x = 0$. The condition for a minimum at $x = 0$ is that

$$\left[\frac{d\,E(N-x, Z+x)}{dx}\right]_{x=0} = 0 \tag{8.10}$$

If we use the Empirical Mass Formula expression for B(N,Z) (Equation 8.5) in Equation 8.8 and we then take the derivative indicated in Equation 8.10, we obtain an equation involving N, Z, and A. Putting in numerical values for all the constants appearing in the equation (m_n,m_p, etc.) we obtain the following equation for β stable nuclei.

$$N - Z = \frac{A^{5/3} - 1.8\,A}{A^{2/3} + 130} \tag{8.11}$$

QUESTION: For the mass number $A = 185$ what are the values of N and Z predicted by Equation 8.11? How does the prediction compare with reality?

ANSWER: From Equation 8.11 we get

$$N - Z = \frac{(185)^{5/3} - 1.8(185)}{(185)^{2/3} + 130} = 35$$

Thus $N + Z = 185$ and $N - Z = 35$. This gives $N = 110$ and $Z = 75$. Experimentally it is found that the only β stable nucleus of mass number 185 is (110, 75) which is ^{185}Re.

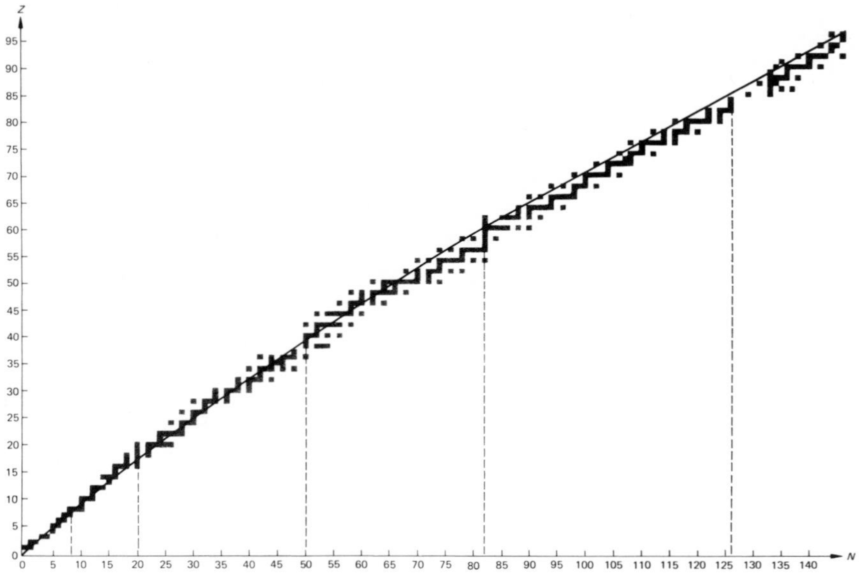

FIGURE 8.14 A comparison of the β-stability curve calculated according to the empirical mass formula and the actual set of β-stable nuclei.

In Figure 8.14 the predicted curve of β stability is superimposed on the chart of β stable nuclei.

For small values of A the β stability curve follows fairly closely the line N = Z, which indicates that the neutron and proton quantum states are occupied equally. This is clear evidence for the charge independence of the nuclear potential. The electrostatic potential which is certainly not charge independent is negligible for small Z, because the basic electrostatic interaction is much weaker than the nuclear interaction. However, as Z increases the electrostatic interaction becomes more and more important. Due to the long range of the Coulomb potential a given proton interacts electrostatically with all other protons in the nucleus. Its nuclear interactions takes place only with its near neighbors. The large number of protons compensates for the intrinsic weakness of the electrostatic interaction, and the electrostatic term in the energy becomes more and more significant. The energy of the proton levels is thus shifted upward, and it is advantageous for the nucleus to take on more neutrons than protons in spite of the fact that the symmetry energy acts to equalize proton and neutron numbers. For nuclei with Z larger than 82 the electrostatic term dominates, and all such nuclei are unstable with respect to α-decay, which is a method by which the nucleus can reduce the proton number and thus lower the electrostatic potential.

8.8. NUCLEAR SHELLS

If we look at Figure 8.15 we see that the discrepancies between the actual binding energies of nuclei and the predictions of the Empirical Mass Formula are not at all random. Particularly beyond A = 50 the actual binding energies are not simply scattered on both sides of the smooth curve. Instead there are certain regions of the curve in which the nuclei are consistently more strongly bound than the simple theory predicts. The underlying cause of these consistent deviations is the existence of a definite shell structure within the nuclear energy levels. Just as the energy levels of atoms are not distributed uniformly but are grouped into shells, with significant gaps between the highest energy level in one shell and the lowest energy level in the next, the energy levels of nuclei come in distinct groups. The common practice of calling a group of such energy levels a nuclear shell can be misleading. There is no spatial separation in the shells with an associated variation in the particle density as a function of the distance from the center of the nucleus. The particle density within a nucleus is quite uniform. It is the energy levels that are grouped together into distinguishable shells. Both theoretical calculations and experimental observation indicate that the shells become filled at the particle numbers 2, 8, 20, 28, 50, 82, and 126. These "magic numbers" are the same for neutrons and protons. Thus when N (or Z) approaches a magic number the binding energy of each further neutron (or proton) is increased until the magic number is reached. Once the magic number is reached the next neutron (or proton) is much more weakly bound.

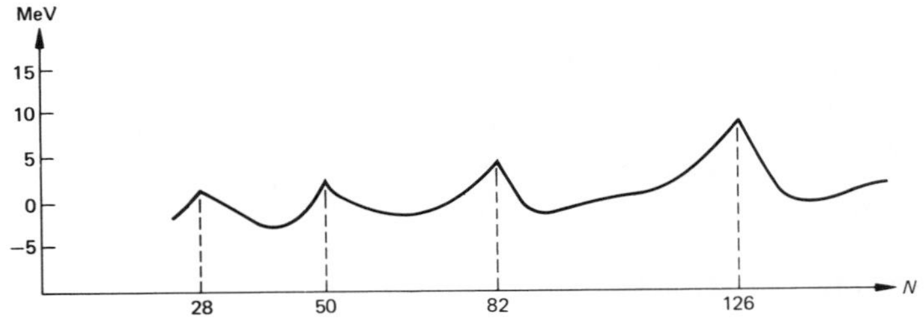

FIGURE 8.15 Actual binding energy of nuclei of neutron number N minus the empirical mass formula prediction of the binding energy. The curve shown gives the average for all nuclei of the given neutron number. At the "magic numbers" 28, 50, 82, and 126 the actual binding energies are larger than those predicted by the empirical mass formula.

In Figure 8.14 the positions at which N or Z are equal to magic numbers are indicated by dotted lines. We should expect the binding energy curve to deviate from the Empirical Mass Formula curve whenever the line of stable nuclei cuts a magic number line. This happens at (N,Z) equal to the following values:

(N,Z)	A
(2,2)	4
(8,8)	16
(20,18)	38
(24,20)	44
(28,24)	52
(34,28)	62
(50,38)	88
(70,50)	120
(82,60)	142
(126,82)	208

Those values of A are indicated by small arrows on Figure 8.4. Except for the case of A $\approx$ 40 the closing of nuclear shells by either neutrons or protons is associated with a significant increase in the binding energy (See Figure 8.15).

QUESTION: How does *carbon-14 dating* work?

ANSWER: The carbon in plants has its source in the carbon dioxide of the atmosphere. It is incorporated into the structure of plant cells by the chemical reactions of photosynthesis. Since animals eat plants or else eat other animals who eat plants the carbon in animal tissue is also atmospheric in origin. The carbon of the atmosphere is composed of two isotopes, ^{12}C and ^{14}C. The ^{14}C nucleus undergoes spontaneous β^--decay with $T_{1/2} = 5570$ years. The atmospheric ^{14}C is replenished by the reaction $^{14}N + n \rightarrow {}^{14}C + p$ where the neutron is produced in the upper atmosphere by incoming cosmic rays. Assuming that the average cosmic ray flux remains steady over a long period of time then the ratio of ^{14}C to ^{12}C will adjust itself so that the decay rate of ^{14}C is balanced by its production rate. Since the production only occurs in the upper atmosphere any carbon sample that was sequestered, such as a buried wooden object, would begin to lose its ^{14}C but not its ^{12}C content (^{12}C is stable). By chemically extracting all the carbon from a bit of the object and measuring the ratio of ^{14}C to ^{12}C in it one can determine the time that has elapsed since that carbon was removed from the atmosphere. Of course, all this depends on the assumption that the cosmic ray flux has remained steady over the last 30,000 or so years (The method is useful for dating objects in an age range of about 1 to 30 thousand years). That assumption must be separately corroborated by comparing carbon-14 dating to other means of dating, such as counting the annual growth rings of old trees. The independent checks have supported the assumption.

8.9. NUCLEAR FISSION AND FUSION

It is common knowledge that vast amounts of energy are released in nuclear reactions. Why this should be expected is clear from Figure 8.4 which gives the binding energy per particle of the most stable nuclei as a function of the mass number of the nucleus. The greatest per particle binding energies occur around the mass numbers between 50 and 60. This range includes the elements chromium, manganese, iron, cobalt, and nickel. The nucleus ^{30}Si contains 16 neutrons and 14 protons. It has a mass of 29.966u. The nucleus ^{60}Ni contains 32 neutrons and 28 protons. It has a mass of 59.915u. Because the binding energy per particle of ^{60}Ni is greater than that of ^{30}Si, the mass of the ^{60}Ni nucleus is less than the combined masses of two ^{30}Si nuclei. If two ^{30}Si nuclei were to combine to form one ^{60}Ni nucleus, the mass difference would be given off in the form of radiant energy. The mass difference, in atomic mass units, is $2\times 29.966\,u - 59.915\,u = .017\,u$. In kilograms the mass difference is

$$.017\,u = (.017\,u)(1.66\times 10^{-27}\,kg/u) = 2.8\times 10^{-29}\,kg$$

This mass difference, converted to energy, would give an energy

$$E = \Delta m c^2 = (2.8\times 10^{-29}\,kg)(9\times 10^{16}\,m^2/s^2)$$
$$= 2.5\times 10^{-12}\,J$$

This may not seem like a great deal of energy; however, one kilogram of silicon contains 2×10^{25} nuclei. If that kilogram of silicon were converted to nickel, the energy released would be 25 trillion joules. One can obtain the same magnitude of energy release by breaking up a very heavy nucleus to form lighter, more tightly bound nuclei. Processes of the first type are called *fusion* reactions. Those of the second type are *fission* reactions. With the large differences that exist in the binding energies of nuclei it is no surprise that highly exothermic (energy releasing) nuclear reactions exist. The question that really needs answering is why these reactions are not taking place spontaneously all the time. Oxygen and hydrogen atoms interact chemically with a much smaller release of energy than is typical of nuclear reactions, and yet having large amounts of oxygen and hydrogen gas mixed together would present a serious health hazard. All that would be needed is a spark. In contrast cadmium 108 is a completely stable nucleus although it contains the same constituents as two chromium 54 nuclei and would release substantial energy if it ever underwent fission to form two of the lighter nuclei. The reason why most nuclei are quite stable in spite of the fact that one can conceive of strongly exothermic reactions involving them is that all the imaginable reactions have the property that large amounts of energy would have to be temporarily supplied to the nucleus in order to get the reaction to occur. Of course, this initiation energy would then be returned in addition to the calculated reaction energy. However, if there is nothing available to supply the initiation energy, then the reaction will never take place, and the nucleus will be effectively stable.

There are two forces at work in nuclei; the nuclear force and the electrostatic force. In both fission and fusion one of these forces is primarily responsible for the energy released (the *reaction energy*) while the other opposes the reaction and is thus responsible for the energy that must be supplied to initiate the reaction (the *initiation energy*). However, the roles played by the two forces are reversed in fission and fusion.

8.10. FUSION

In a fusion reaction the source of the energy released is, according to our liquid drop model of the nucleus, the surface energy term. The combined nucleus formed by the fusion of the two smaller nuclei has a smaller surface area than the two separated nuclei. An equivalent way of stating this fact is to say that the nucleons in the product nucleus have, on an average, more nearest neighbors with which they can form strong nuclear force bonds. It is therefore the nuclear

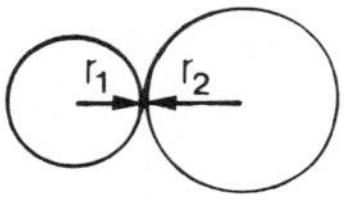

FIGURE 8.16

Because of the short range of the nuclear force, the two fusing nuclei must essentially come into contact in order to react.

force that is the source of the energy released in the reaction. The electrostatic force opposes the fusion reaction. In order for fusion to occur the two smaller nuclei must be brought close enough for the attractive nuclear force to take effect. But both smaller nuclei are positively charged objects and thus repel one another. If the two nuclei have charges Z_1e and Z_2e and radii r_1 and r_2 respectively, then a good estimate of the energy that must be supplied in order to initiate the fusion reaction may be obtained by calculating the potential energy of two charged spheres of those radii just at the touching distance (see Figure 8.16). The potential energy is

$$V = k\frac{Z_1 Z_2 e^2}{r_1 + r_2}$$

QUESTION: What is the estimated initiation energy for the reaction

$$^{30}Si + {}^{30}Si \rightarrow {}^{60}Ni \text{ ?}$$

At what temperature would the average kinetic energy of a silicon atom be equal to that initiation energy?

ANSWER: Using Equation 8.3 we estimate the radius of ^{30}Si as

$$R = (1.2 \times 10^{-15}\,m)(30)^{1/3} = 3.7 \times 10^{-15}\,m$$

The estimated initiation energy is therefore

$$E_{init} = (9 \times 10^9 \frac{J\text{-}m}{C^2})\frac{(14)^2(1.6 \times 10^{-19}\,C)^2}{(2)(3.7 \times 10^{-15}\,m)}$$
$$= 6 \times 10^{-12}\ J$$

At absolute temperature T the average kinetic energy of an atom is $\frac{3}{2}k_BT$. Setting this equal to E_{init} we get the following formula for T

$$T = \frac{2}{3}\frac{E_{init}}{k_B} = \frac{2}{3}\frac{6 \times 10^{-12} J}{1.4 \times 10^{-23} J/K}$$
$$= 3 \times 10^{11}\ K$$

This is about 20,000 times the estimated temperature at the center of the sun. It is about a billion times room temperature which is why this reaction is never seen on earth. Reactions like this do take place, however, in the late stages of the gravitational collapse of stars.

The most potentially useful fusion reaction is the fusion of two deuterium nuclei to form a nucleus of ^4He. Deuterium is the second most common isotope of hydrogen. It is ^2H, which indicates that its nucleus is composed of one proton and one neutron. This is a particularly advantageous reaction from the point of view of energy generation in that the nucleus of ^4He (i.e., the α-particle) is doubly magic ($N = Z = 2$) and therefore very tightly bound while the ^2H nucleus (the deuteron) is quite weakly bound as nuclei go. Thus the reaction energy is large.

The supply of deuterium is effectively unlimited. One in every six thousand hydrogen atoms has a deuteron nucleus. Being twice the mass of a normal hydrogen atom, the deuterium atoms can be easily separated from the hydrogen atoms. Since the reacting nuclei have only one unit of charge the initiation energy is as small as possible. The nuclear force has a range of a couple of fermis. If we take three fermis as the distance between each pair of particles, we should expect deuteron-deuteron reactions if we can bring the protons within nine fermis of one another (See Figure 8.17). The temperature at which the average kinetic energy of the particles would be sufficient to bring them that close is about one billion degrees. This is far too large to be attained in any controlled situation. However, the particles in a gas do not all have the average kinetic energy. Instead there is a broad distribution of velocities among the particles so that even at a much lower temperature, say a few million degrees, a small fraction of the particles will have enough energy to undergo fusion. Also, by a process of quantum mechanical tunnelling, particles which, from a classical point of view, do not have sufficient energy to react, can 'tunnel through' the Coulomb barrier and get sufficiently close for the attractive nuclear force to take effect. The problem of creating and maintaining a controlled fusion reaction thus boils down to the technical problem of maintaining a sufficiently dense sample of deuterium for a long enough time at a few million degrees to allow a reasonable fraction of the deuterium nuclei to fuse. Two quite different methods have shown some promise of solving this thermonuclear confinement problem. One method is called *magnetic confinement*, the other is called *inertial confinement*. Before we describe those methods we should note that what may seem the most obvious method of confining and compressing a hot sample of deuterium simply will not work. That is to compress the deuterium within a strong rigid cylinder. At millions of degrees every known material vaporizes and there is obviously no way of making rigid cylinders out of vapor. It is therefore essential that, while the material is at the very high temperature that is necessary for the fusion reaction to take place, it be kept away from the solid walls of the vessel in which the reaction is occurring.

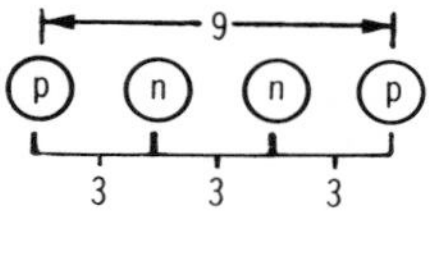

FIGURE 8.17
At a distance of 9 fm between protons the two deuterons can undergo fusion.

8.11. MAGNETIC CONFINEMENT

In magnetic confinement machines the required confining pressures are maintained by magnetic forces. At the high temperatures involved the deuterium gas becomes completely ionized. Such a highly ionized gas, called a *plasma*, reacts strongly with magnetic fields. To understand the effect of a magnetic field on a plasma one should consider the trajectories of the individual charged particles of the plasma in a magnetic field.

In a uniform magnetic field a particle with an initial velocity perpendicular to the field will move in a circle whose radius R can be determined by setting the magnetic force equal to the mass of the particle times its centripetal acceleration.

$$\frac{m v^2}{R} = e\,|\mathbf{v} \times \mathbf{B}|$$

which gives

$$R = \frac{m v}{e B}$$

To estimate the radius of the circle let us take m equal to the deuteron mass, B equal to one Tesla, and $\frac{1}{2} m v^2$ equal to $\frac{3}{2} k T$ where T = 100,000 K. The velocity is then

$$v = \sqrt{3 k_B T/m} = \left[\frac{(3)(1.38 \times 10^{-23}\,\mathrm{J/K})(10^5\,\mathrm{K})}{(3.35 \times 10^{-27}\,\mathrm{kg})}\right]^{1/2}$$
$$= 3.52 \times 10^4\,\mathrm{m/s}$$

With this velocity

$$R = \frac{(3.35 \times 10^{-27}\,\mathrm{kg})(3.52 \times 10^4\,\mathrm{m/s})}{(1.6 \times 10^{-19}\,\mathrm{C})(1\,\mathrm{T})} = .74\,\mathrm{mm}$$

The radius is fairly small in comparison to the typical size of a thermonuclear fusion reaction chamber. The radius of a typical electron's motion in the same field and at the same temperature is much smaller than that of a deuteron. If, as will usually be the case, the initial velocity of the particle is not perpendicular to the magnetic field then the particle will move in a spiral rather than a circle. This is because the magnetic force is always perpendicular to B and therefore the velocity component in the direction of B remains constant. Figure 8.18 shows the paths of positive and negative particles in a uniform magnetic field. It is seen that both types of particles spiral along the field lines, therefore, if it were possible to create a magnetic field whose field lines never intersected the chamber walls we might be able to keep the particles away from the walls by means of magnetic forces. One of the simplest ways of creating such a magnetic field is to wind wire around a large toroidal vessel as shown in Figure 8.19. Running an electric current through the wire will then create an interior

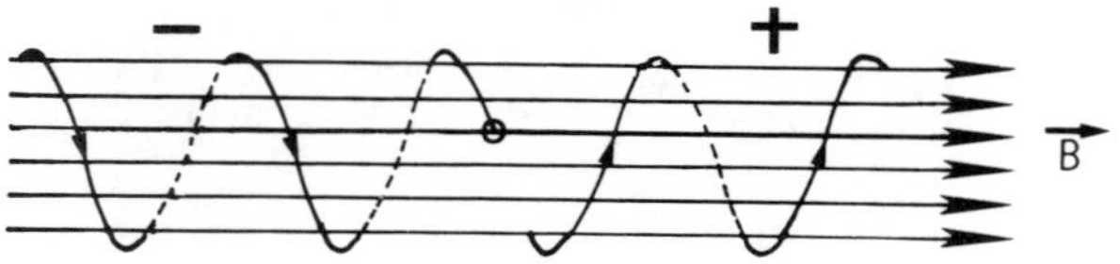

FIGURE 8.18 The trajectories of positive and negative particles in a uniform magnetic field.

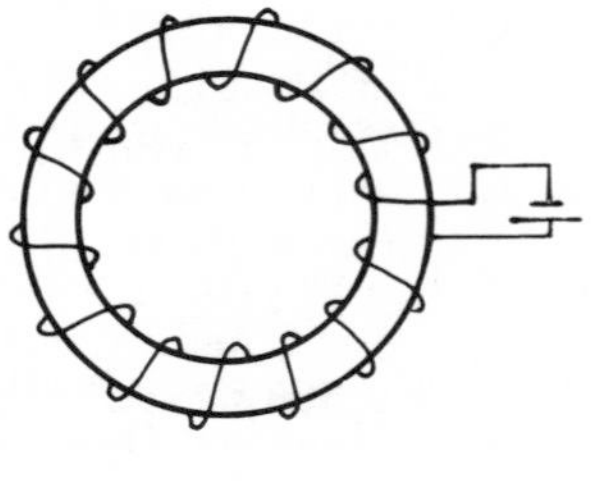

FIGURE 8.19

toroidal magnetic field whose field lines never intersect the vessel walls. This is one of a number of chamber and field configurations used in the attempts to achieve electric power generation by means of controlled nuclear fusion. In order to obtain the densities and temperatures needed for appreciable fusion to take place, the deuterium plasma that is contained in the reaction vessel must be heated and compressed. Both must be accomplished without any material object ever touching the hot plasma. Anything that touches the plasma will immediately vaporize and seriously contaminate the plasma with ionized impurities. One of the ways in which heating is accomplished is shown in Figure 8.20. An iron-core toroidal electromagnet loops the containment vessel. (The other windings around the vessel which were shown in Figure 8.19 should be assumed still to be there although they are not shown in Figure 8.20 in order not to obscure the picture.) An alternating current is passed through the electromagnet, causing an oscillating ring of magnetic flux to link the reaction chamber. According to Maxwell's equation,

$$\int \mathbf{E} \cdot \mathrm{dl} = -\frac{\mathrm{d}\phi}{\mathrm{d}t}$$

a changing magnetic flux generates an electric field. The electric field within the plasma drives the positive ions and the electrons in opposite directions, heating the plasma. If, instead of a single electromagnet, a large number of them are placed at regular intervals around the containment vessel then the electric field they create within the plasma is parallel to the magnetic field created by the windings around the containment vessel. The electric field then exerts no force on the particles in a direction toward the walls. Of course, by heating the plasma, the electric field causes the pressure to increase and that makes it more difficult to contain the plasma by means of the magnetic field.

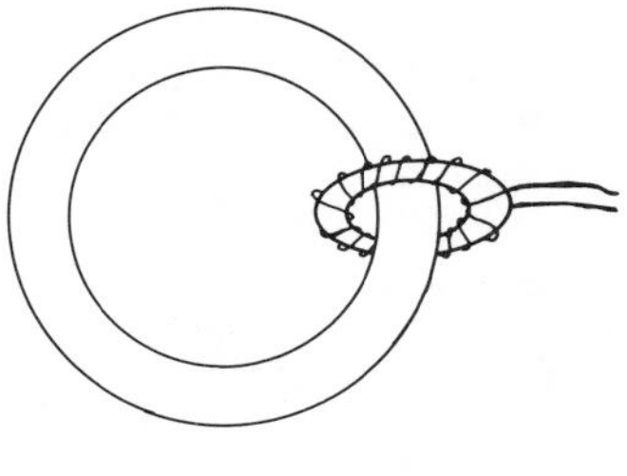

FIGURE 8.20

An alternating current through the coil shown will produce an induced electric field within the containment vessel.

In addition to confining the plasma, the magnetic field within the containment vessel also offers a method of compressing the plasma. Because of their spiral motion the plasma particles may be considered as bound to the field lines they spiral around. If the current in the confining coils (as opposed to the heating coils) is increased so as to increase the magnetic flux density within the vessel, new magnetic field lines enter the vessel from the outside while the field lines within the vessel move toward the center of the vessel (See Figure 8.21). The lines that move toward the center carry the plasma with them, thereby compressing it.

During the last thirty years the temperatures and densities obtainable in magnetically confined plasmas has steadily increased. However there are two

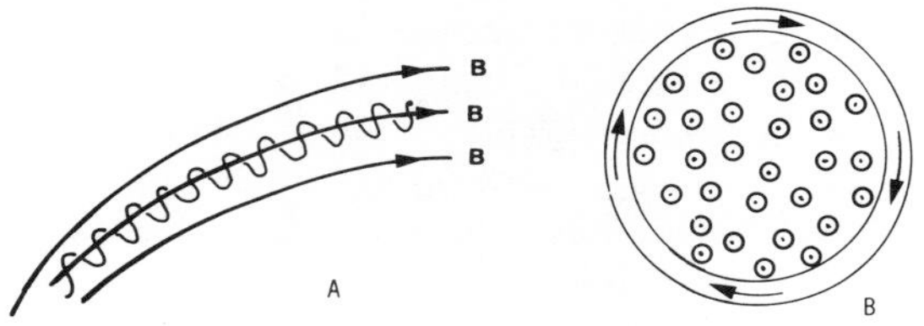

FIGURE 8.21 A. Charged particles, spiralling in the magnetic field can be considered to be attached to the **B** lines they spiral around.
B. As the magnetic field strength is increased new lines are created at the periphery as all the lines move toward the center to make the line density greater.

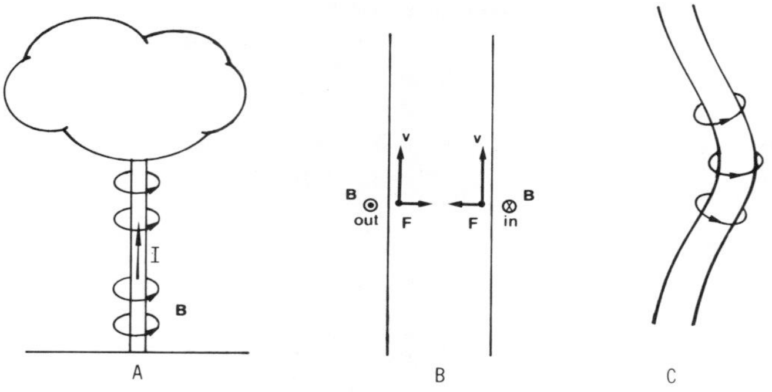

FIGURE 8.22 A. Consider a tube of ionized gas in which there is a strong electric current. (A lightening bolt is a good example.) The current tube is encircled by magnetic field lines.
B. The magnetic force on the moving charged particles in the tube is such as to compress the tube. The magnetic forces act like an external pressure to contain the tube.
C. If the tube develops a small bend the field becomes stronger at the inside of the bend thus increasing the magnetic pressure there and pushing the bend out still further.

problems which become more and more difficult to deal with as the pressure and temperature of the plasma are increased. One of these problems is the diffusion of the plasma toward the vessel walls. The diffusion is caused by particle-particle collisions within the plasma. Because of those collisions the simple spiral curves we have discussed are not the exact trajectories of the plasma particles and it is possible for particles, during a collision, to move from one field line to another one that is closer to the vessel wall. The other serious problem is the existence of *plasma instabilities.* The plasma particles are charged and their movements therefore create large electric currents within the plasma. The plasma currents create their own magnetic field which can disturb the containing field sufficiently to bring the plasma into contact with the walls of the containment vessel. A typical example of a plasma instability is the kink instability illustrated in Figure 8.22.

8.12. INERTIAL CONFINEMENT

In the technique of inertial confinement a small spherical pellet of solid material containing a high density of deuterium is suddenly irradiated on all sides by extremely intense light from a battery of high-powered lasers. The outside shell of the pellet is immediately raised to an extremely high temperature and explodes. The material exploding away from the surface of the pellet applies tremendous reaction forces on the inner material of the pellet, compressing it to high densities and temperatures. If high enough temperatures and densities can be reached this way the fusion of the deuterium in the pellet will produce more energy than was needed to compress the pellet and the device will give a net energy yield.

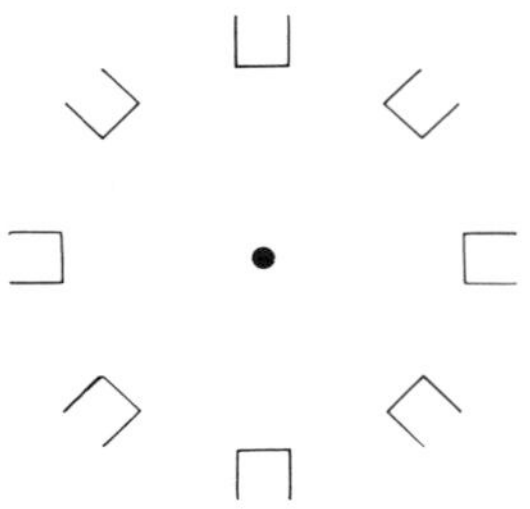

FIGURE 8.23

A battery of high-powered lasers all focused on a small fuel pellet. They can be synchronized to fire at exactly the same time with a very powerful burst of radiation.

8.13. NUCLEAR FISSION

The process of fission is the breaking up of a nucleus into a number of smaller constituents. Although the word is not commonly used in that context, it is clear that α-decay is an example of a fission reaction. The term fission is more often used to describe a process in which a large nucleus breaks up into a few pieces of comparable size.

In order to determine the relative importance of surface energy, electrostatic energy, and symmetry energy in the fission of a large nucleus, we will evaluate the change in each term separately for the case of an even binary fission of ^{232}U. We thus begin with the nucleus (N,Z) = (140,92) and end with two nuclei (70,46). The nucleus (70,46) falls below the curve of β stable nuclei and would therefore be expected to transform by β-decay to the nearest stable species, namely ^{116}Cd which is (68,48). Since β processes have half lives that are much longer than the time involved in the actual fission process, we can ignore the final β-decay since we are interested in the dynamics of fission. Actually this whole process in which the nucleus neatly separated into only two equal pieces is a quite artificial case whose only justification is its simplicity. Table 8.2 shows the value of each term in the empirical mass formula for the initial and final nuclei. Although the total reaction is strongly exothermic, releasing 208 MeV per nucleus of uranium, it is clear that the surface term and the electrostatic term are in strong opposition to one another.

TABLE 8.2

Nucleus	R in fm	$4\pi R^2 b_{sur}$ in MeV	$\frac{3}{5}\frac{kZ^2e^2}{R}$ in MeV	$\frac{(N-Z)^2}{A} b_{sym}$ in MeV
^{232}U, (140, 92)	7.37	610	992	231
^{116}Pd, (70, 46)	5.85	385	312	116
Change (2×2nd row - 1st row)		+160	−368	0

Left to itself the nucleus ^{232}U will undergo spontaneous fission. But it is a rather slow process with a half-life of 72 years. The origin of the relative stability of a nucleus such as this, which has available to it such a strongly exothermic decay reaction is the following fact. For small shape deformation of the ^{232}U nucleus the electrostatic energy decreases while the surface energy increases, *but the surface increases more rapidly than the electrostatic energy decreases.* The net effect is an energy increase, and thus the nucleus is pulled back to its equilibrium shape. It is necessary for a large deformation to occur before the electrostatic term will overcome the surface term and initiate fission (see Figure 8.24). We intended to explain why the nucleus is relatively stable and does not fly apart immediately, but we seem to have overshot our mark. We have apparently shown that unless some external forces create a sufficiently large deformation of the nucleus, it will never undergo fission. But in fact the ^{232}U nucleus does spontaneously decay by fission with the above mentioned half-life. Where does the deformation energy come from? The resolution of the paradox lies in the fact that the nucleus is a quantum mechanical system and cannot be described completely adequately using classical concepts. We will present the quantum mechanical theory of nuclear fission in only enough detail to communicate the way in which the problem we have encountered is resolved.

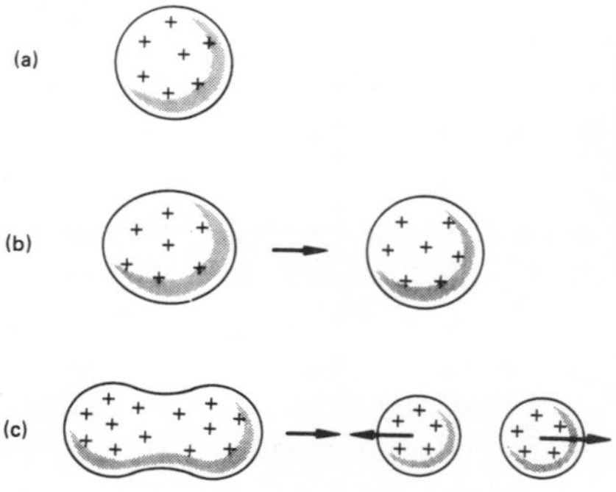

FIGURE 8.24

(a) Equilibrium shape of a large fissionable nucleus.
(b) For small deformations the surface tension pulls the nucleus back to its equilibrium shape.
(c) For large deformations the Coulomb force dominates and the nucleus breaks up.

In Figure 8.25 is shown a sequence of possible shapes of the ^{232}U nucleus that are arranged in order of increasing elongation in the x direction. They are labeled by a parameter x which gives the length of the nucleus in the x direction. The equilibrium value of x is x_c. If we plot the sum of the electrostatic and surface energies of the nucleus (which are the only terms relevant to the fission process) as a function of the parameter x, we get a graph of the form shown in Figure 8.26. As a system satisfying classical dynamics, one can easily picture the nucleus undergoing a kind of oscillatory normal mode vibration between configurations x_b and x_d. Ordinary liquid drops are known to carry out just such vibrations. For the nucleus this mode of vibration is *quantized*. That is, one can write a wave function $\psi(x)$ whose square gives the probability that the nucleus would actually be found with that value of the deformation parameter. We thus have to discuss the question of what would be the solutions of the Schroedinger equation with a potential such as the one shown in Figure 8.26. We will not attack that mathematical problem directly but instead we will first obtain the

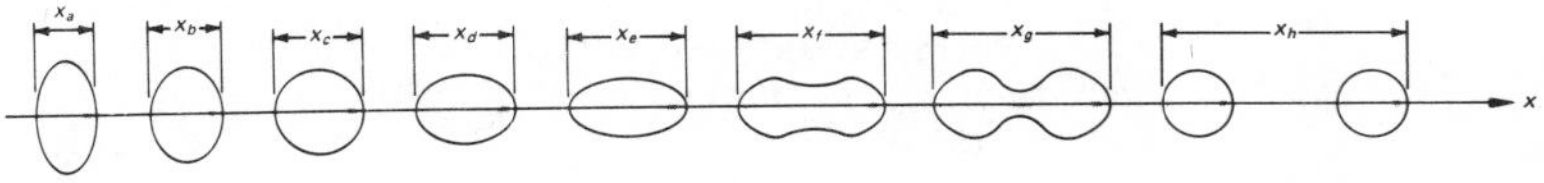

FIGURE 8.25 Various degrees of elongation of a nucleus are described by a deformation parameter x.

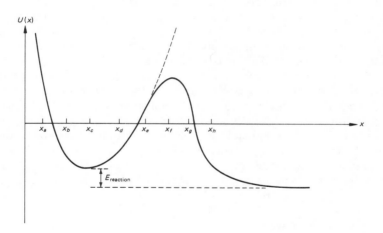

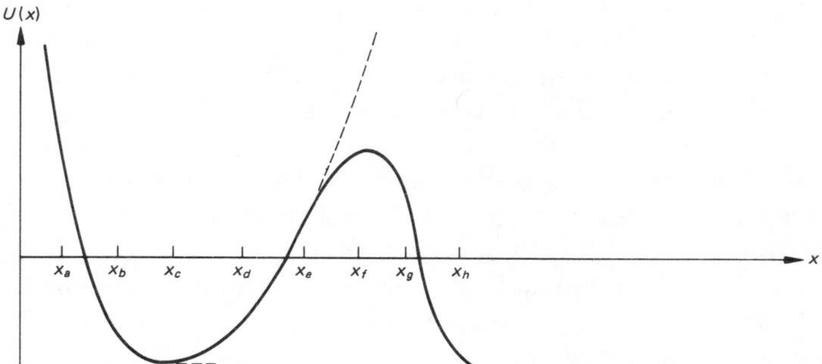

FIGURE 8.26 The "deformation potential" U(x) is the sum of the electrostatic and surface energy terms for various values of the deformation parameter x. It has a local minimum for $x = x_c$ but a true minimum for the separated nucleus at $x = \infty$. The differences between the values of U at these points is the energy released in the fission reaction. The dashed line shows a harmonic oscillator potential that matches U(x) near the local minimum.

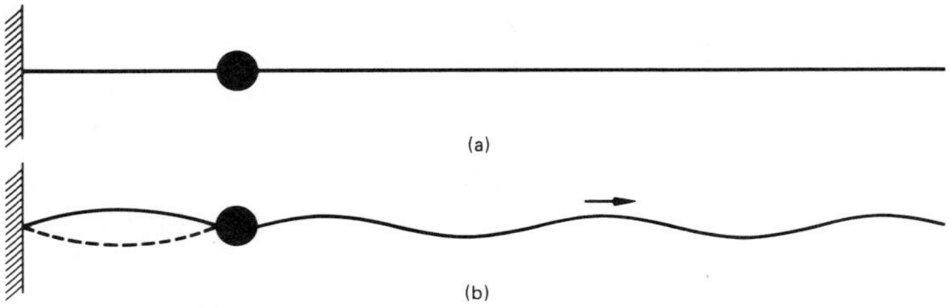

FIGURE 8.27 (a) A string of infinite length has a large mass M attached to it at some distance from the wall.
(b) When the string segment vibrates in its fundamental normal mode, the mass moves slightly, thus causing small-amplitude right-going traveling waves to the right of the mass. Those traveling waves carry energy off to infinity, and thus the amplitude of the normal-mode vibration gradually diminishes.

essential results by using an analogy with a simple string system. We consider the system shown in Figure 8.27. An infinite string is attached to a wall at one end. At some distance from the wall a very heavy mass M is attached to the string. If M were infinite, the string segment would be effectively fixed at both ends and could vibrate in a normal mode as shown. Because the mass M is only large but not infinite, it will actually vibrate with a small amplitude. This will create traveling waves of small amplitude on the string beyond the mass. There is an energy flux associated with the traveling wave to the right of the mass. Since there is no source of energy in the system, the energy being carried away by the traveling wave must be coming from the energy in the normal mode motion of the string segment. Thus the amplitude of the normal mode motion will not remain constant. It will slowly diminish as the energy leaks out of the system to the right. The amplitude of vibration of the mass is proportional to the amplitude of vibration of the string segment. The amplitude of the traveling wave is equal to the amplitude of vibration of the mass. Therefore the amplitude of the traveling wave is proportional to the amplitude of the string segment.

$$A_{tr} = \alpha A_{vib} \tag{8.12}$$

Squaring Equation 8.12, we get

$$A_{tr}^2 = \alpha^2 A_{vib}^2$$

But the total energy of vibration of the string segment is proportional to the square of the amplitude of vibration

$$E \propto A_{vib}^2$$

and the rate at which energy is being carried away by the traveling wave is proportional to the square of A_{tr}.

$$\frac{dE}{dt} \propto -A_{tr}^2$$

Thus

$$\frac{dE}{dt} = -\lambda E \tag{8.13}$$

where λ is some constant of proportionality which we will not bother to calculate. This equation gives an exponential decay in the energy of the string segment.

$$E(t) = E(0)\, e^{-\lambda t} \tag{8.14}$$

Equation 8.14 is similar to equation 8.5 that describes radioactive decay. This is no accident. As we will see when we draw the analogy between the string system and the fission decay of a nucleus, it is exactly the same mathematical phenomenon that leads to the law of radioactive decay.

Let us now return to our consideration of the Schroedinger equation for a system with the potential shown in Figure 8.26. Near the local minimum of the potential, at $x = x_c$, the potential V(x) can be accurately approximated by a harmonic oscillator potential. We would therefore expect that the Schroedinger equation with the potential V(x) would have a bound state solution that is similar to the ground state wave function of a harmonic oscillator. This is almost true. The ground state wave function of the system is similar to that of harmonic oscillator near $x = x_c$. However, for large x the potential V(x) approaches zero and thus the wave function must approach a plane wave, which is the only solution of Schroedinger's equation with zero potential. What happens is that the small "tail" of the harmonic oscillator wave function connects smoothly to a very small amplitude outgoing wave (See Figure 8.28). Near the local minimum the wave function oscillates harmonically in time with a factor $e^{-i\omega t}$ where $\hbar\omega$ is very close to the ground state energy of the harmonic oscillator. However, the

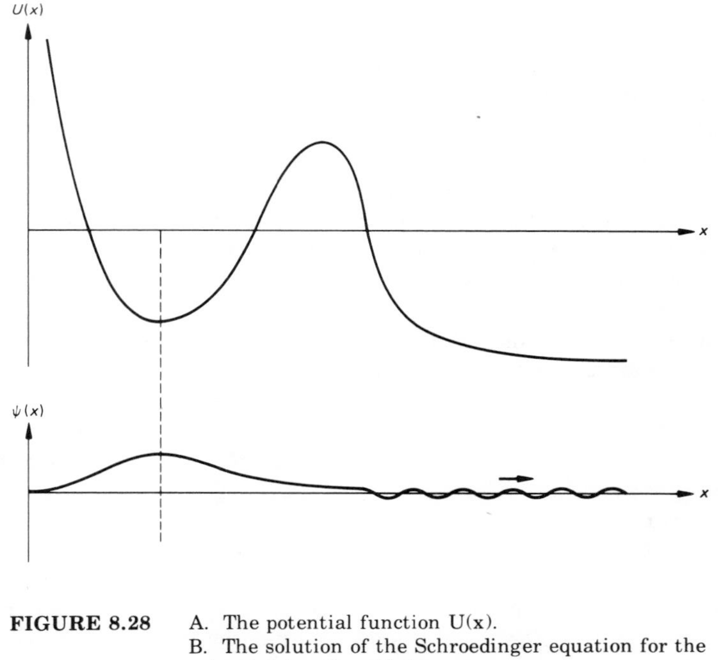

FIGURE 8.28 A. The potential function U(x).
B. The solution of the Schroedinger equation for the potential function U(x).

amplitude gradually diminishes in such a way that the probability of finding the system with a deformation parameter near the local minimum decreases exponentially in time. The traveling wave is associated with a flux. When one refers to Figure 8.25 for the interpretation of the deformation parameter (in particular the right-most figure labeled x_h) one realizes that the flux is a flux of the fission decay products. As time goes on it becomes less and less likely that the nucleus will be found intact and more and more likely that it will have undergone fission.

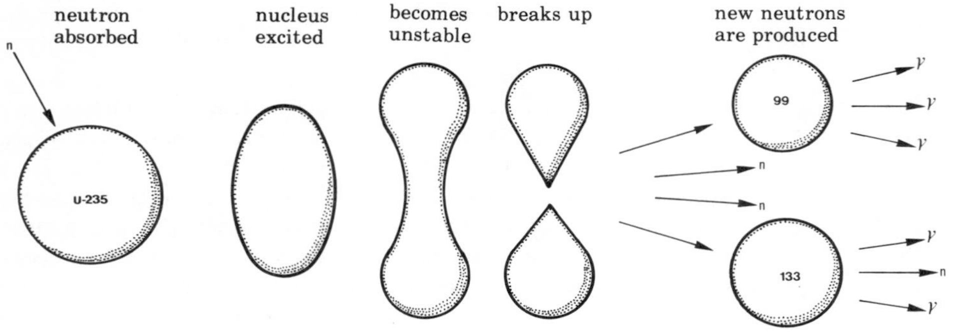

FIGURE 8.29 The chain reaction. The neutron-induced fission of a ^{235}U nucleus creates a number of fragments, some of which boil off excess neutrons. Those neutrons can then induce further fission reactions, causing the whole process to be amplified.

8.14. NEUTRON INDUCED FISSION

In applications of the fission reaction to power generation and nuclear weapons, one does not take a sample of spontaneously radioactive nuclei and simply wait for them to disintegrate at their own pace. Rather one induces fission in the nuclei by supplying the needed deformation energy by means of a neutron capture reaction. The basic process can be understood in terms of the ideas we have already developed. Suppose a low energy neutron approaches a large unstable nucleus. The neutron, being uncharged, will not be repelled by the electric charge of the nucleus. When it comes within a few fermis of the nuclear surface, it will come under the influence of the attractive nuclear potential and be drawn into the nucleus with tremendous force. The energy of the incoming neutron will be quickly transmitted to the nucleus as a whole, and the new nucleus will be left in a highly excited state. Some of the excitation energy will go into the deformation vibrational modes, driving them beyond the range of stability and causing fission of the nucleus. We have discussed only the simple case of binary fission in which the parent nucleus breaks up into two pieces. More commonly a large number of pieces result, including some single neutrons. Those emitted neutrons are then free to move to other nuclei causing them to undergo fission and give rise to still more free neutrons. If this process continues undiminished or grows, we say that a *chain reaction* has been initiated. Whether a chain reaction is sustained depends on the balance between the rate at which new neutrons are being created and the rate at which neutrons are being lost through the surface of the sample of fissionable material. That in turn depends upon the surface to volume ratio of the sample. With increasing sample size the surface to volume ratio decreases. Thus there is a certain sample size, referred to as the *critical mass*, for which a chain reaction will sustain itself or grow.

8.15. THE CONTROLLED FISSION REACTOR

In a controlled fission reactor the nuclear reaction rate of a large sample of nuclear fuel containing fissionable material is controlled by inserting long rods of neutron absorbing material (*control rods*) into the fuel sample. If the control rods are inserted all the way they absorb a large fraction of the emitted neutrons and the fission rate is much reduced. If the control rods are pulled out, those neutrons are free to cause fission reactions, and thus the fission rate increases. However, the concentration of fissionable material in the fuel elements is kept low enough so that even without control rods the reaction rate would not grow without any bound. For the types and amounts of fuel used in nuclear power generating reactors the reaction rates that would occur if all control were lost are sufficient to produce enormous amounts of heat energy but are not comparable to those occurring in nuclear explosions.

8.16. THE YUKAWA THEORY OF THE NUCLEAR FORCE

In 1935 the Japanese physicist, Hedeki Yukawa, presented a theory of the nucleon-nucleon force. His analysis used a form of relativistic quantum theory, called *quantum field theory.* In the next chapter we will discuss the basic ideas of quantum field theory, but only in a somewhat sketchy and qualitative way. Therefore we will not be able to repeat the fairly complex mathematical argument given by Yukawa. The outlines of the Yukawa theory are as follows. One assumes that nucleons are capable of emitting and absorbing some kind of massive particle, now called a *pion* or a π-meson, in the same way that electrons can emit or absorb photons. (When Yukawa's theory was published the pion had not yet been detected. Thus the predicted existence of a new class of particles was a very bold conjecture.) Using the principles of quantum mechanics one can show that the possibility of one nucleon emitting a meson which is absorbed by a second nucleon has the effect of producing an attractive force between the two nucleons. Furthermore, the detailed equations lead to a relation between the mass of the particle transferred and the form of the attractive force. Yukawa showed that the exchange of a particle of mass m would lead to an attractive nucleon-nucleon potential of the form

$$V(r) = -g^2 \frac{e^{-r/d}}{r} \tag{8.15}$$

with $d = \hbar/mc$. g is the *interaction strength.* It plays the same role in Yukawa's theory as e, the electric charge, plays in electromagnetic theory.

A potential function of this form is now called a *Yukawa potential.* d is the range of the potential. Beyond a distance d the exponential function in the Yukawa potential becomes so small that the potential is completely negligible. The range of the potential is proportional to the inverse of the mass of the particle transferred. Thus heavy particles lead to very short-range interactions. If the exchanged particle has a mass of zero, then $d = \infty$ and the Yukawa potential has the same form as a Coulomb potential. However, it differs from the

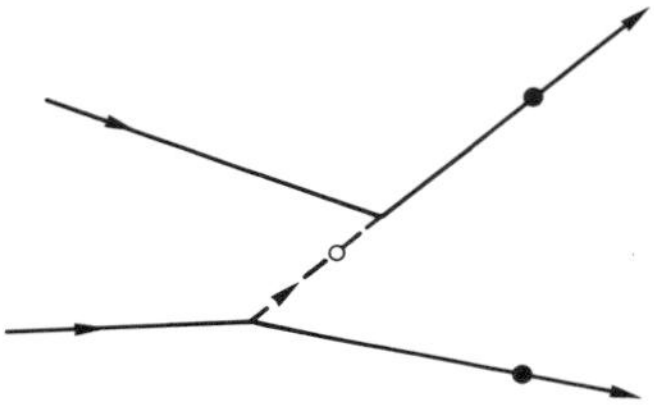

FIGURE 8.30
The transfer of a massive particle has the effect of producing an attractive force between the two nucleons.

electromagnetic Coulomb potential in one fundamental way. The minus sign in Equation 8.15 shows that the Yukawa force between two identical particles is an attractive force. The electromagnetic interaction between two like particles leads to a repulsive force. This disagreement between the electromagnetic force and the zero-mass Yukawa force can be traced to the fact that the Yukawa theory assumes that the particle transferred (usually called the *force carrier*) is a particle of zero spin. The Yukawa theory could be modified to describe the exchange of a particle of spin $\hbar$ and it then leads to a repulsive force between like particles. In fact, if the force carrier is assumed to be a spin-one particle of zero mass, like the photon, one obtains electromagnetic theory exactly. This description is reversing the historical order of things. The relationship between photon transfer and the Coulomb interaction was already known at the time of Yukawa's work. His contribution was to show that the exchange of a spin-zero particle of finite mass would lead to a short-range attractive nucleon-nucleon potential.

We know that the range of the nuclear force is about one fermi. Thus one would expect the force carrier to have a mass of about

$$m = \frac{\hbar}{c\,d} = \frac{1.05 \times 10^{-34}\,\text{J-s}}{(3 \times 10^{8}\,\text{m/s})(10^{-15}\,\text{m})} = 1.7 \times 10^{-28}\,\text{kg}$$

This estimate is confirmed by experiment. Pions have zero spin and come in three different types. One type, called the pi-zero (written π^0) has an electric charge of zero. The other types, the π^+ and π^-, have charges of $+e$ and $-e$ respectively. The masses of the pions are

$$m_{\pi} = m_{\pi} = 2.49 \times 10^{-28}\,\text{kg} \quad \text{and} \quad m_{\pi^0} = 2.41 \times 10^{-28}\,\text{kg}$$

Actually, Yukawa's analysis makes a number of approximations which cause it to be valid only for distances of order d or larger, but in that range the Yukawa form of nucleon-nucleon potential has been well confirmed.

When Yukawa's work was published, no particles with masses between the electron mass and the proton mass were known. No accelerators existed at that time which had sufficient energy to create such particles. The predicted particles were searched for in cosmic rays, the steady stream of high-energy particles hitting the earth from outer space. Particles with about the right mass were soon discovered. However, the newly discovered particles were muons. The muon has a mass only slightly smaller than the pion mass, but it is a totally different type of particle. The muon is basically a heavier version of the electron and plays no role at all in the strong nucleon-nucleon force. The closeness of the muon and pion masses is a pure accident. There was a period of some confusion before pions were discovered and clearly differentiated from muons.

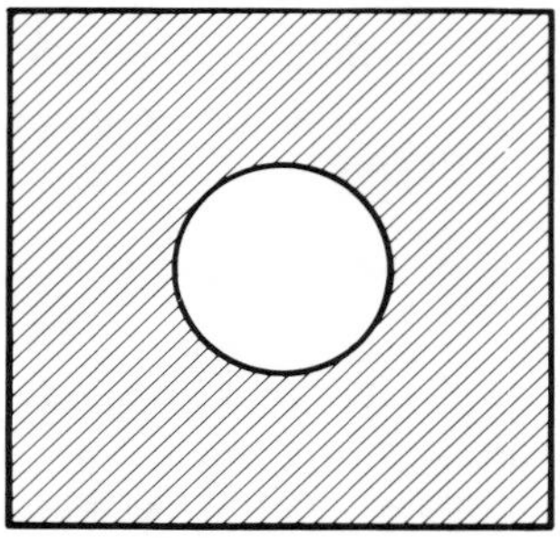

FIGURE 8.31
A spherical step potential. Within the sphere $V=0$. Within the shaded region $V=E+\Delta E$.

Although we cannot carry out a complete derivation of the Yukawa potential we can obtain some idea of the source of the relationship between the mass of the exchanged particle and the range of the force by treating the pion exchange as a quantum mechanical tunneling process. In Chapter 5 we considered the problem of a one-dimensional particle of energy E striking a potential step of height $V = E+\Delta E$. Since the potential in the step region is greater than the particle's energy by an amount ΔE the particle could not, according to *classical* physics, enter the potential step region at all. However, according to the Schroedinger equation, the wave function actually extends into the classically forbidden region, where it has the form $u = c\,e^{-\lambda x}$. According to Equation 5.19 on page 133, λ is related to ΔE by

$$\lambda = \frac{\sqrt{2m\,\Delta E}}{\hbar} \tag{8.16}$$

Let us now consider a similar problem in three dimensions. We picture a particle of energy E surrounded by a spherical potential step of height $V = E+\Delta E$ (See Figure 8.31). Using the Schroedinger equation one can find that the wave function in the surrounding, classically forbidden, region has the form

$$u(r) = c\,\frac{e^{-\lambda r}}{r} \tag{8.17}$$

where λ is still given by Equation 8.16.

Now we turn our attention to a system of one neutron. If the difference between the neutron mass and the proton mass were about 100 times larger than it actually is the neutron would be capable of decaying by the process

$$n \rightarrow p + \pi^-$$

With the actual masses such a process would violate classical energy conservation by an amount

$$\Delta E = m_\pi c^2 - (m_n - m_p)c^2 \approx m_\pi c^2 \tag{8.18}$$

The process is therefore forbidden by the requirement of energy conservation. In spite of that fact, if one were to shoot another high-energy particle, as an experimental probe, near the neutron there is a finite probability that the particle would encounter, not a neutron, but a proton and a π^-, separated by some distance r. In other words, the wave function of an isolated neutron contains a certain component that describes the system as being in a state of a proton and a π^- separated a distance r. (It also contains a component describing a separated neutron plus a π^0.) The wave function for finding the system in this classically forbidden state is approximately of the form of Equation 8.17 where λ is given by

Equations 8.16 and 8.18. The potential felt by a second nucleon, capable of absorbing the π^-, at a distance r from the neutron is proportional to the π^- wave function at that location. If we compare the Yukawa potential, given in Equation 8.15, with the π^- wave function, given in Equation 8.17, we see that

$$d = \frac{1}{\lambda} = \frac{\hbar}{\sqrt{2m_\pi(m_\pi c^2)}} = \frac{\hbar}{\sqrt{2}m_\pi c} \tag{8.19}$$

The erroneous factor of $\sqrt{2}$ can be traced to the fact that we have used nonrelativistic formulas in a problem that should be done completely relativistically. The main point of this calculation is to show that processes, such as pion transfer, that involve intermediate states that are forbidden by energy conservation have a probability that decays exponentially with distance and that the exponent is related to the amount of energy violation in the intermediate state.

Since Yukawa's work it has become clear that all particle-particle interactions are associated with the exchange of particles and that the range of the interaction is inversely proportional to the mass of the transferred particle. The weak interaction, for example, is not really very weak. It is actually only very short-ranged because of the very high masses of the force carriers associated with the interaction. The weak interaction force carriers are two particles, called the Z and W particles.

SUMMARY

The nucleus of an atom consists of two types of *nucleons*; *protons*, which are positively charged, and *neutrons*, which are uncharged.

The nuclear force is the same for protons and neutrons. It is therefore said to be *charge independent*.

The nuclear force is very strong but very short-ranged. Its range is about $10^{-15}\,\mathrm{m} = 1\,\mathrm{fm}$.

The structure of a nucleus is qualitatively similar to the structure of a liquid drop. The particle density in the nuclear fluid is about 1.5×10^{44} particles/m^3.

The *mass number* of a nucleus is $A = N+Z$ where N is the neutron number and Z the proton number.

Nuclei with the same number of protons (and thus the same chemical characteristics) but different mass numbers are called different *isotopes* of the same element.

The mass of the nucleus (N,Z) is related to its *binding energy* by

$$M(N,Z) = N\,m_n + Z\,m_p - \frac{B(N,Z)}{c^2}$$

The binding energies of nuclei are fairly well predicted by the *empirical mass formula*.

$$B(N,Z) = A\,b_{vol} - 4\pi R^2 b_{sur} - \tfrac{3}{5}\frac{k Z^2 e^2}{R} - \frac{(N-Z)^2}{A} b_{sym}$$

where the values of b_{vol}, b_{sur}, and b_{sym} are given in Equation 8.4. The empirical mass formula can be written in the form

$$B = [2.5\,A - 2.6\,A^{2/3} - .115\,Z^2/A^{1/3} - 3.7\,(N-Z)^2/A]\times 10^{-12}\,\mathrm{J}$$

The three most important processes of nuclear transformation are known as α-decay, β-decay, and γ-decay. α-decay is the emission of a ^4He nucleus, which is called an α-particle. β-decay is a process in which a neutron changes into a proton plus an electron and the electron is emitted along with a massless antineutrino. In another type of β-decay, called *electron capture*, a proton plus one of the atom's electrons combine and are transformed into a neutron along with the emission of a neutrino. γ-decay is the emission of a high-energy photon as the nucleus makes a transition from one quantum state to another of lower energy.

The law of *radioactive decay* states that for all spontaneous decay processes, the rate at which decays occur in a large sample of nuclei is proportional to the number of undecayed nuclei still left. Mathematically this says that

$$\frac{dN}{dt} = -\lambda N(t)$$

The number of undecayed nuclei at time t is

$$N(t) = N_o e^{-\lambda t}$$

where N_o is the number at time zero. The time for half of the original nuclei to decay is called the *half-life* and is given, in terms of the *decay constant* λ, by

$$T_{1/2} = \frac{\ln 2}{\lambda}$$

A nuclear γ-decay with a half-life of $T_{1/2}$ has a *natural line width* of $\Delta\omega = 1/T_{1/2}$.

Nuclear emission and absorption lines are usually wider than the natural line width due to *thermal Doppler broadening* and the *nuclear recoil effect.* For a γ-ray of angular frequency ω emitted by a nucleus of mass m at a temperature T the Doppler and recoil broadening effects can be estimated by the formulas

$$\frac{\Delta\omega}{\omega} \approx \frac{\sqrt{k_B T/m}}{c} \qquad \text{(Doppler)}$$

$$\frac{\Delta\omega}{\omega} \approx \frac{\hbar\omega}{2mc^2} \qquad \text{(Recoil)}$$

The *Mossbauer effect* is the emission and absorption of nuclear γ-rays with no recoil or Doppler broadening.

Of all the possible combinations of integers, N and Z, only a small fraction actually describe stable nuclei. The stable nuclei all lie close to the curve given by the equation

$$N - Z = \frac{A^{5/3} - 1.8\,A}{A^{2/3} + 130}$$

The energy levels of nuclei are grouped into *shells* similar to atomic shells. The shell closings occur when N or Z is equal to one of the *magic numbers* 2, 8, 20, 28, 50, 82, or 126.

Nuclear fission is a process in which a large nucleus breaks up into a number of smaller pieces with the release of nuclear energy.

Nuclear fusion is a process in which two smaller nuclei fuse, with the release of energy, to form a larger, more tightly bound nucleus.

The *Yukawa theory* describes the nuclear force as being due to the transfer of spin-zero particles, called pions. According to the Yukawa theory the nuclear potential for $r > d$ should be given by the *Yukawa potential*

$$V(r) = -g^2 \frac{e^{-r/d}}{r}$$

where the range of the potential, d, is related to the pion mass by $d = \hbar/mc$.

PROBLEMS

8.1 Assuming uniform density, calculate the diameter of the ^{209}Bi nucleus.

8.2* The nuclear force becomes strongly repulsive at an internucleon distance of about .5 fm. If we construct a sphere about the center of each nucleon of diameter .5 fm and call it the repulsive core of the nucleon then, in a large nucleus, what fraction of the nuclear volume is occupied by the repulsive cores of the nucleons?

8.3 Estimate the number of nucleons that are within a distance of 1 fm from the surface of a nucleus of mass number A. Evaluate your estimate for the cases of ^{12}C and ^{238}U.

8.4 The ^{16}O nucleus has a mass of 15.9906 u. What is its binding energy?

8.5** In the gravitational collapse of the white dwarf to form a neutron star the nucleus (N,Z) swallows up Z electrons to become a pure neutron nucleus (A,0). These neutron nuclei then fuse to form the giant neutron nucleus which is the neutron star. Considered as a nuclear reaction, the reaction $(N,Z) + Z e^- \rightarrow (A,0)$ is highly endothermic (energy absorbing). The energy to drive the reaction comes from the gravitational pressure. The energy absorbed from the gravitational field when the atom collapses is $E = pV$ where p is the pressure and V is the original volume of the atom. Taking the radius of the Fe atom as 1 Å, estimate the pressure required to drive this reaction for the nucleus ^{56}Fe.

8.6* Calculate the electrostatic force on a proton at the surface of a nucleus (N,Z).

8.7* Using the empirical mass formula, estimate the mass of the ^{232}Th nucleus. Compare your result with the actual mass.

8.8 Using the empirical mass formula, calculate the binding energy of ^{12}C.

8.9** Show that, when Equation 8.2 and 8.3 are combined, one obtains the empirical mass formula in the form

$$B = a A - b A^{2/3} - c Z^2 A^{-1/3} - d (N-Z)^2 A^{-1}$$

and evaluate the coefficients a, b, c, and d.

8.10* Use the empirical mass formula to estimate the excitation energy (the energy above the ground state energy) that the ^{235}U nucleus would be left with after the slow neutron capture reaction $^{234}U + n \rightarrow {}^{235}U$.

8.11* Calculate separately the electrostatic, surface, and symmetry energies of each of the following three nuclei: ^{9}Be, ^{59}Co, and ^{209}Bi.

8.12* The nuclei ^{38}Ca and ^{38}Ar have $(N,Z) = (18,20)$ and $(20,18)$ respectively. They are called *mirror nuclei.* According to the empirical mass formula, the mass difference of two mirror nuclei is attributable entirely to the difference in electrostatic energy. All other terms in the formula have the same values for both nuclei. Use the empirical mass formula to estimate the difference in electrostatic energy for these two nuclei and compare it with the actual difference in their binding energies.

8.13** Using the analysis that goes with Figure 8.6 and the numerical value of b_{sym}, estimate the spacing between nucleon energy levels in ^{78}Se. Compare your result with the level spacing shown in Figure 8.7.

8.14 What is the wavelength of the γ-ray emitted when the ^{78}Se nucleus decays from its excited state at 613.2 keV to its ground state? (See Figure 8.7).

8.15* Suppose a particular nucleus had a probability of .001 of emitting an α-particle during any second and also a probability of .002 of emitting a positron during any second. What would be the half-life of the nucleus (i.e., the time required for half the nuclei to decay by any means)?

8.16 ^{32}P decays by β^--decay with a half-life of 14.2 days. What is its decay constant? What is the resulting nucleus? How long would it take for .9 of the original ^{32}P to have undergone decay?

8.17* According to Equation 8.11 what should be the β stable isotope of mass number 80?

8.18* Calculate four well spaced points on the curve of β-stable nuclei using Equation 8.11 and draw the curve.

8.19** Carry out the detailed analysis leading from Equation 8.10 to Equation 8.11.

8.20* Estimate the initiation energy for the reaction $^{24}Mg + {}^{4}He \rightarrow {}^{28}Si$.

8.21* $^{235}U \rightarrow {}^{124}Sn + 2\,{}^{48}Ca + {}^{4}He + 11\,n$. What is the energy released by the reaction?

8.22* Using the actual masses of the nuclei involved, calculate the total energy that would be released by the fusion of one kilogram of tritium (^{3}H) with an equal number of protons to form ^{4}He nuclei.

8.23* Make a rough graph of the ratio of electrostatic energy to surface energy of stable nuclei from A = 4 to A = 238.

CHAPTER NINE

HIGH ENERGY PHYSICS

9.1. INTRODUCTION

In 1680, in a book called *Optiks*, Newton described the structure of matter in the following words:

> Now the smallest Particles of Matter may cohere by the strongest attractions, and compose bigger Particles of weaker Virtue; and many of these may cohere and compose bigger Particles whose Virtue is still weaker, and so on for divers Successions, until the Progression end in the biggest Particles on which the Operations in Chemistry, and the Colours of natural Bodies depend, and which by cohering compose Bodies of sensible Magnitude. There are therefore Agents in Nature able to make the Particles of Bodies stick together by very strong Attractions. And it is the Business of experimental Philosophy to find them out.

During the three centuries since that time there certainly have been diverse successions of experimental philosophers engaged in the problem of finding out the most fundamental agents in nature. The above words of Newton, which include an amazingly accurate general description of our current picture of the structure of matter, clearly separate the problem into two parts. On one hand we are to discover the most fundamental *particles* of which matter is composed. On the other hand we are to determine the nature of the *forces* that cause those particles to cohere. The portion of physics that deals with the fundamental particles and their interactions has come to be called *high-energy physics*, because, in order to investigate very small things one must use very small wavelengths*, and particles of very small wavelength have very high energy. After a brief discussion of the experimental techniques used in high-energy physics we will present the currently accepted picture of the fundamental structure of matter. The presentation will be along the lines suggested by Newton. Namely, a description of the fundamental particles involved, followed by a description of their interactions.

---

* Large wavelength radiation is simply diffracted around small structures and fails to reveal any of their details.

9.2. PARTICLE PRODUCTION

Most high-energy experiments fit nicely into the pattern we called the standard quantum mechanical experiment. They include a device for creating a beam of particles of known type and energy. They include an interaction region in which the particles in the beam can interact with those in the target material (or sometimes in a second particle beam). And finally, they include an array of detectors for detecting and making measurements on the particles that were scattered or created in the interaction region. The devices used for the production of the particle beam can be broken down into five groups:

1. Natural sources.
2. High-voltage machines (primarily Van de Graaff machines).
3. Cyclotrons.
4. Synchrotrons.
5. Linear accelerators.

Natural sources were very important to high-energy research before the development of particle accelerators but are used much less now. They are of two types; *cosmic rays* and *naturally radioactive isotopes.* Cosmic rays are a steady stream of particles that strike the earth from outer space. The cosmic-ray particles detected at sea level consist of *primary particles,* whose sources are somewhere in space, and *secondary particles,* which are particles produced by the collision of primary particles with atoms in the atmosphere. The primaries consist of about 91% protons, 8% α-particles, and 1% heavier nuclei. Considering these as the nuclei of hydrogen, helium, and heavier elements, we find that the fractional distribution of various elements in cosmic-ray primaries agrees fairly well with the distribution of elements in the universe at large, including the distribution of the elements in stars.

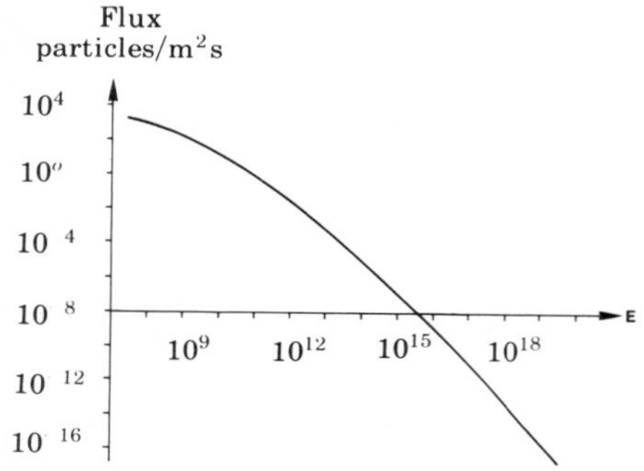

FIGURE 9.1

The energy distribution of primary cosmic ray particles. The graph shows the flux, above the atmosphere, of cosmic-ray primaries of energy E and higher, as a function of E. It is averaged over all latitudes. For low-energy charged primaries the deflection in the earth's magnetic field causes a strong dependence of the flux on geomagnetic latitude; the flux is greatest near the magnetic poles and weakest near the magnetic equator.

The energy distribution of primary cosmic-ray particles is shown in Figure 9.1. As can be seen from that figure, the distribution extends to extremely high energies. Primaries have been detected with energies of the order of 10^{20} eV. For very high-energy primaries what is actually detected is not the original particle but the extensive shower of secondary particles it produces by collisions with nuclei in the upper atmosphere. The energy of the primary is then determined by estimating the total energy of the particles in the shower (See Figure 9.2). The sources of these high-energy cosmic-ray particles are thought to be collapsed or

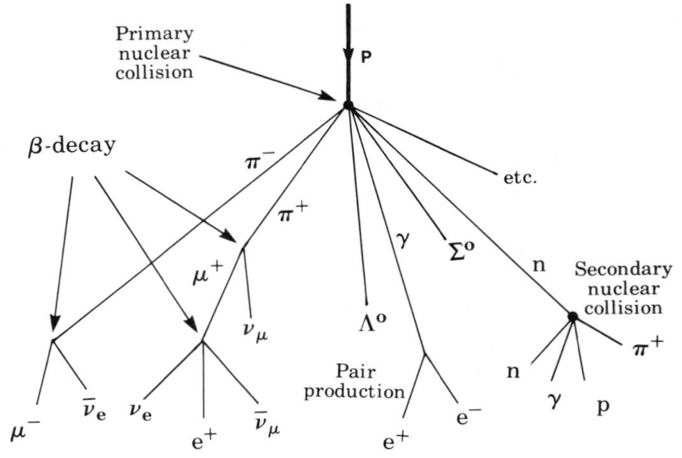

FIGURE 9.2 The structure of a cosmic-ray air shower. The incoming primary collides with a nucleus (usually of nitrogen or oxygen) in the upper atmosphere. The collision products undergo further collisions and decays at lower altitudes. Very high energy primaries can lead to massive particle showers at sea level that extend over many square kilometers and involve very large numbers of particles.

collapsing stars. That is, stars that have become, or are in the process of becoming, neutron stars or black holes. Supporting this view is the observation of one astronomical object (called SS 433) that appears to have the following structure. It is a pair of stars, orbiting one another. One member of the pair is a giant star while the other is a collapsed star. The collapsed star is emitting two oppositely-directed jets of particles. The speed of the particles in the jets, measured by the Doppler shift of their emission lines, is about one fourth the speed of light. A widely accepted theoretical model of SS 433 explains the jets as the result of material being continuously drawn from the larger star by the collapsed companion. As the material approaches the collapsed star it is accelerated toward the speed of light by the enormous gravitational field of the collapsed star. This stream of infalling particles acts as the energy source for the jets of outgoing particles. A very strong magnetic field associated with the collapsed star forces the outgoing particles to be emitted in the direction of the two magnetic poles, thus producing two narrow, oppositely-directed jets. Of course, there is a large distance between the energy of a particle of speed $c/4$ and the extremely high energies of some cosmic-ray particles. There are no presently known astronomical objects that produce particles in the very high energy portion of the cosmic-ray spectrum, and therefore the theory that collapsed stars act as the sources of most cosmic-ray particles has not yet been verified.

In the period before World War II the study of cosmic rays supplied much of the experimental data for high energy physics. It was in cosmic rays that positrons, muons, pions, lambdas, and sigmas were first detected and studied. However, the fact that the flight paths and energies of incident cosmic-ray particles are completely uncontrollable and unpredictable made life very difficult for the pioneer high-energy physicists. Because the atmosphere screens out most cosmic-ray particles and severely reduces the energy of those that get through, most experiments were conducted above the atmosphere, using high-altitude balloons. The balloons usually had only enough lift capacity to carry the experimental equipment without the experimenter. The experimenter would operate the equipment from the ground by radio, while someone else chased the balloon along back-country roads with a car. (The winds were not much more predictable than the particles.) The inconvenience of this mode of experimental research provided strong motivation for the development of particle accelerators.

The other natural sources of high-energy particles are radioactive isotopes. These can be chosen to provide steady sources of α-particles, γ-rays, or electrons. While naturally radioactive isotopes are readily available and more convenient to use than cosmic rays, they have one crucial deficiency. The energies of the particles emitted by radioactive substances are no more than a few MeV. Such energies, although adequate for some nuclear physics experiments, are too small for studying most processes of interest to the high-energy physicist.

High-voltage machines: High-voltage machines are devices for producing a large electric potential difference between two metal electrodes within an evacuated container. Charged particles (usually electrons) from a source at one electrode are accelerated by the electric field between the electrodes. If the

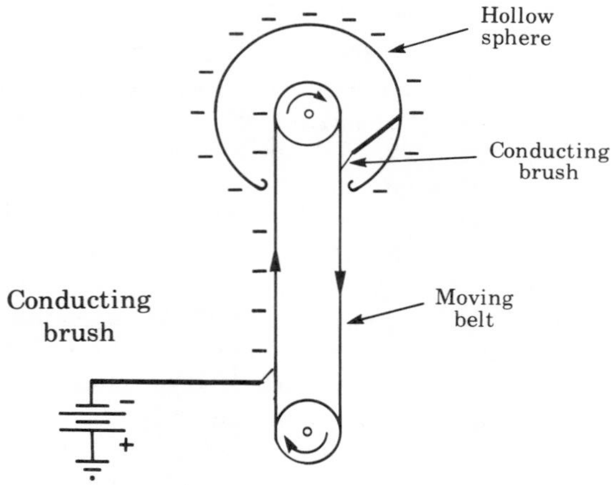

FIGURE 9.3 A Van de Graaff generator. Negative charge is carried up the moving belt and conducted by the metal brush to the outside of the large hollow sphere.

particles have negligible initial energy then their energy, in electron-volts, when they reach the second electrode, will be equal to the potential difference between the electrodes.

The most important high voltage machine in present use is the *Van de Graaff accelerator.* The operation of a Van de Graaff machine depends on the fact that excess charge will always move to the surface of a conductor, regardless of how much charge already resides on that surface. In a Van de Graaff machine, charge is physically transported into a hollow conductor as surface charge on a moving belt (See Figure 9.3). As charge continues to collect on the outer surface of the hollow conductor its potential increases. There is a certain amount of unavoidable leakage of charge from the hollow conductor. This leakage current increases with increasing potential. When the leakage current is equal to the current transported by the belt the Van de Graaff machine has reached its maximum potential. For large modern Van de Graaffs the maximum achievable particle energies are some tens of MeV. Thus, because of its limited energy the machine is used primarily in nuclear physics research where its high beam current and precise beam energy make it very useful.

Cyclotrons: The operation of a cyclotron depends on two physical principles. (1) Charged particles within a hollow conductor feel no electrostatic forces, regardless of the potential difference between the conductor and other conductors outside it. (We are assuming that the fields created by the charged particles within the hollow conductor are negligible.) (2) A charged particle, moving in a uniform magnetic field, with an initial velocity that is perpendicular to the field, will move in a circle whose radius is proportional to the particle's speed.

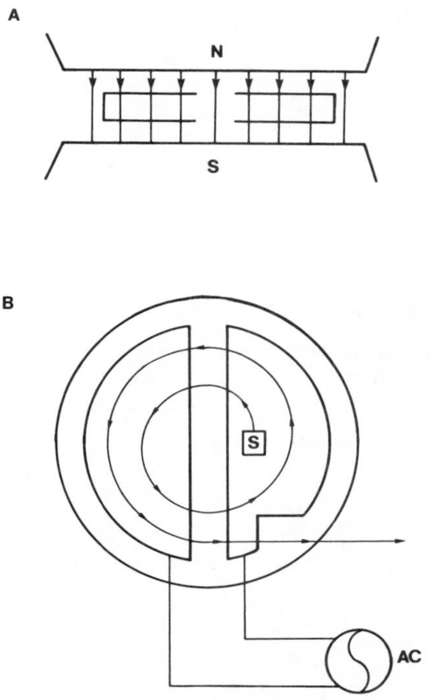

FIGURE 9.4

A. The two dees are located between the pole pieces of a large magnet.

B. The trajectory of a particle as it is accelerated within the cyclotron. The particle originates at the source (S) with low energy and is accelerated each time it passes from one dee to the other.

Therefore, the time it takes the particle to traverse the circle, given by $T = 2\pi r/v$, is independent of the particle's speed. A simple calculation, made by setting the mass of the particle times its centripetal acceleration (v^2/r) equal to the magnetic force on the particle (qvB) gives the radius of its circular orbit as $r = mv/qB$. Thus the frequency with which it traverses its orbit is given by $\nu_c = 1/T = qB/2\pi m$. ν_c is called the *cyclotron frequency* of the particle.

A cyclotron is made by splitting a hollow metal cylinder, of large diameter but low height, so as to form two *dees* (See Figure 9.4). The dees are placed in a uniform magnetic field and connected to a high-voltage AC generator operating at the cyclotron frequency. A steady source of low-energy particles is placed inside one of the dees as shown. The alternating generator creates a strong alternating electric field across the gap between the dees but no field within the dees. Consider a particle from the source that enters the gap just as the field is in a direction to accelerate it. The particle will be accelerated across the gap and enter the other dee with a greater speed than it had when it left the first. It will move in a circular orbit within the second dee and emerge from it, into the gap, at a time equal to one half a cyclotron period. But by that time the electric field across the gap will have reversed and so the particle will again be accelerated as it crosses the gap. Clearly, the same thing will happen at each crossing of the gap and the particle will travel on a spiral path of increasing radius and energy until it reaches the outer edge of the dees. There an exit window is located and the beam of high-energy particles is drawn off into an experimental area.

Unfortunately, relativistic effects spoil the nice agreement between the orbital frequency and the electric field frequency. In Problem 2.15 the reader was asked to show that a particle of mass m and charge q, moving in a uniform magnetic field, will travel on a circular orbit with an angular frequency ω that is related to the radius R of the circular orbit by

$$\omega = \frac{(qB/m)}{\sqrt{1 + (qBR/mc)^2}} \qquad (9.1)$$

Thus the orbital angular frequency is independent of R only when $qBR/mc << 1$. If we consider the typical case of a proton in a magnetic field of 1 Tesla then this inequality becomes

$$R << \frac{m_p c}{eB} = \frac{(1.67\times 10^{-27})(3\times 10^8)}{(1.6\times 10^{-19})(1)} \approx 3\,\mathrm{m}$$

Thus a proton that has been accelerated to an orbital radius of about 30 cm will have an orbital frequency that differs significantly from the frequency of the alternating electric field between the dees. It will therefore get out of synchronization with the electric field and will no longer be accelerated when it passes through the gap. A proton of orbital radius 30 cm in a field of 1T has a kinetic energy of 4.3 MeV. Therefore, in order to accelerate protons in a cyclotron to energies of more than a few MeV something must be done about the loss of synchronization due to relativistic effects. One has no way of changing the charge or mass of a proton. The only controllable variables in the system are the frequency of the alternating field on the dees and the strength of the magnetic field. This suggests two different ways of solving the sychronization problem.

One can vary the frequency of the alternating voltage as the particles are accelerated and thereby keep the electric field in phase with the particle motion. This is the method used in the *synchrocyclotron*, sometimes called the *frequency modulated cyclotron*. Such machines can accelerate protons up to relativistic energies of about 1.7 GeV. The rest energy of a proton is .94 GeV. Therefore, the speed of a 1.7 GeV proton is definitely relativistic. The disadvantage of the synchrocyclotron is that the beam current is very much reduced, in comparison with a nonrelativistic cyclotron, by this mode of operation. In order to maintain synchronization one must adjust the frequency of the alternating voltage as a particular packet of particles is accelerated to larger orbits. But this gets the alternating voltage out of synchronization with any particles in smaller orbits, which are thus not accelerated. One has to follow one bunch of particles at a time from the center to the outer edge. In contrast, the nonrelativistic cyclotron supplies a bunch of particles to the experimental beam with each cycle of the alternating voltage.

The other way of solving the synchronization problem is to make the magnetic field increase with increasing R so as to keep ω , given in Equation 9.1, constant (See Problem 9.1). This is the method used in the *isochronous cyclotron.* In comparison with a synchrocyclotron, an isochronous cyclotron has the advantage of having a much greater beam current. However technical problems that are too complicated to discuss here have limited the output energy of isochronous cyclotrons to less than 100 MeV for protons. Thus, with their large beam current, they are excellent tools for nuclear physics research, but have limited value for research on elementary particle processes which mostly take place at higher energies.

Synchrotrons: In a cyclotron the radius of the particle's orbit increases as its energy increases. Therefore, it is necessary to maintain a strong magnetic field over a large circular area. This makes the cyclotron design quite impractical for obtaining particle energies of more than a few GeV. For example, a proton of energy 100 GeV, in a magnetic field of 2 Tesla (which is a very strong field to maintain over a large area), has an orbit with a diameter of about .3 km (See Problem 9.2). It is obviously unfeasible to build a solid magnet of that size.

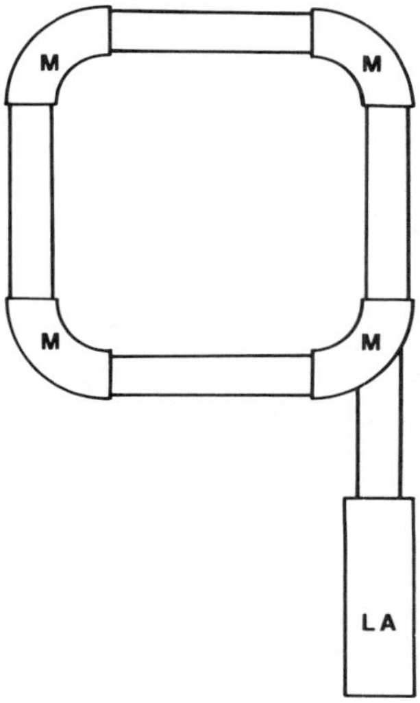

FIGURE 9.5

A synchrotron. The particles are preaccelerated in a small linear accelerator (LA) or cyclotron and then enter the beam tube. They are kept in the beam tube by the bending magnets (M) as they traverse the circuit many times during their acceleration. The accelerating mechanism and the exit tube are not shown in the picture.

A synchrotron is a machine in which the particles are kept on a single orbit as they are accelerated. The size of the orbit can be made much larger than would be practical in a cyclotron because the magnetic field needs to be maintained only along that single orbit (See Figure 9.5). In Chapter 6 we have shown that the radius of a particle's orbit is related to the particle's momentum by

$$r = p/qB \tag{9.2}$$

Thus, in order to maintain a fixed trajectory as the particle's momentum increases during acceleration, the magnetic field strength in the curved portions of the path must be increased along with the momentum. Particles are always preaccelerated by a linear accelerator or cyclotron to a speed close to c before being injected into a synchrotron for final acceleration.

Linear Accelerators: In a linear accelerator no attempt is made to confine the accelerated particles to a circular or closed orbit. The particles travel in a straight line down a long tube while they are being accelerated. The earliest versions had the tube broken into sections, separated by gaps. An alternating voltage was applied across adjacent sections as shown in Figure 9.6(A). The particles were accelerated as they passed from one section to the next. Since a particle's velocity increases as it moves down the tube, the sections must have

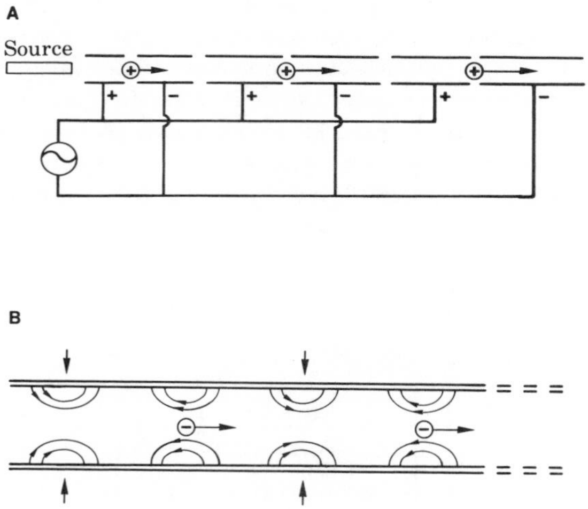

FIGURE 9.6 A. A 'drift tube' linear accelerator. The AC voltage is timed so that the particles are accelerated each time they pass from one tube section to the next.
B. In a modern linear accelerator electrons are accelerated by electromagnetic waves that travel with them so that the electrons remain always in a region of backward-pointing electric field and are therefore accelerated continuously down the tube. Notice that positrons could also be accelerated in the same machine by 'bunching' them at the locations of the arrows where the electric field is forward-pointing. This fact is important for the operation of the Linear Collider to be described later.

increasing lengths in order to have the particle appear at each gap in phase with the alternating voltage. In more recent linear accelerators the tube is continuous and the particles are accelerated by traveling electromagnetic waves (microwaves) that are created and amplified by microwave generators distributed along the length of the tube (See Figure 9.6(B)). In free space the electric and magnetic fields in a traveling wave are always perpendicular to the direction of travel, but this is not so for waves traveling within a conducting *waveguide*. A circular waveguide can carry electromagnetic waves that have a strong longitudinal component near the center of the waveguide. In a linear accelerator the particles are preaccelerated before being injected into the accelerating tube. The speed of the traveling wave is matched to the particle velocity so that the particle sits in a region of forward pointing electric field all during its trip down the tube. (For electrons the electric field is made backward pointing so that the force points forward.)

The linear accelerator design is particularly well-adapted to accelerating electrons. High-energy electrons, moving in a curved path, radiate energy in the form of electromagnetic waves very rapidly. Actually, all charged particles radiate electromagnetic waves if they travel on curved paths. But, in order to obtain a prescribed energy, electrons, because of their very small mass, must travel at a greater speed than heavier particles. The rate of energy loss due to radiation increases rapidly as the speed of a charged particle, moving on a curved path, approaches c. For this reason, radiation loss is a much more serious problem in circular electron accelerators than it is in circular proton accelerators of equal particle energy. Particles being accelerated in a longitudinal, rather than a transverse direction have much smaller radiation loss.

The largest of the linear electron accelerators is located at the Stanford Linear Accelerator Center (SLAC) in California. It has a length of approximately two miles, and is capable of accelerating electrons to an energy of over 27 GeV. That is 54,000 times their rest energy, $m_e c^2$. At that energy $v = (1 - 1.7 \times 10^{-10})c$.

9.3. COLLIDING BEAM EXPERIMENTS

A stationary particle has a relativistic energy $E = mc^2$. The same particle, viewed in a frame moving at velocity $v = \beta c$ with respect to the particle's rest frame, has an energy $E' = mc^2/(1-\beta^2)^{1/2}$. If β is close to one then E' will be much greater than E. That is, the *energy* of a particle or a system of particles depends upon the frame of the observer. Therefore, if no information is given about the inertial frame in which the process is to be viewed, then a statement such as: "2 GeV of energy is required for the reaction, $e^- + e^+ \rightarrow p + \bar{p}$." is really meaningless. In fact, such statements are made by physicists all the time. When they are made there is an assumption, usually unstated, that the process is to be viewed in the *center-of-mass frame*. That is, the frame in which the total momentum of the system is zero. It can be shown that it is in the center-of-mass frame (usually abbreviated as the cm frame) that the energy required for any particular reaction is a minimum. Most high-energy collision experiments are carried out and observed in a frame of reference, called the lab frame, in which one of the colliding particles is part of stationary target while the other is contained in the incoming beam. We will soon show that the ratio of the energy of the pair of colliding particles in the cm frame to their energy in the lab frame becomes very small as both energies increase. Therefore, to carry out a reaction that requires 2

GeV in the cm frame we may need a particle beam of many times that energy in the lab frame. The laws of relativistic dynamics make high-energy collision experiments, carried out in the lab frame, very inefficient. For example, the $e^- + e^+ \rightarrow p + \bar{p}$ experiment mentioned above, which needs 2 GeV of center-of-mass energy in order to create the proton-antiproton pair, could not be done with a 27 GeV positron beam from the SLAC linear accelerator. In fact, it would be beyond the capacity of a 1000 GeV positron accelerator!

Before we discuss the colliding beam machines that have been designed to avoid this problem let us convince ourselves that the relativistic transformation laws do lead to this unpleasant result. We view two particles of equal mass, m, in the cm frame. Their velocities are $\pm v_{cm}$. Using Equation 2.14, we calculate the ratio of their energy in this frame to their rest energy, $2mc^2$.

$$\frac{E_{cm}}{2mc^2} = \frac{1}{\sqrt{1-\beta_{cm}^2}} \tag{9.3}$$

where $\beta_{cm} = v_{cm}/c$. The lab frame is the frame in which one of the particles is stationary. Therefore the lab frame moves with velocity v_{cm} with respect to the cm frame. In the lab frame the beam particle will have a velocity v^*, given by the velocity addition law (Equation 1.22).

$$v^* = \frac{2v_{cm}}{1+\beta_{cm}^2} \tag{9.4}$$

Letting $\beta^* = v^*/c$, a little algebra gives

$$1-\beta^{*2} = \frac{(1-\beta_{cm}^2)^2}{(1+\beta_{cm}^2)^2} \tag{9.5}$$

In the lab frame the energy of the system is

$$E_{lab} = \frac{mc^2}{\sqrt{1-\beta^{*2}}} + mc^2 \tag{9.6}$$

Using Equation 9.5 we get

$$\frac{E_{lab}}{2mc^2} = \frac{1}{2}\left(\frac{1+\beta_{cm}^2}{1-\beta_{cm}^2} + 1\right)$$

$$= \frac{1}{1-\beta_{cm}^2} = \left(\frac{E_{cm}}{2mc^2}\right)^2 \tag{9.7}$$

For the $e^- + e^+ \rightarrow p + \bar{p}$ reaction

$$\frac{E_{cm}}{2mc^2} = \frac{2\text{GeV}}{1\text{MeV}} = 2000$$

Therefore the beam energy needed in the lab frame would be

$$E_{lab} = (2000)^2(1\,\text{MeV}) = 4000\,\text{GeV}$$

In order to have 2GeV available in the cm frame we must supply 4000 GeV in the lab frame. No accelerator, either existing or projected, can produce 4000 GeV electrons. It is clear that something must be done to eliminate the lab frame to cm frame transformation. Machines have now been designed to eliminate that transformation by producing a beam of particles moving in one direction and

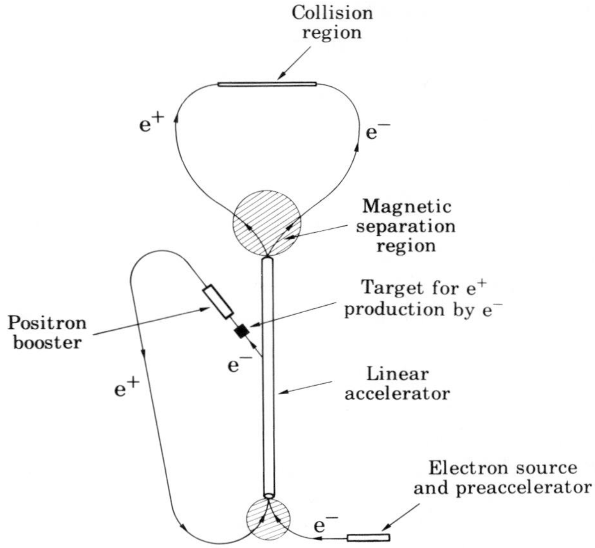

FIGURE 9.7 The plan of the Linear Collider, presently (1983) under construction at SLAC. Intermittent bunches of the electrons are pulled out of the accelerating tube to the left and used to create a positron (e^+) beam. The positrons are then preaccelerated and then directed into the accelerating tube to be accelerated to full energy (see note to Figure 9.6). At the end of the tube electrons and positrons are deflected in different directions by the separation magnet (**B** into paper). They are then directed into two opposing beams that collide 'head on' in the collision region.

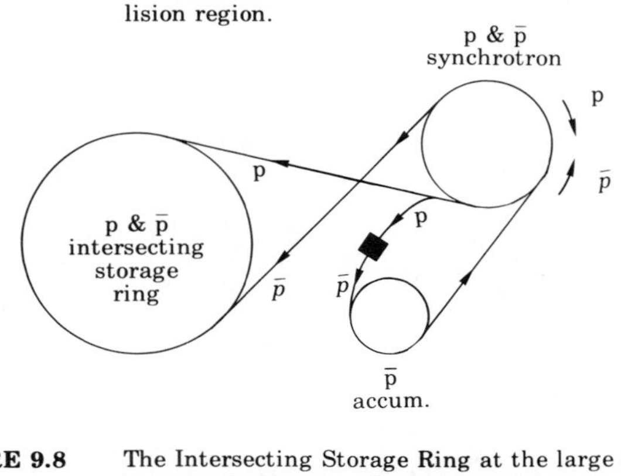

FIGURE 9.8 The Intersecting Storage Ring at the large CERN laboratory in Geneva. The protons (p) and antiprotons ($\bar{p}$) travel in opposite directions along slightly separated figure 8 paths between the large storage ring and the synchrotron in which they are accelerated. The $\bar{p}$ beam is created by drawing off a fraction of the proton beam into a target (solid black square) where antiprotons are created by collisions. The antiprotons are then brought up to speed in the $\bar{p}$ accumulator and directed into the main figure 8 path. In the storage ring the two beams are made to intersect in order to obtain $p + \bar{p}$ collisions.

another beam of their antiparticles moving in the opposite direction with equal speed. If the beams are made to intersect at some point then one obtains collision events in which the cm frame and the lab frame coincide. Two examples of such *colliding beam machines* are shown in Figures 9.7 and 9.8.

9.4. PARTICLE DETECTORS

In order to obtain useful scientific information from the collisions of high-energy particles the particles leaving the collision site must be detected and identified. If possible, their momenta should be accurately measured. A wide variety of particle detection and measurement devices have been designed for these purposes. We will consider only a few illustrative examples of them.

All present particle detection methods depend in some way upon the interaction of charged particles with matter. They therefore do not directly detect electrically neutral particles. The existence and properties of the neutral particles are inferred from their interaction with the directly-detected charged particles. For example, the sudden appearance of a pair of charged particle tracks in a bubble chamber might be clearly identifiable as being due to the decay of a neutral Λ^o into a π^- and a proton as in Figure 9.9.

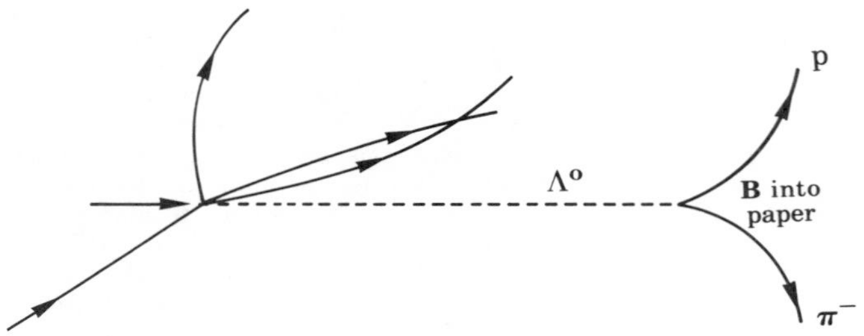

FIGURE 9.9 The identification of a neutral (and therefore invisible) Λ^o particle by means of the visible tracks of charged particles. In a bubble chamber two tracks (on the far right) can be identified as being due to a proton (p) and a pion (π^-). The total momentum of the pair of charged particles can be determined by the curvature of their tracks (a magnetic field permeates the chamber). The total momentum vector is seen to point away from the site of a charged particle collision (indicated by the arrow). This shows that a neutral particle must have traveled along the unseen (dotted line) path and spontaneously decayed at the point of origin of the p and π^-.

Geiger Counters and Proportional Counters: One of the simplest charged-particle detectors is the Geiger counter. It consists of a gas-filled metal cylinder with nonmetal caps. A fine metal wire is stretched along the axis of the cylinder and is maintained at a positive potential with respect to the cylinder. Any charged particle moving through the gas in the cylinder will dislodge electrons

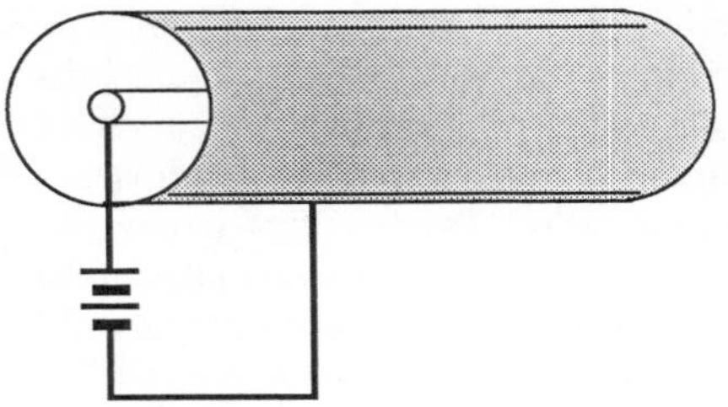

FIGURE 9.10
The geometry of a Geiger or proportional counter.

from the gas atoms. The electrons will be accelerated toward the wire by the electric field. As they get close to the wire, where the field is very strong, the electrons gain enough energy to knock more electrons off the molecules. Those are also accelerated and create still more free electrons. If the electric field is strong enough, an *electron avalanche* is created and a substantial pulse of electric current can be obtained from what started out as a very small number of electrons. A proportional counter has the same basic design as a Geiger counter. The distinction between the two is that the Geiger counter, which operates with a larger potential difference, gives a larger pulse whose size is independent of the number of electrons initially created by the moving charged particle. The proportional counter gives a smaller pulse, but one that is approximately proportional to the number of initial ions created by the charged particle. In high-energy experiments proportional counters are almost always used in preference to Geiger counters. The smallness of the pulse is no significant problem because electronic amplification is easy, and the extra information regarding the ionizing power of the initial charged particle can be very useful.

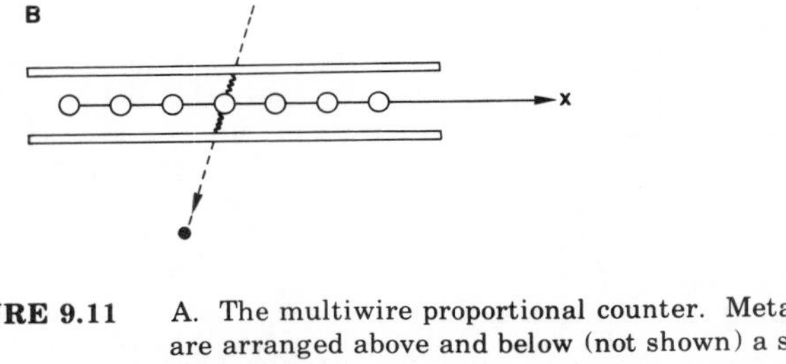

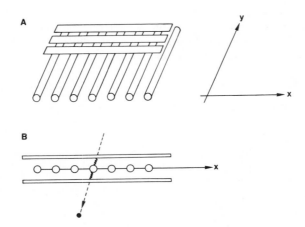

FIGURE 9.11 A. The multiwire proportional counter. Metal strips are arranged above and below (not shown) a series of parallel wires.
B. A charged particle trajectory creates a temporary conducting path between a particular wire and a particular pair of strips.

Multiwire Counters: The major deficiency of the Geiger or proportional counter is its inability to define the location of the ionizing particle's trajectory. For the usual cylindrical counter geometry the signal produced by a particle is independent of the position at which the particle passed through the counter. This lack of spatial resolution is eliminated in a device called a *multiwire proportional counter* (See Figure 9.11). In the multiwire counter a series of parallel wires are suspended between two series of metal strips. The strips are oriented perpendicular to the wires. The localized ionization track of a particle passing through the counter will create a short-lived current pulse in a particular wire and a particular strip. With realistic geometries the x and y coordinates of the particle track can be determined within a fraction of a millimeter.

Cloud Chambers and Bubble Chambers: The best device for studying a particular collision in complete detail is the *bubble chamber.* The bubble chamber is similar in design and operation to an earlier device, called the *cloud chamber,* which it has completely supplanted because of its much shorter cycle time and its greater convenience. A cloud chamber contains a volume of moist air that has been cooled below its dew point and is thus *supersaturated.* A sample of supersaturated air is unstable. Left to itself it will eventually separate into a combination of liquid droplets and saturated air (i.e. air at 100% humidity). However, in the absence of dust particles or other things that can act as centers of droplet formation, the process of droplet production can be very slow. A fast-moving charged particle, passing through such supersaturated air, will create ionized air molecules by collisions along its trajectory. Because the water molecule is a strongly polar molecule, the newly-created ions attract surrounding water molecules and thus act as excellent condensation sites for droplet formation. A visible track of tiny droplets is created, showing the path of the ionizing particle. If the cloud chamber is placed in a region of uniform magnetic field, then the momentum of the ionizing particle can be inferred from the curvature of its track.

The bubble chamber differs from the cloud chamber in two ways. (1) The saturated air is replaced by a liquid, and (2) the condensation process is replaced by the boiling process. A rigid chamber is completely filled with a liquid that is kept at a pressure of about 10 atmospheres and at a temperature that lies

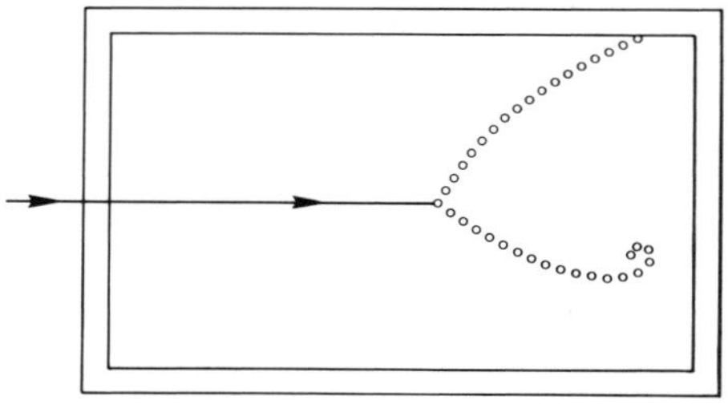

FIGURE 9.12 A neutral particle, coming into the cloud chamber from the left, decays into two charged particles that are seen as a series of droplets. A magnetic field, pointing into the paper, causes the paths of the charged particles to be curved.

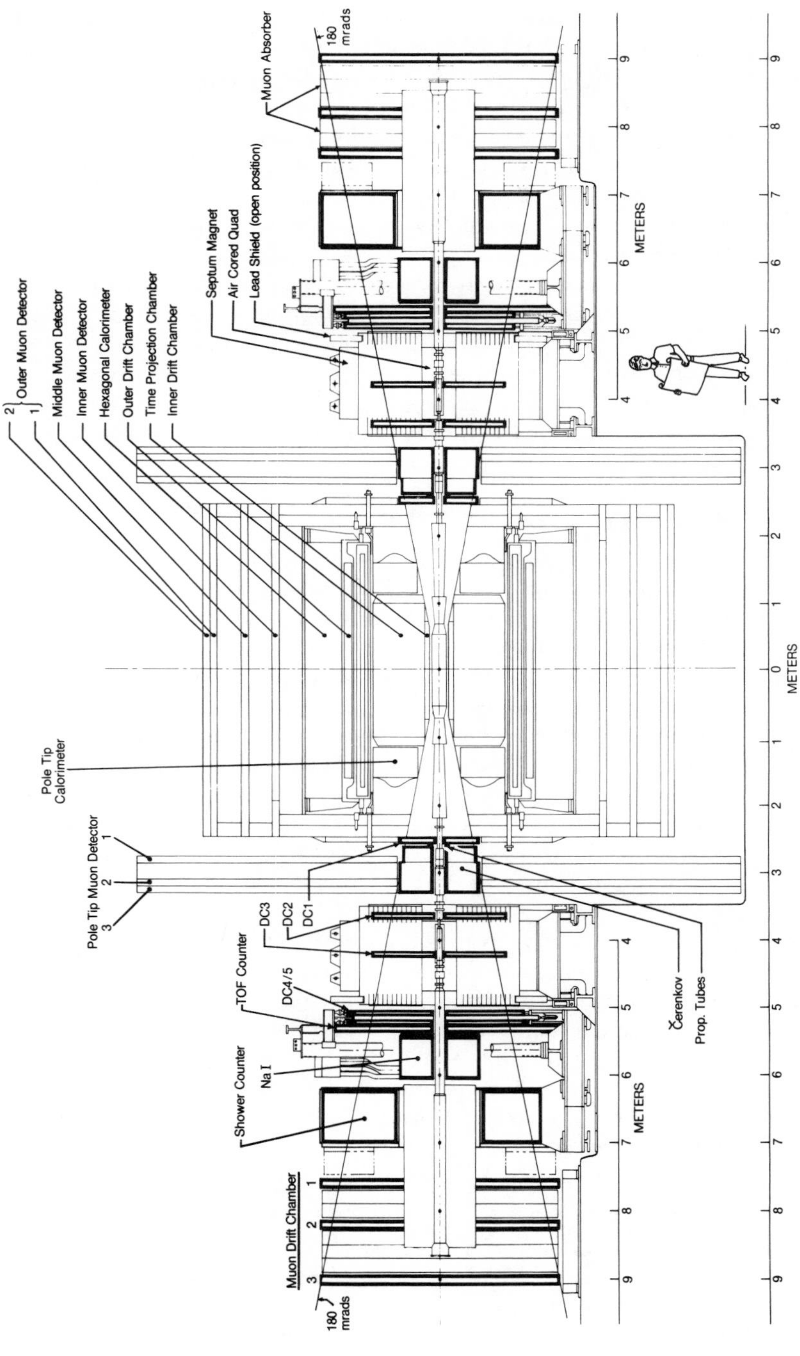

FIGURE 9.13 An example of a large, complex particle detection system; the Time Projection Chamber, recently completed at SLAC. The system of counters and measuring devices allows an essentially complete reconstruction of collision events that take place within the chamber.

somewhere between the boiling temperature at 1 atmosphere and the boiling temperature at 10 atmospheres. By means of a movable piston, the pressure is suddenly reduced and the liquid finds itself in an unstable, superheated state. Boiling commences by bubble formation at any nucleation site within the liquid. If very clean liquid is used, and the liquid is exposed to high-energy ionizing particles, then the nucleation sites will lie along the particle trajectories which will therefore appear as sequences of small bubbles. The bubble tracks are quickly photographed using more than one camera so that a complete, 3-dimensional reconstruction of any collision process that has taken place within the bubble chamber can later be created. As soon as the picture is taken the liquid is brought back up to high pressure, causing the bubbles to quickly disappear. Since ionization centers in a liquid quickly decay, by means of ion recombination and diffusion, the traces of the old tracks are gone within a fraction of a second and the chamber is ready to record new events. Thus the bubble chamber is well-matched to most particle accelerators which produce bunches of high-energy particles at a rate of about one bunch per second.

9.5. THE STRUCTURE OF MATTER

The currently accepted picture of the structure of matter is a hierarchical one. At the base of the hierarchy is a small number of *fundamental particles.* All of the fundamental particles have certain common experimental and theoretical characteristics. In all experiments they appear to be point particles with no signs of internal structure. No finite size can be assigned to them. The meaning of these statements can be clarified by considering the characteristics of a proton; a particle that is definitely not considered to be a fundamental particle. The proton has an electric charge e. By observing how protons scatter electrons from a beam of electrons with known velocity it is possible to determine just how that charge is distributed within the proton. We find that the electric charge of the proton is spread over a region of about 1 fermi (10^{-15}m). In contrast, when electrons are scattered by other electrons the observed scattering pattern agrees precisely with the scattering pattern computed by assuming that the charges of the electrons are concentrated into geometric points. There is no observed charge spread or size. This is no proof that the electron is truly a geometric point. It only indicates that any internal structure of the electron is on a distance scale that is too small to be detected using current experimental techniques. When the electron energy in an electron-proton scattering experiment is made sufficiently high (a few GeV) even more interesting details of the proton's structure are revealed. The high-energy electron-proton scattering pattern agrees very well with what would be expected when electrons are scattered by a bound system of three point particles, two of which have charge $\frac{2}{3}$e while the third has charge $-\frac{1}{3}$e. (Notice that $\frac{2}{3}e+\frac{2}{3}e-\frac{1}{3}e = e$, which is the proton charge.) Thus the proton exhibits both size and internal structure. In the hierarchical scheme there are many levels of structure. For example, the helium atom is a structure composed of two electrons and an α-particle. The α-particle is itself a structure composed of two protons and two neutrons. The protons and neutrons are each structures composed of three particles called *quarks.* According to the currently accepted theory this worlds-within-worlds-within-worlds sequence stops at the level of the quarks. They are not structures composed of still lower level particles. The quarks and electrons are both basement level particles. There are no lower levels. Of course if you are not entirely convinced that this is really the ultimate, final level of structure you are not alone in your skepticism. However, each generation's scientists must work with the experimental and theoretical tools available to them. It is quite possible that future physicists will probe more deeply and find structure in what presently seems structureless.

As mentioned in Chapter 7, all particles are either *fermions* (Fermi-Dirac particles) or *bosons* (Bose-Einstein particles). The characteristic of being a fermion or a boson is called the *statistics* of the particle. This unfortunate and slightly absurd terminology has its origin in the fact that the statistical properties of collections of particles are strongly affected by whether the particles are fermions or bosons. We will look at the fermions first.

9.6. THE FUNDAMENTAL FERMIONS

All the fundamental fermions have an intrinsic spin of $\hbar/2$. They are classified in a sequence of *families*. Each family contains four fermions. The four fermions in a given family are further divided into two groups; a pair of *leptons* and a pair of *quarks*. The pair of leptons has one member of finite mass and charge −e. (The electron is that member of the first family.) The other lepton is an uncharged, zero-mass particle. (For the first family it is the electron neutrino.) The two quarks in a family are both massive. One of them has a charge of $\frac{2}{3}$e. The other has a charge of $-\frac{1}{3}$e. Table 9.1 lists the members of the first three families and some of their properties. No other families are presently known but there is no good reason to assume that further families do not exist. As of this writing one member of the third family, namely the top quark, has not been observed*.

TABLE 9.1
THE FUNDAMENTAL FERMIONS

FAMILY		LEPTONS		QUARKS	
First	Symbol	e^-	ν_e	u	d
	Name	electron	electron neutrino	up quark	down quark
	Charge	−e	0	$+\frac{2}{3}$e	$-\frac{1}{3}$e
	Mass*	.511 MeV	0	~.35 GeV	~.35 GeV
Second	Symbol	μ^-	ν_μ	c	s
	Name	muon	muon neutrino	charm quark	strange quark
	Charge	−e	0	$+\frac{2}{3}$e	$-\frac{1}{3}$e
	Mass	105.6 MeV	0	~ 1.8 GeV	~ .55 GeV
Third	Symbol	τ^-	ν_τ	t	b
	Name	tau	tau neutrino	top quark	bottom quark
	Charge	−e	0	$+\frac{2}{3}$e	$-\frac{1}{3}$e
	Mass	1.87 GeV	0	?	~ 4.5 GeV

*The table gives mc^2, the energy equivalent to the rest mass.

*During the typesetting of this chapter one group of experimenters has announced a possible detection of the top quark, including a rough determination of its mass.

For each of the fundamental fermions there exists an antiparticle with the same mass and spin. For the charged fermions the antiparticle is distinguished from the corresponding particle by having the opposite value of the electric charge. The uncharged neutrinos differ from their antiparticles by having opposite values of helicity. For each neutrino the component of spin in the direction of the particle's motion is $-\hbar/2$ while for it's antiparticle it is $\hbar/2$.

The property of a quark of being a certain type, such as an up-quark, a down-quark, or a charm-quark is called its *flavor.* This use of the word, flavor, is obviously unrelated to the ordinary use of the word. It was intended to be humorous. By the year 2000 the joke may become a little stale, but the terminology has become so standardized that there is no way of changing it now.

9.7. THE FUNDAMENTAL BOSONS

In listing the fundamental bosons we will not include the *graviton,* a massless particle of spin $2\hbar$ expected to be associated with the gravitational field. Although the gravitational force between large objects is noticeable every time we sit down the gravitational interaction between elementary particles is so extremely weak that no quantum gravitational effects have ever been detected and it is extremely unlikely that the theoretically conjectured graviton will be observed experimentally any time in the near future. At present, gravity is treated as a strictly macroscopic effect, and although there is a lot of work being done on quantum theories of gravity there is currently no acceptable way of incorporating gravity into the scheme of elementary particle structure and interactions. When we neglect the graviton the set of fundamental bosons has only four members. They are (1) the *photon,* associated with electromagnetic interactions, (2) the Z^0 and W bosons, associated with the weak interactions, and the *gluon,* associated with the strong interactions. A list of some of their properties is given in Table 9.2. As can be seen there, the bosons are all uncharged except for the W^- and its antiparticle, the W^+. They all have spin $\hbar$.

TABLE 9.2
THE FUNDAMENTAL BOSONS

SYMBOL(NAME)	CHARGE	SPIN	MASS	ANTIPARTICLE
γ (photon)	0	$\hbar$	0	none*
Z^0 (Z-zero)	0	$\hbar$	$\approx$ 96 GeV	none*
W^- (W-minus)	$-e$	$\hbar$	$\approx$ 81 GeV	W^+
G (gluon)	0	$\hbar$	0	$\overline{G}$

*This means that the particle and its antiparticle are identical.

9.8. QUANTUM CHROMODYNAMICS

In addition to having a mass, an electric charge, and a spin angular momentum, the quarks and gluons have another physical property, called their *color.* The word color is used to describe this new property because there are certain similarities between the way particles with different values of the color property combine to form structures with zero color and the way colored lights

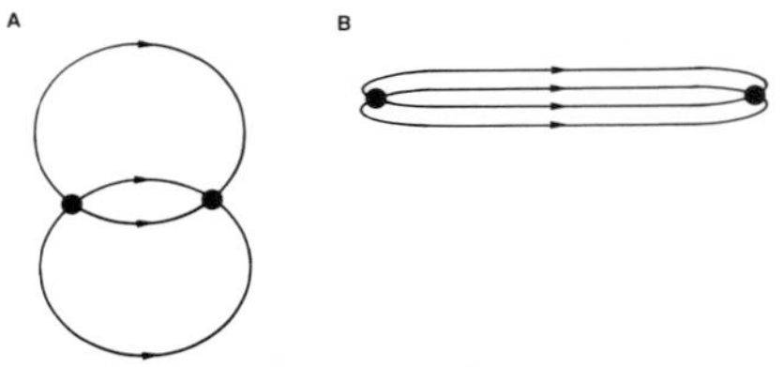

FIGURE 9.14
A. The field pattern of an electric dipole.
B. The color field pattern of a separated quark and antiquark. Actually the color field is a many-component field that cannot be represented by a single vector at each point so that this diagram is only intended to be suggestive.

can be combined to form white (colorless) light. The color property has no relationship at all to the visual properties of the particles; that is, to the way in which they scatter light. In fact, the property of quarks and gluons called color has no simple relationship to any large-scale observable phenomena. It is best to regard color as a new type of charge, similar to, but different from, electric charge. The total color of every structure above the level of single quarks and gluons is exactly zero. This causes color to be completely unobservable on a macroscopic scale. This should be compared with the properties of electric charge. Because of the attraction of unlike charges and the repulsion of like charges the numbers of positive and negative charges in any macroscopic body are always very close to equal. But they are not exactly equal, and the relatively small residual electric charge leads to easily observable physical effects. Also, for microscopic amounts of matter, such as atoms or molecules, the electric charge imbalance can be a large fraction of the total number of charges in the object. In contrast, all isolated structures, regardless of whether they are microscopic, like protons and mesons, or macroscopic, like rocks and pencils, contain exactly zero total color charge. In fact, according to the equations that describe the color field* any isolated object with a net color charge would have infinite energy. We will not try to present any of the mathematical theory of the color field equations. The quantum theory of quarks, gluons, and the color field is called *quantum chromodynamics* (chromo means color). It has many resemblances to quantum electrodynamics, which is the theory of electrons, photons, and the electromagnetic field. However the field equations of chromodynamics are not linear equations, as Maxwell's equations are, and they are therefore immensely more difficult to solve.

Without giving any mathematical details we can show, in a pictorial way, why all isolated systems have zero color. We picture an isolated system composed of two particles of equal but opposite color. In order to obtain a system of nonzero color we need only move one of the particles to infinity. For small separations the color field pattern resembles the electric field pattern of an electric dipole (See Figure 9.14). However, for larger separations the field lines do not spread out in space as the electric dipole field lines would but remain concentrated in a narrow tube connecting the positive and negative color charges. There is an energy density associated with the color field just as there is with an electric field. For large separations of the opposite color charges the total field energy clearly becomes proportional to the separation, s. The work necessary to increase the separation an amount ds would then be fds where f is the energy per unit length in the long tubelike section of the field pattern. But the work necessary to move a particle a distance ds is always given by the force on the particle times ds. Thus, for large separations the force between the two particles is independent of distance. It is just equal to the energy per unit length in the long field tube which will stretch from one particle to the other, no matter how far they are separated. It would therefore take an infinite amount of work to separate them completely. This property of being impossible to be completely separated is called *quark confinement.* It is the reason why particles that are composed of quarks, such as protons, are never broken up into isolated quarks during collisions with other particles, no matter how energetic the collisions are.

This discussion, involving equal and opposite color charges seems to indicate that color charge is very similar to electric charge. Actually there is a major fundamental difference in the two kinds of charge. The electric charge of any

---

*The color field is a field generated by color charge in a way that is similar to the way the electromagnetic field is generated by electric charge. However, the detailed equations relating the color field to the color charge are not the same as Maxwell's equations.

object is given by a single number, q, that can be positive or negative. It is a scalar charge in the sense that all quantities that can be expressed as single numbers are called scalars. In contrast, the color charge of any object can only be defined by giving three separate numbers. We will write them as q_R, q_B, and q_G for red charge, blue charge, and green charge respectively. Thus each of the before mentioned quarks can have three different color charge states. That is; there is a red up-quark, a blue up-quark, and a green up-quark. (The same is true for the down-quark, the charm-quark, etc.) The antiquarks have opposite color charges (antired, antiblue, and antigreen). The color charges q_R, q_B, and q_G are defined as positive for the quarks. They are therefore negative for the antiquarks.

There is only one way to produce a structure with zero electric charge; namely to have it contain as many positive electric charges as negative electric charges. In contrast, there are two ways of constructing an object of zero color charge. One way is to cancel each positive red charge with a negative red charge and do the same for the blue charges and the green charges. For example, a colorless π^+ can be made with a red u (an up-quark) and an antired $\bar{d}$ (an anti-down-quark). The other way is to combine equal amounts of the three color charges. It is the similarity between this way of adding equal amounts of the three different color charges to obtain a net color charge of zero and the way that lights of the three primary colors can give colorless light that has led to the use of the term color to describe this characteristic of quarks.

9.9. QUARK STRUCTURE OF THE HADRONS

We have already mentioned that all allowable structures must have zero color charge. By combining that rule with the two different ways of producing colorless structures we can understand many of the properties of the known strongly-interacting particles. The observed strongly-interacting particles are called *hadrons.* The set of hadrons is composed of a subset of quark-antiquark structures, called mesons, and a subset of three-quark structures, called baryons. Tables 9.3 and 9.4 show the quark structure of some of the hadrons. The

TABLE 9.3
QUARK STRUCTURE OF SELECTED BARYONS

SYMBOL & NAME	QUARK CONTENT	QUARK SPINS*	CHARGE (in e)	BARYON SPIN	mc^2 (GeV)
Proton[p]	uud	↑↑↓	$\frac{2}{3}+\frac{2}{3}-\frac{1}{3}=1$	$\hbar/2$	.938
Neutron[n]	ddu	↑↑↓	$-\frac{1}{3}-\frac{1}{3}+\frac{2}{3}=0$	$\hbar/2$	.940
Sigma minus[Σ^-]	dds	↑↑↓	$-\frac{1}{3}-\frac{1}{3}-\frac{1}{3}=-1$	$\hbar/2$	1.197
Sigma zero[Σ^o]	uds	↑↑↓	$\frac{2}{3}-\frac{1}{3}-\frac{1}{3}=0$	$\hbar/2$	1.192
Lambda zero[Λ^o]	uds	↑↑↓	$\frac{2}{3}-\frac{1}{3}-\frac{1}{3}=0$	$\hbar/2$	1.116
Sigma plus[Σ^+]	uus	↑↑↓	$\frac{2}{3}+\frac{2}{3}-\frac{1}{3}=1$	$\hbar/2$	1.189
Xi minus[Ξ^-]	dss	↑↑↓	$-\frac{1}{3}-\frac{1}{3}-\frac{1}{3}=-1$	$\hbar/2$	1.321
Xi zero[Ξ^o]	uss	↑↑↓	$\frac{2}{3}-\frac{1}{3}-\frac{1}{3}=0$	$\hbar/2$	1.315
Omega minus[Ω^-]	sss	↑↑↑	$-\frac{1}{3}-\frac{1}{3}-\frac{1}{3}=-1$	$3\hbar/2$	1.672

* ↑↑↑ means that all quark spins are parallel. ↑↑↓ means that one of them is antiparallel.

TABLE 9.4
QUARK STRUCTURE OF SELECTED MESONS

NAME AND SYMBOL	QUARK CONTENT	QUARK SPINS	CHARGE (in e)*	MESON SPIN	MESON MASS
Pi minus[π^-]	$\bar{u}d$	↑↓	$-\frac{2}{3}-\frac{1}{3}=-1$	0	139.6 MeV
Pi plus[π^+]	$u\bar{d}$	↑↓	$\frac{2}{3}+\frac{1}{3}=1$	0	139.6 MeV
Pi zero[π^0]	$d\bar{d}+u\bar{u}$	↑↓	0	0	135.0 MeV
Rho minus[ρ^-]	$\bar{u}d$	↑↑	$-\frac{2}{3}-\frac{1}{3}=-1$	$\hbar$	769 MeV
Rho plus[ρ^+]	$u\bar{d}$	↑↑	$\frac{2}{3}+\frac{1}{3}=1$	$\hbar$	769 Mev
Omega[ω]	$d\bar{d}+u\bar{u}$	↑↑	0	$\hbar$	783 MeV
K minus[K^-]	$\bar{u}s$	↑↓	$-\frac{2}{3}-\frac{1}{3}=-1$	0	494 MeV
K plus[K^+]	$u\bar{s}$	↑↓	$\frac{2}{3}+\frac{1}{3}=1$	0	494 MeV
K zero[K^0]	$d\bar{s}$	↑↓	$-\frac{1}{3}+\frac{1}{3}=0$	0	498 MeV
K zero bar[$\bar{K}^0$]	$\bar{d}s$	↑↓	$\frac{1}{3}-\frac{1}{3}=0$	0	498 MeV
Phi[ϕ]	$s\bar{s}$	↑↑	0	$\hbar$	1.02 GeV
Psi[ψ]	$c\bar{c}$	↑↑	0	$\hbar$	3.10 Gev

structure of the antiparticle to any hadron can be obtained by simply replacing each quark in the table by the corresponding antiquark. Both the particle and antiparticle are listed in the table for the mesons but not for the baryons. The antiparticle always has the same mass as the particle but opposite charge.

Before the quark theory of elementary particles existed a baryon was defined as a particle which, left to itself, would eventually decay to a proton plus some leptons and photons. An isolated antibaryon would decay into an antiproton plus leptons and photons. Because of charge conservation the set of particles into which a given baryon decays must naturally have a total charge equal to the charge of the original baryon. In contrast with a baryon, a meson would eventually decay completely into leptons and photons. The fact that baryons never decayed into lighter particles, such as leptons and photons, even though such decays were not prohibited by charge or energy conservation was described (but not really explained) by a *baryon conservation law.* It was known that, if to any isolated system one assigned a number, called the *baryon number*, and defined as the number of baryons minus the number of antibaryons in the system, then throughout all the future transformations of that system (beta decays, alpha decays, etc.) the baryon number would never change. Baryon number, like electric charge, was an exactly conserved quantity.

Recently a number of *Grand Unified Theories* havn devised which describe deep fundamental connections among the weak, electromagnetic, and strong interactions. A common feature of all the grand unified theories is that they predict very slight violations of baryon conservation. In particular, they predict that a proton, left to itself, will decay into leptons and photons. The half-life for proton decay is different in different theories, but always very large. Typical values for the proton decay half-life are $T_{1/2} = 10^{31}$ years. Since the estimated age of the universe is about 2×10^{10} years it is obvious that proton decay at this extremely slow rate would be difficult to observe. However, recent careful experiments have been making life very difficult for those theories that predict proton decay. Theories with $T_{1/2} \approx 10^{30}$ years have already been made untenable. (As of this writing there is one experiment in which proton decay seems to have been detected, but the results of that experiment seem to be contradicted by other experiments.) Within the quark picture of matter, baryon conservation is a logical consequence of quark conservation. If we assign a *quark number* of $+1$ to each of the quarks and -1 to each of the antiquarks and ignore (without prejudice) the currently speculative grand unified theories, then all allowed processes, regardless of whether they are caused by electromagnetic, weak, or strong interactions, conserve the total quark number.

9.10. INTERACTIONS IN QCD

Quantum chromodynamics (abbreviated as QCD) is not only a theory of the structure of matter but includes a detailed description of the interactions among all of the elementary particles. According to QCD the strong, electromagnetic, and weak interactions are all associated with the exchange of fundamental bosons. The electromagnetic interaction between particles is described by the emission and absorption of photons. Strong interaction processes involve the transfer of gluons. Weak interaction processes can be separated into two classes. Neutral weak interaction processes are those that involve the exchange of Z^0 bosons. Charged (or sometimes called charge-changing) weak interaction processes involve the exchange of W^- or W^+ bosons.

It is convenient to pictorially represent the various interaction processes by means of a set of pictures, called *Feynman diagrams.* Although we will only use the Feynman diagrams to communicate qualitative features of the interaction processes, their more important use is as part of a detailed computational scheme for predicting the rates and other quantitative characteristics of the processes. That computational scheme was first devised by Richard Feynman.

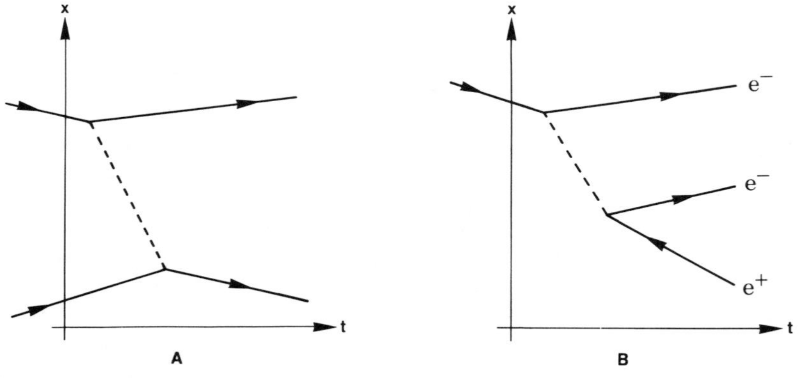

FIGURE 9.15 A. A Feynman diagram describing the transfer of a photon from one electron to another. The trajectories of the electrons are indicated by solid lines with arrows.
B. A Feynman diagram showing the creation of an $e^- e^+$ pair. The event depicted here could not really take place by itself because it does not conserve momentum and energy.

In drawing a Feynman diagram, the spacetime trajectories of particles are represented by lines in an x-t plane. (We will draw the t axis horizontal and the x axis vertical.) Fermions are represented by solid lines; bosons by dashed or wavy lines. (We will use dashed lines.) For example, the emission of a photon by one electron and its absorption by another would be represented by the diagram in Figure 9.15(A). The arrows in that diagram indicate the direction of motion of the electrons. For reasons we will give shortly, the trajectories of positrons and other antiparticles are indicated by solid lines with the arrow directions reversed. The diagram of Figure 9.15(B) therefore describes the emission of a photon by an electron and the later conversion of the photon's energy and momentum into an electron-positron pair. The two diagrams in Figure 9.15 are simply related. If we take the positron line in diagram (B) and pivot it clockwise about the vertex we can convert that diagram to diagram (A). This pictorial relationship is also reflected in the mathematical equations describing these processes. The probability of the process shown in diagram (A) (photon transfer between two electrons) depends upon the spacetime trajectory of the incoming and outgoing electrons. It was shown by Feynman that, if we take the formula that gives the probability of process (A) and, in that formula, replace the spacetime trajectory of the electron by one that describes an electron *moving backwards in time* toward the vertex then one obtains exactly the formula for the probability of the pair creation process shown in diagram (B). In general, in all calculations, antiparticles can be treated as if they were the corresponding particles moving backwards in time! At the time it was made, this discovery led to a small burst of philosophical speculation regarding its deeper meaning. At present it is simply considered to be a very appealing symmetry in the fundamental equations of physics. It certainly does not mean that we should expect to meet antipeople who walk backwards from the grave to the womb.

Before we draw the relevant Feynman diagrams for the weak, electromagnetic, and strong interactions we will list some of their general properties in a single table. This will facilitate a comparison of the three fundamental interactions. (If we had included gravity there would be four fundamental interactions.)

Electromagnetic interactions: Any process involving only electromagnetic forces can be described by a Feynman diagram that is made up of simple *electromagnetic vertex diagrams* like that shown in Figure 9.16(A). The vertex has an incoming and an outgoing charged fermion line and a photon line. Figure 9.16(B) shows a typical complex process, made up of elementary vertices. In that diagram an electron interacts electromagnetically with a proton and shortly thereafter produces an electron-positron pair.

Neutral weak interactions: All weak interaction processes can also be represented by diagrams composed of simple weak interaction vertices. For neutral weak processes the photon is replaced by the Z^0 boson. The fermions no longer need to be electrically charged, and therefore all fermions are affected by neutral weak interactions. The complete set of neutral weak interaction vertices, involving the fermions of the first two families, is shown in Figure 9.17(A) (There are similar vertices for the third family). Part (B) of that figure shows a more complex process that involves neutral weak interactions. The overall process is one in which a muon neutrino knocks a π^- out of a neutron, leaving a proton. The detailed process is analyzed as follows. The neutrino emits a high-energy Z^0 which is absorbed by a down quark in the neutron. As the colored quark separates from the other quark constituents of the neutron the energy in the

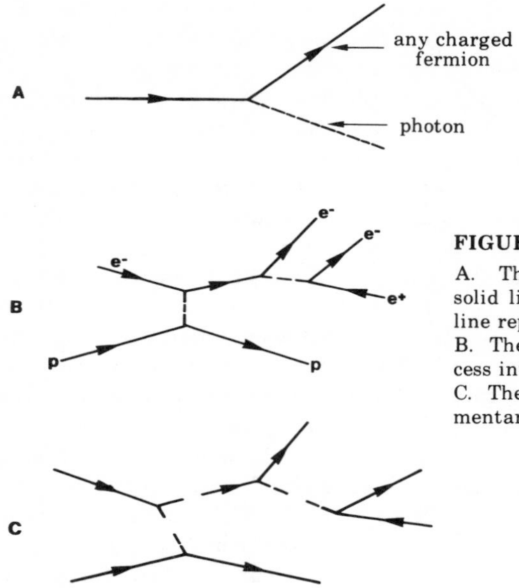

FIGURE 9.16

A. The elementary electromagnetic vertex. The solid line can be any charged fermion. The dashed line represents a photon.
B. The Feynman diagram for a pair production process involving a proton.
C. The diagram shown in B is composed of four elementary vertices.

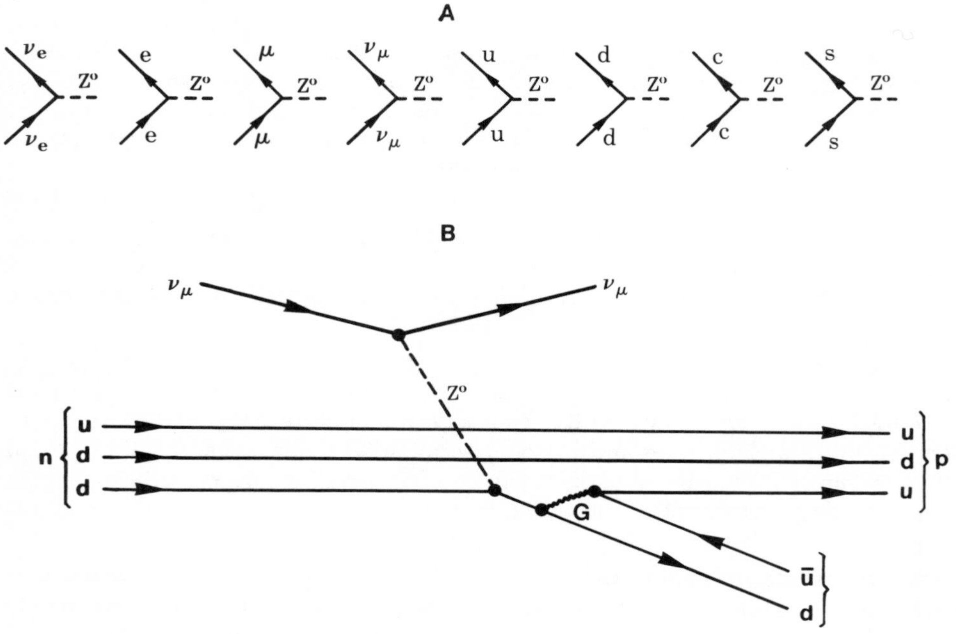

FIGURE 9.17 A. The set of neutral weak vertices involving fermions of the first two families.
B. A process involving neutral weak vertices. A neutron (u d d) and a high-energy μ-neutrino approach one another. The ν_μ emits a Z^o which is absorbed by a d-quark in the neutron. The momentum transferred to the quark causes it to separate from the other quarks in the neutron. As it pulls away the energy of the color field increases until there is sufficient energy available to create a quark-antiquark pair by gluon exchange. Two colorless particles (a π^- and a proton) emerge from the site of the reaction.

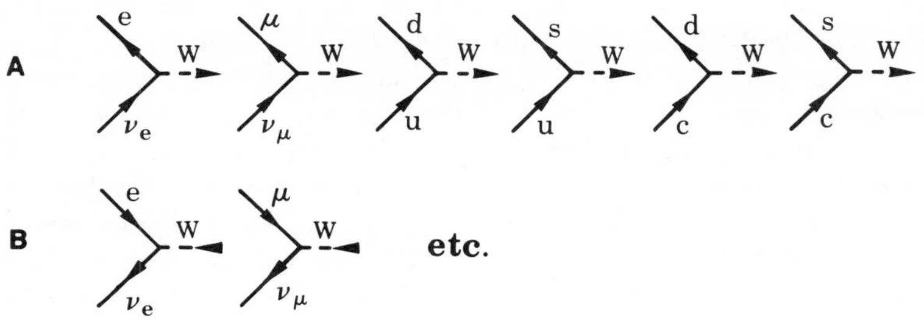

FIGURE 9.18 A. Half of the charged weak interaction vertices involving fermions of the first two families. In these processes the W carries electric charge away from the fermion.
B. The other half of the charged weak vertices are drawn by reversing the arrows on the first half. The incoming fermion becomes the outgoing fermion and vice versa. The W now carries charge to the fermion.

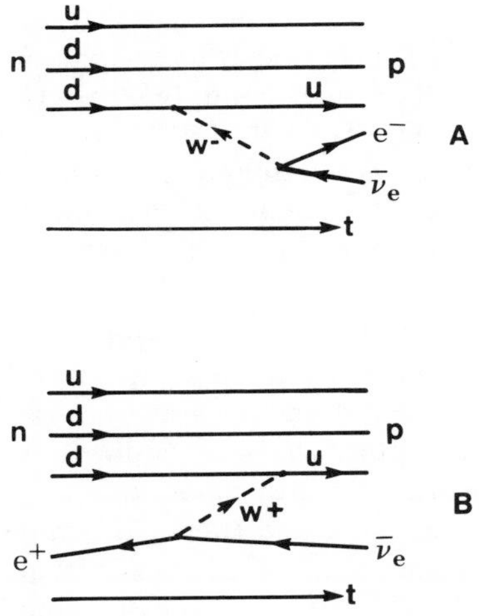

FIGURE 9.19

Two processes involving the charged weak interactions. A. An incoming neutron (u d d) emits a negatively charged W^-. This adds charge to the fermion line, which changes from a d (charge $-\frac{1}{3}e$) to a u (charge $+\frac{2}{3}e$). The W^- decomposes into an e and a ν_e.
B. A different but closely related process. An incoming positron emits a W^+ and changes to an electron-neutrino. The W^+ is absorbed by a d-quark, which changes the neutron into a proton.

color field rises until it becomes energetically advantageous to create a u-$\bar{u}$ pair, giving two colorless outgoing particles. The final pair creation takes place via the strong interaction (i.e. gluon exchange).

Charged weak interactions: Figure 9.18 shows the set of charged weak interaction vertices for processes involving fermions of the first two families. We have added an arrow to the boson line in order to indicate the direction in which electric charge is being carried. Two processes involving charged weak interaction vertices are shown in Figure 9.19. The triplet of quark lines represent the constituents of the proton or neutron. The constant gluon exchanges going on among the three quarks, which yield the binding force of the hadrons, are not shown in the diagrams. Their existence actually creates great mathematical difficulties in the calculation of the probability of the processes described by these diagrams. The process shown in (A) is the ordinary β-decay of a neutron. The overall process described by part (B) of that figure is the absorption of a positron by a neutron, producing a proton plus an electron-antineutrino. From an experimental point of view, the two processes seem very different. The first is a spontaneous decay process, while the second, which requires an incoming positron beam, is a type of collision process. However, the basic vertices involved in both processes are the same and therefore the rates of the two processes are closely related (in a way that we unfortunately cannot discuss because we have not presented the detailed computational method associated with the Feynman diagrams). In process (A) charge is transferred by the W from the later vertex to the earlier one. Therefore the W would be interpreted as a W^-, being emitted by the neutron. In the second process charge is transferred by the W boson from the earlier vertex to the later one. Therefore, in that case the W boson line in the diagram corresponds to a W^+.

Previously we mentioned that the recently-proposed grand unified theories attempt to unify the weak, electromagnetic, and strong interactions. They propose to derive all three types of effects from a single field equation in much the same way as magnetic and electric effects both follow from Maxwell's equations. Although the ultimate fate of the grand unified theories is still rather uncertain, there exists a somewhat less ambitious unified description of weak and electromagnetic interactions that has been spectacularly successful. In the unified electro-weak theory, formulated by Sheldon Glashow, Abdus Salam, and Steven Weinberg, the existence of the neutral weak interaction effects and the Z^o boson is required by the field equations. Since the theory was proposed before any neutral weak interaction effects had been observed or even strongly suspected the subsequent confirmation of the effects and the detection of the Z^o boson, with exactly the properties predicted by the theory, was a convincing confirmation of the theory. Because of its mathematical complexity, we shall not present any of the unified electro-weak theory but will instead treat the electromagnetic and weak interactions as unrelated effects.

Strong interactions: The strong interaction results from the transfer of gluons. The source of the gluon field is color charge. Therefore none of the uncolored particles feel the strong interactions. The only colored particles are quarks and gluons. The fact that gluons are colored means that there is a direct strong interaction between two gluons. Gluons can absorb other gluons or spontaneously create new gluons. It is like having an electromagnetic field theory in which the electric field itself is charged and therefore acts as a source of itself. In such a theory, just calculating the electric field due to a point charge would be challenging. The charge density giving rise to the field would be that of the point charge plus the charge density of the field itself. Therefore one would have to calculate the unknown charge density and the field simultaneously. Fortunately this does not happen for electric fields but the self-interaction effect, caused by the fact that the field quanta themselves are charged, creates very great obstacles to finding detailed solutions of the field equations for the color field.

The quarks each carry one unit of color charge (either red, green, or blue) while their antiparticles carry negative color charges (usually referred to as antired, antigreen, and antiblue). As we mentioned before, any flavor of quark can come in any color. The emission or absorption of a gluon by a quark may or may not cause a change in the quark's color. Since all components of the color charge are conserved, one can see by looking at the strong interaction vertices

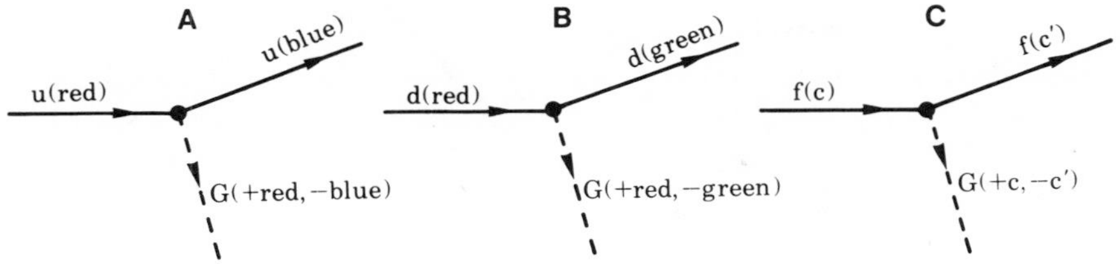

FIGURE 9.20 A. Two elementary quark vertices. The gluon, which carries one color and one anticolor, may change the color but not the flavor of the quark.
B. The general case. A quark of any color or flavor can emit or absorb a gluon and may or may not thereby change its color but not its flavor.

shown in Figure 9.20 that the gluons must carry a positive charge of one color and a negative charge of another (or possibly the same) color. For example a gluon may be (red,antigreen) or (blue,antired), etc.. The flavor of a quark is unaffected by absorption or emission of a gluon. That is, an up-quark remains an up-quark, a down-quark remains a down-quark, etc. after absorbing or emitting a gluon.

9.11. STRONG, ELECTROMAGNETIC, AND WEAK DECAYS

The reader might very reasonably question the wisdom of introducing all these Feynman diagrams without the detailed calculational scheme that goes with them. What useful information do the pictures contain? The answer to that question is that it is possible, by using the Feynman diagrams, to determine what physical processes are possible at all and, for those that are possible, to estimate very roughly their rate or probability of occurrence. To illustrate the method we will consider the case of spontaneous decay processes.

1. A process, such as the decay of a neutron into an electron-positron pair, which cannot be drawn as a Feynman diagram using the elementary vertices given above *never happens.*

That the process $n \rightarrow e^+ + e^-$ cannot be described by any Feynman diagram can easily be confirmed. A diagram describing that process would have the general structure shown in Figure 9.21. Since every elementary vertex involving a quark has both an incoming and an outgoing quark line it is clear that there is no way of completing the diagram without introducing other incoming or outgoing particles.

We are not suggesting here that the possibility of drawing a Feynman diagram for a process is the *only* requirement for a process to be possible. In particular, any possible process must satisfy the energy and momentum conservation laws in order to have a nonzero probability of occurence. If we had the detailed calculational scheme available we would find that the energy and momentum conservation requirements are built into the theory. Without it we must separately check that it is possible for the process to satisfy energy and momentum conservation. For example, it is easy to draw a Feynman diagram for the process $p \rightarrow n + e^+ + \nu_e$ but the sum of the masses of the products in this reaction is greater than the mass of the proton. Therefore the reaction is impossible.

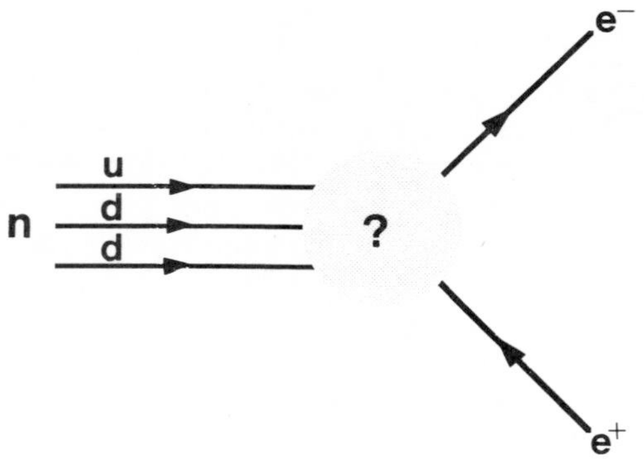

FIGURE 9.21 An incomplete Feynman diagram for the process $n \rightarrow e^+ + e^-$. A study of all the possible elementary vertices reveals that there is no way of completing it.

When it *is* possible to describe a decay process by means of a Feynman diagram the rate of the resulting process depends very much on the kinds of vertices that must be used in the diagram. The strong-interaction vertices (that is, the gluon emission or absorption vertices) describe processes that occur much more rapidly than the processes described by electromagnetic or weak-interaction diagrams. In general, one can state that:

2. Any decay process that can be described by a Feynman diagram using only strong-interaction vertices will occur with a half-life of order 10^{-23} seconds.

To appreciate how very short a time 10^{-23} seconds is, one need only ask how far any meson which can decay by the strong interaction will travel before it decays, assuming that it is traveling at relativistic speed. A meson has a diameter of about 1 fermi. At a speed of $c/3$ it would travel just one diameter in 10^{-23} seconds. A particle that cannot move more than one diameter without falling apart is hardly a particle at all. In fact, all the mesonlike or baryonlike quark structures that have strong decay modes are referred to as meson- or baryon-*resonances* rather than particles. A resonance is an unstable structure that just hangs together for an instant. Its fleeting existence is detected by a tendency of the particles that compose the structure to act together as a group in processes involving them. A good example is the rho-meson resonance. A rho decays by the strong interactions into two pions. It has a lifetime of only about 10^{-23} seconds. However, when strongly-interacting particles collide there is a strong tendency for pions to be emitted in pairs, where the pair of pions, in their center-of-mass frame, have just enough energy to make up a rho meson. The probability of such a pair of pions being emitted together is very much larger than one would expect if one assumed that the pions were acting independently. What is actually happening is that, in the process of collision, the strongly-interacting particles emit a rho meson which immediately decays into the observed pion pair.

Decay processes that can be described by Feynman diagrams using only gluon-quark vertices are called strong decays. Those that must be described by diagrams that include photon-fermion vertices are called electromagnetic decays. Those that require $W^{\pm}$ or Z^{0} vertices in their Feynman diagrams are called weak decays. The decay rates and half-lives associated with the three types of processes are very different.

3. $T_{1/2}$(strong) << $T_{1/2}$(electromagnetic) << $T_{1/2}$(weak)

Typically electromagnetic decays involving a single photon are about a thousand times slower than strong decays while those involving two photons are a million times slower than strong decays Weak decays are another million times slower than two-photon electromagnetic decays.

The decay of the pions provides a good example of the use of these ideas. The π^0 decays (about 99% of the time) into two photons. The π^0 has a half-life of 5.8×10^{-17}s. The π^+ decays (more than 99% of the time) into a μ^+ and a ν_μ. The π^+ decay has a half- life of 1.8×10^{-8}s. Both pions are composed of a quark and an antiquark. Since their structures are so similar why does the π^+ have a lifetime that is larger than the lifetime of the π^0 by a factor of 3×10^8? Also, since quarks and antiquarks are strongly-interacting particles, why do the pions not decay by the strong interactions with a lifetime of order 10^{-23}s? Even the short-lived π^0 seems to last six million times as long as it should. We can answer these questions by looking at a detailed quark description of the decay modes of the two pions. (The π^- decays to a μ^- and a ν_μ by a process that is essentially identical to the π^+ decay.) Let us consider the case of the π^+ first.

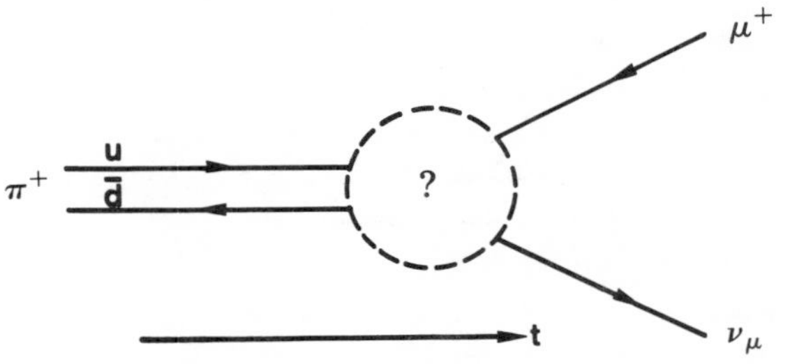

FIGURE 9.22 An incomplete diagram for the process $\pi^+ \rightarrow \mu^+ + \bar{\nu}_\mu$.

In Figure 9.22 we show a portion of a Feynman diagram describing the decay of a π^+. The two quark lines represent the u and the $\bar{d}$ that compose the pion. The fermion lines on the right represent the decay products. What diagram can connect the two pieces? First we notice that, since gluon emission or absorption changes the color but not the flavor of a quark, no combination of gluon vertices will eliminate either the u or the $\bar{d}$ line. Photon vertices also leave quark flavor unchanged and therefore they too are of no value in completing the diagram. We must introduce at least one weak-interaction vertex. As soon as we look at the W vertices in Figure 9.18 it is easy to see how the diagram may be completed as simply as possible. The result is shown in Figure 9.23 (Note that the $\bar{d}$ line is a backwards-pointing d line). The decay proceeds through an intermediate state consisting of a W^+. That fact explains why the decay is so strongly suppressed. The pion has a mass m_π. Thus the maximum energy available to the system is $m_\pi c^2$. But the energy needed to create a W^+ is $m_W c^2$, where $m_W >> m_\pi$. It is clear that the intermediate state would constitute a gross violation of classical energy conservation and that therefore the overall process can only proceed through the quantum mechanical tunneling effect. There is more than enough energy in a pion to create a μ^+ and a ν_μ. Thus the system must 'tunnel through'

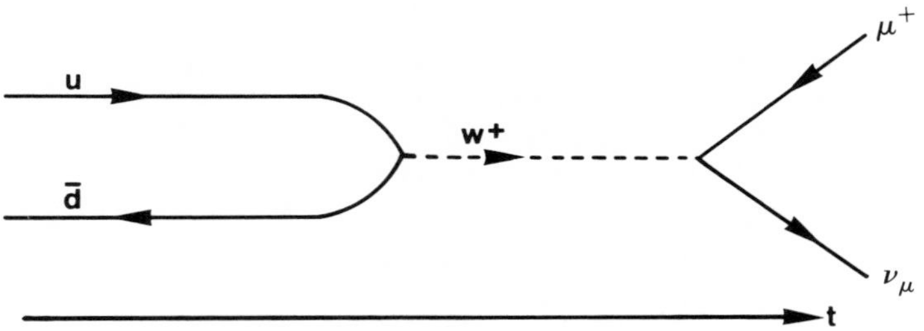

FIGURE 9.23 The completed diagram for the weak decay of the π^+.

the classically forbidden intermediate state in order to reach the classically allowed final state. The tunneling probability is so small that the pion gets to live to the ripe old age of 10^{-8}s rather than dying in its infancy at 10^{-23}s.

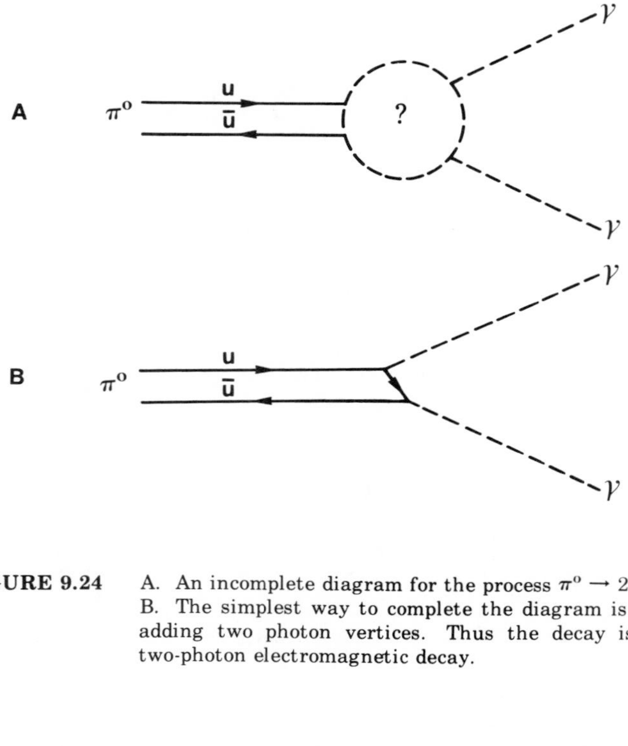

FIGURE 9.24 A. An incomplete diagram for the process $\pi^0 \rightarrow 2\gamma$. B. The simplest way to complete the diagram is by adding two photon vertices. Thus the decay is a two-photon electromagnetic decay.

As can be seen in Figure 9.24, the decay of the π^0 is a purely electromagnetic decay. There is no massive intermediate state to suppress the process. It is therefore much more rapid than the π^+ decay. But there is still no way of describing π^0 decay using only strong-interaction vertices, and thus the π^0 lifetime has an intermediate value typical of purely electromagnetic processes.

A good example of a purely strong decay process is the decay of the ρ^+, mentioned before (The ρ^- has a similar decay). The ρ^+ decays, almost all of the

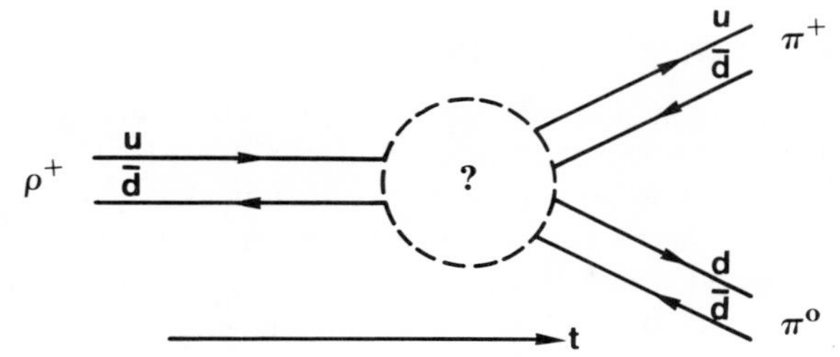

FIGURE 9.25 An incomplete diagram for the process $\rho^+ \rightarrow \pi^+ + \pi^o$.

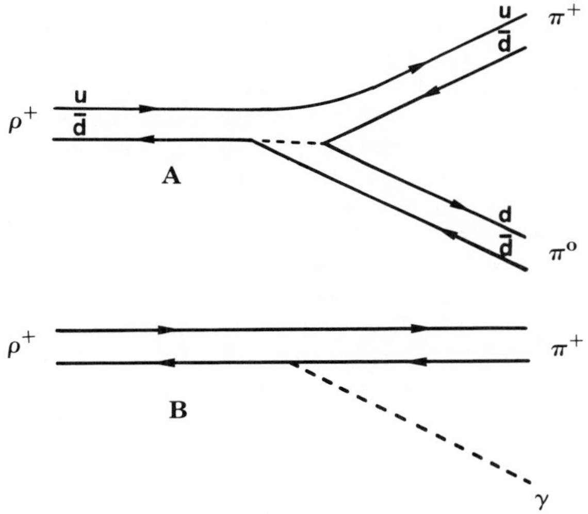

FIGURE 9.26 A. The completed diagram for the process $\rho^+ \rightarrow \pi^+ + \pi^o$.
B. A purely electromagnetic decay of the ρ^+.

time, by the process $\rho^+ \rightarrow \pi^+ + \pi^o$. In Figure 9.25 the quark structure of the initial and final particles is shown. It is not difficult to complete the diagram with gluon vertices alone (See Figure 9.26(A)). Since the ρ^+ has the same quark content as the π^+ it is also possible to describe a decay process that does not include the creation of a π^o (See Figure 9.26(B)). However, this process must take place by photon emission, since gluons, being colored, cannot act as isolated final-state particles. The electromagnetic decay does occur but, because it is an electromagnetic decay it is intrinsically weaker than the strong decay. The observed ratio of the rate of the $\rho^+ \rightarrow \pi^+ + \gamma$ decay to the rate of the $\rho^+ \rightarrow \pi^+ + \pi^o$ is called the *branching ratio* of the two processes. The branching ratio for these decay processes is .00044, which shows that the gluon-mediated process strongly dominates.

9.12. QUANTUM FIELD THEORY

In earlier chapters we have tried to show that there is really no conflict between quantum mechanics and classical mechanics. In the overall scheme of physics, quantum mechanics is the more fundamental and more general theory. Classical mechanics can be derived from quantum mechanics as an approximation that is quite accurate within a limited range; the range of large, heavy objects. Well, the quantum theory presented in those chapters is itself an approximation. What was presented there is nonrelativistic quantum mechanics. It is valid only for systems in which the particles move at speeds much smaller than c. When the velocity of a particle approaches c, then the particle's kinetic energy becomes larger than its rest energy, mc^2 and completely new processes, such as electron-positron pair production become possible. These processes are not accounted for at all by the Schroedinger equation. The Schroedinger equation is fundamentally nonrelativistic because it assumes that the interactions between particles can be described by a potential energy function. But forces that are derivable from a potential are communicated instantaneously. To see this let us imagine two particles with a simple Coulomb force (which is derivable from a potential) between them. The force on one particle would be proportional to $1/r^2$ where r is the distance between the particles. If that distance were increased by moving one particle the force on the other particle would decrease immediately. There would be no time interval required for the effect of moving one particle to propagate to the other. Such a thing does not happen in reality. For real charged particles the simple Coulomb potential can only be used when the particles are stationary or when their motions are slow enough that the finite time for the propagation of effects is negligible. Whenever the relative particle velocities are close to c then the full time-dependent electromagnetic field must be calculated in order to predict the forces on the particles. When both relativistic and quantum effects are important, as they would be for small systems of high-velocity particles, then the dynamics of the field must be treated quantum mechanically. This leads to a mathematical structure called *quantum field theory.* We will only give a superficial description of quantum field theory and show some of the new and interesting things described by it.

The essential trick in applying the quantum theory to a continuous field is to convert the field into a collection of harmonic oscillators. That can be done by focusing our attention on the normal modes of the field. As an example, let us consider a uniform stretched string. That is about the simplest possible "field". It is considered a field because it is a continuous system. The position of a string at any instant cannot be defined by a finite set of coordinates. It is a one-dimensional field which makes it easier to visualize and analyze than three-dimensional fields such as the electric field or the magnetic field. For a string of length L the normal modes are standing waves whose shapes are given by sine functions of the form

$$y(x,t) = A_k(t) \sin kx \tag{9.8}$$

In order for the vibration amplitude to be zero at both ends the possible values of the wave vector, k, are restricted to those of the form $k = K\pi/L$ where $K = 1, 2, \cdots$. The wave function y(x,t) must satisfy the *wave equation*

$$\frac{\partial^2 y}{\partial t^2} - v^2 \frac{\partial^2 y}{\partial x^2} = 0 \tag{9.9}$$

where v is the velocity of waves on the string. Inserting the normal mode solution into the wave equation shows that the normal mode amplitude must satisfy the equation of motion of a harmonic oscillator of angular frequency $\omega_k = vk$. That is

$$\frac{d^2 A_k(t)}{dt^2} = -\omega_k^2 A_k(t) \tag{9.10}$$

Thus the amplitude function for each of the normal modes behaves like a classical harmonic oscillator with the frequency appropriate to that normal mode.

Now we have to take a conceptual leap and simply state that when the string is treated as a quantum mechanical system each of the normal modes becomes a quantum harmonic oscillator. The system is therefore equivalent to a collection of quantum harmonic oscillators. The kth oscillator has an angular frequency $\omega_k = K\pi v/L$. Each one of those quantum harmonic oscillators has the possible energies given by the harmonic oscillator energy spectrum. For instance, the kth oscillator has the energy levels $E_k = \frac{1}{2}\hbar\omega_k, \frac{3}{2}\hbar\omega_k, \frac{5}{2}\hbar\omega_k, \cdots$ or, in general

$$E_k = (n_k + \tfrac{1}{2})\hbar\omega_k \quad \text{with } n_k = 0, 1, 2, \cdots \tag{9.11}$$

Because different normal modes do not interact the possible values for the total energy of the string are given by summing the energies of all the normal modes.

$$E = \sum_k (n_k + \tfrac{1}{2})\hbar\omega_k \tag{9.12}$$

The lowest possible energy is obtained by choosing all the integers n_k to be zero. This is equivalent to assuming that each normal mode harmonic oscillator is in its ground state. Therefore the ground state energy of the system is

$$E_o = \sum_k \tfrac{1}{2}\hbar\omega_k \tag{9.13}$$

We can now see that a serious problem has developed. There are an infinite number of normal modes, one for each value of $k = K\pi/L$. This means that the value of E_o is infinite! In quantum field theory little embarrassments of this type are called *divergences.* They occur in almost every calculation. The ground state energy divergence can be easily eliminated in the string system we are considering. It makes no sense to talk about a transverse wave in a string if the wavelength of the wave is smaller than the thickness of the string. When the wavelength is very small the detailed internal structure of the string becomes important. The divergence caused by the infinite sequence of waves of smaller and smaller wavelength is only a consequence of our oversimplified picture of the string. The actual string can be shown to have finite ground state energy. Since the actual ground state energy is finite we can eliminate the infinite constant by shifting the zero point of our energy scale by an amount equal to the ground state energy. Then we would know that the ground state energy on this new energy scale is exactly zero. We would therefore be justified in simply subtracting the apparently infinite constant E_o from all our energy expressions. This process of shifting the energy zero point so as to eliminate a divergent constant energy is called *energy renormalization.* In quantum field theories involving massive charged particles there are also divergent constants that must be subtracted from the mass of the particle (*mass renormalization*) and the charge of the particle (*charge renormalization*). If, after these renormalization procedures have been carried out, the theory predicts finite values for all experimentally observable quantities then the theory is called *renormalizable.* A renormalizable theory is considered to be an acceptable theory since all of its predictions can be compared with experiments. Of course it would be still better if the divergences never appeared in the first place so that no infinite subtractions were necessary.

QUESTION: Are all the renormalizations used in quantum field theory based on oversimplified descriptions of systems that actually have microscopic internal structure? In other words, can all the apparently infinite constants that are subtracted be shown to be actually finite?

ANSWER: Not at all. In the string system the existence of microscopic string structure means that waves of very short wavelength do not satisfy the simple wave equation whose normal mode solutions lead to the energy divergence. In contrast, electromagnetic waves show no signs of deviation from the standard wave equation down to the smallest observed wavelengths. In the cases of mass and charge renormalization it is an *experimental* fact that m_e and e are actually finite but there is no clear explanation of why the calculated values for these constants always contain divergences. In practice the divergent expressions for m_e and e are simply replaced with the finite experimental values. After this very dubious procedure is carried out all other predictions of the theory are finite and in almost miraculously close agreement with experimental results.

The ground state energy, on our *renormalized energy scale* is exactly zero. Using the same scale, Equation 9.13, which describes the energy spectrum of the system becomes

$$E = \sum_k n_k \hbar \omega_k$$

9.13. THREE-DIMENSIONAL WAVES

For waves in a three-dimensional field the same analysis can be used. However, the wave vector **k** is now a three-dimensional vector. Thus the energy spectrum for such a system (on a renormalized energy scale) is given by

$$E = \sum_{\mathbf{k}} n_{\mathbf{k}} \hbar \omega_{\mathbf{k}} \qquad (9.14)$$

But waves also carry momentum. The reader is probably familiar with the fact that electromagnetic waves, in being absorbed by an object, transmit momentum to that object. Sound waves also carry both energy and momentum. Since the field can carry momentum, there is meaning in the following question. How much momentum is contained in the system when it is in the quantum state described by a particular set of quantum numbers, $n_{\mathbf{k}}$? The calculation that must be done to answer that question is much too difficult for us to present; however, the answer it gives is very simple. The total momentum carried by the collection of quantized waves is given by the formula

$$\mathbf{P} = \sum_{\mathbf{k}} n_{\mathbf{k}} \hbar \mathbf{k} \qquad (9.15)$$

In other words, a quantized traveling wave of wave vector **k** (and wavelength $\lambda = 2\pi/k$), excited to the nth quantum level, carries an amount of energy $n\hbar\omega_{\mathbf{k}}$ and an amount of momentum $n\hbar\mathbf{k}$.

9.14. THE PARTICLE INTERPRETATION

Suppose the three-dimensional wave field we are discussing is an electromagnetic field. Where are the photons? How do particles fit into the picture of quantized harmonic waves? One of the most remarkable things about quantum field theory is that the relationship between particles and fields does not have to be put into the theory. It emerges as a natural consequence of the theory. In fact, the theory explains the very existence of discrete particles. The following argument will show how this happens.

We imagine our system of quantized waves replaced by a system of a few noninteracting particles. Some of the particles may have exactly the same value of momentum. Suppose the number of particles that have momentum $\mathbf{p} = \hbar\mathbf{k}$ is called $n_{\mathbf{k}}$. Naturally, since $n_{\mathbf{k}}$ is the number of particles with a certain value of $\mathbf{p}$ it must be a nonnegative integer. It would make no sense at all to say that 1.27 of the particles had zero momentum. It would be possible for none of the particles to have zero momentum, or one of them, or two of them, etc. but the number of particles with any characteristic must, if the phrase is to have any meaning, be an integer. The total momentum of the system of particles can be obtained by adding the momenta of all the particles. We get

$$\mathbf{P} = \sum n_{\mathbf{k}} \hbar\mathbf{k} \tag{9.16}$$

The energy of any one of the particles depends on its momentum. If we include the rest energy of the particle in our definition of the energy then, for a particle of momentum $\mathbf{p}$, the energy is $E(\mathbf{p}) = c(m^2c^2 + p^2)^{1/2}$. For a zero mass particle the relationship is $E(\mathbf{p}) = cp$. In either case, if we define the quantity $\omega_{\mathbf{k}}$ by

$$\omega_{\mathbf{k}} = E(\hbar\mathbf{k})/\hbar \tag{9.17}$$

then we can write the total energy of the system of particles as

$$E = \sum_{\mathbf{k}} n_{\mathbf{k}} \hbar\omega_{\mathbf{k}} \tag{9.18}$$

It is obvious that these two equations, for the total momentum and energy of a system of particles, are identical in form to Equations 9.15 and 9.14, which give the momentum and energy of a system of quantized waves. In other words, the two equations can be given one reasonable interpretation if we use a quantized field picture of the world and another reasonable interpretation if we use a discrete particle picture of the world. Within the context of one picture the integer $n_{\mathbf{k}}$ is the excitation level of a quantized harmonic oscillator. Within the framework of the other picture it is the number of particles with momentum $\hbar\mathbf{k}$. If the various fields in nature did not give rise to harmonic oscillator energy spectra then the particle interpretation would be impossible; there would be nothing in the possible states of the universe that one could identify as the numbers of electrons, photons, and so forth.

Of course, the fact that two equations are identical does not imply that the things that the equations describe are identical. By using the symbols F, m, and a for voltage, resistance and electrical current we can make Ohm's law come out as $F = ma$. That would not prove that voltage is a force and current an acceleration. In order to show that two concepts can be identified one must show that there is an exact correspondence in all aspects of the two things. For example, we can identify the points of a two-dimensional plane with all possible pairs of real numbers (x,y), as we do in analytic geometry, because we can give a geometrical interpretation to *every* arithmetic operation with pairs of numbers and an arithmetic interpretation to *every* theorem in plane geometry.

With the tiny bit of quantum field theory we have presented above we cannot reasonably attempt to prove the detailed equivalence of the quantized wave field interpretation and the discrete particle interpretation of the equations of quantum field theory. The reader will have to take it on faith that there are two different but equivalent descriptions or pictures of every possible physical process.

In the wave or *field picture* the world is composed of a certain number of continuous fields (quark fields, electron fields, etc.). The normal mode oscillations of the fields are traveling waves. The fields are quantized so that any particular normal mode can be excited only to certain discrete excitation levels $n_{\mathbf{k}}$. For fermion fields it can be shown that it is a consequence of the detailed field equations that the excitation level $n_{\mathbf{k}}$ can have only the two values 0 or 1. For boson fields, such as the electromagnetic field (or the quantized string waves that we considered above) the excitation level of a normal mode can be any nonnegative integer. The wave fields interact with one another. The interactions between the fields are described by certain definite, known terms in the mathematical equations for the fields. Because of the interactions the normal mode oscillators for the various fields can make transitions and thereby exchange energy and momentum. The total energy and momentum in all the fields remains constant but a particular normal mode of the electromagnetic field, for instance, may make a transition from a higher energy state to a lower energy state while, simultaneously, the electron field normal modes change in such a way that the electron field has absorbed the energy lost by the electromagnetic field.

In the *particle picture* the world is composed of empty space with discrete particles in it. The field picture statement that "the electromagnetic normal mode of wave vector **k** is excited to the level $n_{\mathbf{k}} = 3$" is, in the particle picture, described by the statement that "there are 3 photons in the world with wave vector **k**". The process that was before described by saying that the electromagnetic field made a transition in which it transferred energy to the normal modes of the electron field is now described by saying that an electron absorbed a photon and thereby increased its energy. There are both field picture and particle picture descriptions of every possible occurrence. For particular processes the field description seems natural and the particle description seems forced or far-fetched. For other processes just the reverse is true. There is no possibility of determining which of the pictures is 'right'. They agree exactly in their predictions. They use the same equations; only the words are different. Niels Bohr described them as two *complimentary* aspects or descriptions of nature. It is possible that at some future time the theory will be modified in such a way that one of the pictures becomes completely unreasonable when applied to some, as yet unknown, phenomena. That picture will then have to be rejected. That has not yet happened.

There has been a centuries-old competition between continuous field descriptions of nature and discrete particle descriptions. According to quantum field theory, that competition has resulted in a stalemate.

SUMMARY

Cosmic rays are a shower of particles that originate in outer space. The *primary* particles strike nuclei in the atmosphere, and those collisions create *secondary* particles. The primaries consist of protons (91%), α-particles (8%), and heavier nuclei (1%).

Naturally radioactive isotopes are substances that spontaneously disintegrate and emit α-particles, γ-rays, or other types of particles.

In the *Van de Graaff accelerator* charge is transported on a moving belt to the inside of a conducting sphere. A conducting brush allows the charge to flow to the outside of the sphere where it collects, creating a large potential difference between the sphere and the ground.

In a uniform magnetic field **B** a particle of charge q and mass m moves in a circle if its initial velocity is perpendicular to **B**. If $v << c$ then the frequency of the motion is independent of v and is equal to the *cyclotron frequency*

$$\nu_c = qB/2\pi m$$

In a *cyclotron* the particles move in a circular orbit while they are inside one of the *dees*. As they go from one dee to the other they are accelerated by a voltage between the dees that oscillates with the cyclotron frequency.

As v approaches c the frequency of the particle motion decreases and something must be done to keep the particles in synchronization with the alternating voltage. In a *synchrocyclotron* the frequency of the alternating voltage is varied as a packet of particles is accelerated to relativistic energy. In an *isochronous* cyclotron the magnetic field is made stronger near the edge of the magnet than in the center. Since the cyclotron frequency increases with increasing field strength, this can counteract the relativistic decrease of frequency.

In a *linear accelerator*, particles (usually electrons or positrons) are accelerated by a traveling electric field as they move down a long straight tube.

The *lab frame* is a frame in which one particle (the target particle) is stationary while the other particle (the beam particle) is moving. For the case of equal mass particles the energy of the pair of particles in the lab frame is related to their energy in the center-of-mass frame by

$$\frac{E_{lab}}{2mc^2} = \left(\frac{E_{cm}}{2mc^2}\right)^2$$

For high energy processes this means that E_{lab} must be much larger than E_{cm}.

In a *colliding beam machine*, particle beams of equal but opposite velocity are made to intersect. The frame of the laboratory then coincides with the cm frame so that the full energy of the beam particles is available for reactions.

A *Geiger* or *proportional counter* is made by suspending a fine wire down the center of a gas-filled metal cylinder. Charged particles, passing through the gas, knock electrons off the gas atoms. These electrons accelerate as they fall toward the center wire, which is kept at a positive potential. The accelerating electrons knock out other electrons causing an *electron avalanche* which appears as a current pulse in the counter. The current pulse in a proportional counter is proportional to the number of electrons originally produced by the charged particle.

In a *multiwire counter* the location of the ionizing particle can be accurately determined.

In a *bubble chamber* bubble tracks are created by charged particles moving through superheated liquid. The bubble tracks are photographed to determine the trajectories of the charged particles.

The *fundamental fermions* all have spin $\hbar/2$. Each *family* consists of 2 *leptons* and 2 *quarks*. One lepton is massive (e^-, μ^-, τ^- for the 1st, 2nd, and 3rd families) and the other is a massless neutrino (ν_e, ν_μ, and ν_τ). One quark has charge $+\frac{2}{3}$e (u, c, and t quarks for the 1st, 2nd, and 3rd families) and the other has charge $-\frac{1}{3}$e (d, s, and b quarks).

Quarks and gluons have a property called *color* or *color charge*. The color charge has three independent components, called red, blue, and green. Each type of quark can have red, blue, or green color. The antiquarks have antired, antiblue, or antigreen color.

All isolated structures have zero net color charge. A colorless object can be made either by combining equal amounts of red, blue, and green or by combining any color with its anticolor.

The *quark number* of a system is equal to the number of quarks minus the number of antiquarks in it. The *baryon number* is one third the quark number. The quark number is conserved by all interactions.

The *flavor* of a quark is its type (up, down, etc.).

The *fundamental bosons* include the photon, the Z^0, the $W^\pm$ bosons, the gluon, and possibly, the graviton. The photon, Z^0, W, and gluon all have spin $\hbar$.

The observed strongly-interacting particles are called *hadrons* (quarks are not directly observed). The hadrons consist of one group of particles called *mesons* and one group called *baryons.*

A meson is composed of one quark and one antiquark of opposite colors.

A baryon is composed of three quarks of three different colors.

The quantum theory of quarks and gluons is called *quantum chromodynamics.*

The basic processes of high-energy physics (particle scattering, $e^+ e^-$ creation, etc.) can be represented by *Feynman diagrams.*

The Feynman diagrams are made up of *elementary vertices.* An elementary vertex describes the emission or absorption of a fundamental boson by a fundamental fermion (We are ignoring gluon-gluon vertices).

The fermions are represented in Feynman diagrams by solid lines with arrows pointing forward in time. Their antiparticles are represented by solid lines with backward-pointing arrows.

The electromagnetic vertex describes the emission or absorption of a photon by a charged fermion.

The neutral weak vertex describes the emission or absorption of a Z^o boson by a fermion. The fermion need not be electrically charged.

The charged weak vertex describes the emission or absorption of a Z^+ or Z^- boson by a fermion. Since the boson carries away electric charge the fermion must change its flavor in the process.

The strong interaction vertex describes the emission or absorption of a gluon by a quark. The color but not the flavor of the quark may be changed in the process.

Physical processes that cannot be described by Feynman diagrams composed of these elementary vertices do not occur at all.

Physical processes that require weak interaction vertices in their Feynman diagrams occur very slowly unless they involve incoming particles with very large cm energies (typical $T_{1/2} \approx 10^{-11}$ s).

Physical processes that require electromagnetic vertices (but not weak vertices) in their Feynman diagrams occur much more rapidly than weak-interaction processes but much more slowly than strong-interaction processes (typical $T_{1/2} \approx 10^{-20}$s for one-photon decays and Tsub½ $\approx 10^{-17}$s for two-photon decays).

Physical processes that can be described by Feynman diagrams using only strong-interaction vertices proceed very rapidly (typical $T_{1/2} \approx 10^{-23}$s).

Quantum field theory is the quantum mechanical analysis of continuous fields. The field is first described in terms of its normal modes. There is a normal mode for each value of the wave vector $\mathbf{k}$. The normal mode acts like a harmonic oscillator with energy values (after renormalization) $n_{\mathbf{k}} \hbar \omega_{\mathbf{k}}$, where $n_{\mathbf{k}} = 0, 1, 2, \cdots$.

The total energy and momentum of the field is given by

$$E = \sum_{\mathbf{k}} n_{\mathbf{k}} \hbar \omega_{\mathbf{k}}$$

$$\mathbf{P} = \sum_{\mathbf{k}} n_{\mathbf{k}} \hbar \mathbf{k}$$

In the *particle interpretation of the theory* $n_{\mathbf{k}}$ is interpreted as the number of particles that have momentum $\hbar \mathbf{k}$ and energy $\hbar \omega_{\mathbf{k}}$.

PROBLEMS

9.1* In an isochronous cyclotron the strength of the magnetic field increases toward the edge of the magnet. For such a cyclotron, show that the condition required to keep the particle frequency constant is that the ratio of the magnetic field at radius R to the magnetic field near the center is

$$\frac{B}{B_o} = \frac{1}{1-\alpha R^2}$$

and derive a formula for the constant α in terms of q, m, c, and B_o.

9.2* Using Equation 9.2, determine the radius of the circular orbit of a 100 GeV proton in a uniform magnetic field of 2 Tesla.

9.3* A proton is accelerated in a cyclotron. The magnet pole has a diameter of 1 meter and the magnetic field strength is 2 T. The proton begins with negligible velocity and is accelerated by a potential difference of 100 kV each time it crosses the gap between the dees. How much time does the proton spend in the cyclotron?

9.4* Derive Equation 9.5.

9.5 The linear accelerator at SLAC is about 2 miles long and produces electrons of energy 27 GeV. What is the average electric field at the location of the electrons on their trip down the tube? Express the answer in volts/cm.

9.6* (a) How much energy would be needed *in the lab frame* to produce the ψ particle ($m_\psi c^2 = 3.10\,\text{GeV}$) using an antiproton beam and a proton target? The reaction is simply $\bar{p} + p \to \psi$.

(b) How is the answer to part (a) changed if you use a positron beam on an electron target?

9.7** A π^o of energy .5 GeV decays in flight by the reaction $\pi^o \to 2\gamma$. What are the maximum and minimum possible energies of the emitted photons (measured in the lab frame)?

9.8** A proton in a bubble chamber follows the spiral path shown in Figure P.9.8. The magnetic field strength is 4 T. What is its energy?

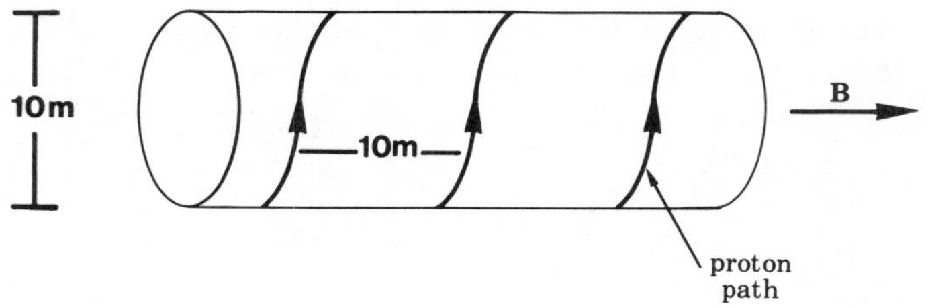

FIGURE P.9.8

9.9* Consider a macroscopic spinning object with linear momentum **P** and angular momentum **L** pointing in the same direction.

(a) Show that there is another inertial frame in which **P** and **L** are antiparallel.

(b) Show that the argument used in part (a) cannot be used for a neutrino. Therefore it is at least *possible* for neutrinos to have negative helicity in *every* Lorentz frame.

9.10* The reaction $\pi^+ + \pi^- \rightarrow 2\gamma$ is seen commonly, but the reaction $\Sigma^+ + \Sigma^- \rightarrow 2\gamma$ never occurs. Explain these facts, using the quark structure of the particles involved.

9.11* Consider the reaction $\pi^- + p \rightarrow n + \pi^0$.

(a) Draw a Feynman diagram for the reaction.

(b) Assume that the incoming π^- and p have negligible velocities. What is the velocity of the outgoing neutron? (Measuring the neutron velocity in this reaction is an excellent way of determining the mass difference between the π^0 and the π^0.)

For each of the reactions listed below:

(a) Determine whether the reaction is possible at all.

(b) If the reaction is possible, draw the simplest Feynman diagram for it and state whether it proceeds by the strong, electromagnetic, or weak interaction.

(c) For those reactions with two incoming particles, determine the *threshold energy* of the reaction in the cm system.
Note: The *kinetic energy* of the incoming particles is the sum of their relativistic energies minus their rest energies. The threshold energy is the smallest kinetic energy that will allow the reaction to proceed.

(d) For the decay processes, determine whether there is any *much faster* decay process available to the particle.

9.12	$\Lambda^0 \rightarrow p + \pi^-$	9.13	$\Lambda^0 \rightarrow p + e^- + \bar{\nu}_e$
9.14	$\Xi^0 \rightarrow \Lambda^0 + \gamma$	9.15	$\phi \rightarrow e^+ + e^-$
9.16	$\Sigma^+ + \Sigma^- \rightarrow \Sigma^0 + \bar{\Sigma}^0$	9.17	$\Sigma^+ + \Sigma^- \rightarrow 2\Sigma^0$
9.18	$\Sigma^+ + \Sigma^- \rightarrow 2\gamma$	9.19	$\Xi^- \rightarrow \Lambda^0 + e^- + \bar{\nu}_e$
9.20	$\pi^- + K^+ \rightarrow 2\gamma$	9.21	$\pi^- + \bar{\Lambda}^0 \rightarrow \bar{p} + \pi^+ + K^-$
9.22	$\Omega^- \rightarrow \Lambda^0 + K^-$	9.23	$\pi^0 \rightarrow \nu_\mu + \bar{\nu}_\mu$
9.24	$n \rightarrow p + \mu^- + \bar{\nu}_\mu$		

CHAPTER TEN

THE QUANTUM THEORY OF SOLIDS

10.1. AMORPHOUS AND CRYSTALLINE SOLIDS

Although the solid state is often considered as a single state of matter there are really two very different kinds of solids; *amorphous solids* and *crystalline solids*. On a macroscopic scale an amorphous solid is very much like a liquid that has been cooled sufficiently for the interatomic bonds to prevent the liquid from flowing although no fundamental change in the microscopic structure of the material has taken place. In contrast a crystalline solid is totally different from a liquid. In a crystalline solid the constituent atoms or molecules arrange themselves in a geometrically ordered crystal lattice. The geometrical pattern is

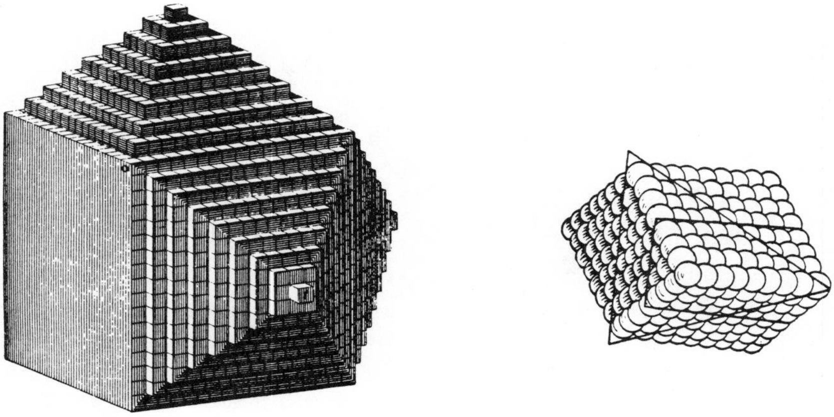

FIGURE 10.1 Two illustrations from important early treatises, showing the relationship between cleavage planes and underlying structure.
A. From *Traité de Cristallographie*, by R. J. Haÿ (1801).
B. From *Traité de Lumière*, by C. Huyghens (1690).

usually nearly perfect throughout regions that are macroscopic in size and therefore contain vast number of atoms. In fact if the solid is crystallized with some care a single crystal may result in which one microscopic pattern is repeated, with only occasional and localized defects, throughout the whole sample. For the rest of the chapter we will study crystalline solids exclusively. Their regular geometrical arrangement makes them much easier to analyze than the randomly arranged amorphous solids.

The ordered geometrical structure of crystals has easily observable effects on some of their macroscopic properties. The most striking effect is the existence of cleavage planes along which the crystal tends to split, creating the smooth facets of individual crystal grains. Another important although less obvious consequence of the ordered internal structure is a directionality of the physical parameters of most crystals. Things such as heat flow or electrical conduction take place at different rates in different directions in many crystals. When this happens the crystal is said to be *anisotropic.*

10.2. CHEMICAL BONDING IN SOLIDS

It is the chemical bonds between neighboring atoms or molecules that create and maintain the spatial structure of a crystal. Therefore solids are separated into five broad classes according to the type of chemical bonding that is most important in their structure. This classification is not completely sharp in that many cases exist in which two or more different types of bonds play an important role in a single crystal. In Table 10.1 the five bonding types are listed with a fairly pure example of each type given. The melting temperatures of the examples given are typical for crystals of those types. It is obvious that the covalent bonds that maintain the structure of diamond are enormously stronger than the weak Van der Waals forces that are responsible for the crystallization of neon.

TABLE 10.1

BOND TYPE	EXAMPLE	MELTING POINT
Covalent	Diamond	4100 K
Ionic	NaCl	1074 K
Metallic	Na	371 K
Hydrogen	H_2O	273 K
Van der Waals	Ne	24.3 K

Covalent: We have already discussed the nature of the covalent bond in Chapter 7. Good examples of covalently bonded crystals are carbon (diamond), germanium, and silicon. They are all tetravalent elements with a crystal structure that places four nearest neighbors adjacent to each atom. The four adjacent atoms are located at the vertices of a tetrahedron. This gives the ideal spatial arrangement for covalent bonds in those elements.

Ionic: Ionic crystals are likely to occur in substances made up of two elements of very different electronegativity such as alkali-halogen salts. Within the crystal electrons are transferred from the atoms of one element to the atoms of the other. This results in a crystal structure composed of alternating positive and negative ions. The primary source of the bonding energy is the simple Coulomb attraction between the unlike charges. Ionic crystals are soluble in water because the water molecules, which themselves have an electric dipole charge distribution, can weaken the bonds between oppositely charged ions by interposing their own electric charges between the unlike charges of the ions.

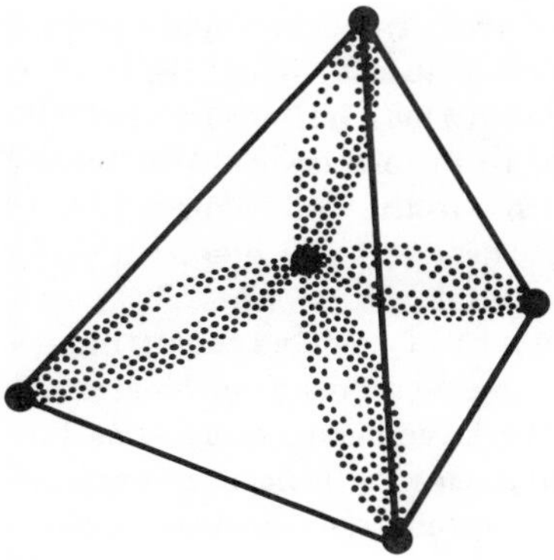

FIGURE 10.2

The tetrahedral structure of a covalently bonded crystal.

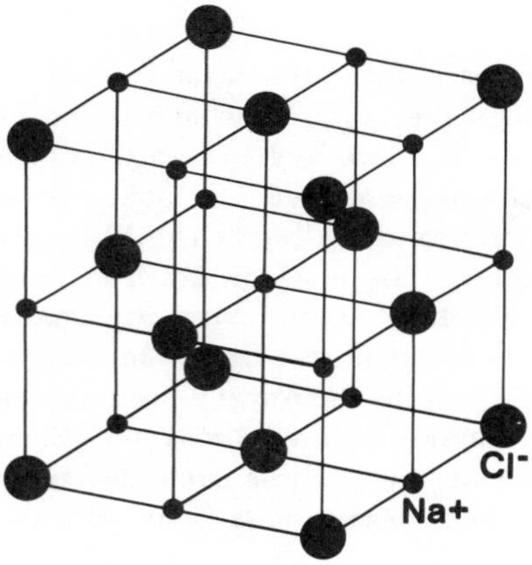

FIGURE 10.3

The crystal structure of common salt (NaCl). The Cl^- ions are represented by large dots; the Na^+ ions by small ones.

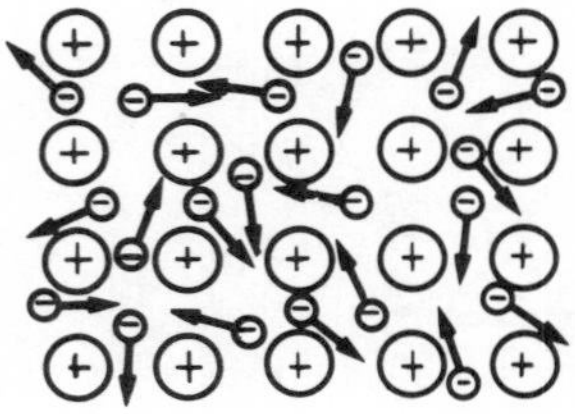

FIGURE 10.4

Metallic crystals are composed of a rigid lattice of positive ions plus a 'fluid' of conduction electrons. The ions are made up of the atomic nuclei surrounded by the tightly-bound inner shell electrons. The inner-shell electrons do not contribute to the electrical conductivity of the metal.

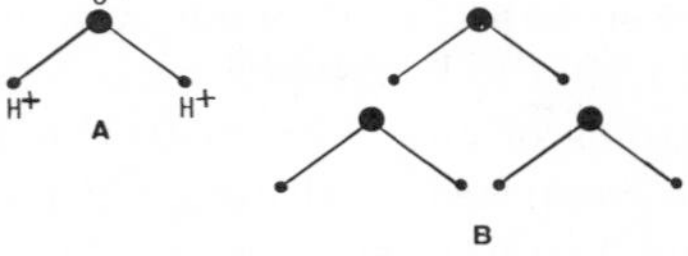

FIGURE 10.5

Hydrogen bonding in ice. The water molecule (A) has a negatively-charged oxygen atom and two positively-charged hydrogens in a fixed spatial arrangement. In the ice crystal (B) the molecules arrange themselves so that positive ions are close to negative ones, producing electrostatic bonding.

Metallic: The metallic bond is quite different from the other molecular bonds in that it is not really a bond between two particular atoms. It occurs in crystals composed of atoms that have only a few electrons in their outermost shell. When an assemblage of such atoms is brought together to form a crystal the outermost electrons, which are only weakly attached to the individual atoms, can move freely from atom to atom within the crystal. These mobile electrons go into electronic quantum states that are spread throughout the crystal. The resulting structure can be pictured as an ordered arrangement of positive ions (the atomic nuclei surrounded by their inner shell electrons) permeated by an electron fluid, composed of the original outer shell electrons. We have seen before that forcing an electron to remain within a small volume increases its kinetic energy. When the outer shell electrons (*the conduction electrons*) spread out through the sample there is a corresponding decrease in their kinetic energies, leading to a reduction in the total energy of the system. The delocalization of the conduction electrons and their corresponding mobility account for the high electrical conductivity of metals.

Hydrogen: Hydrogen bonding occurs in crystals composed of molecules containing hydrogen and some strongly electronegative atom, such as oxygen or fluorine. The water crystal (ice) is a good example. The two hydrogen atoms in a single water molecule are covalently bonded to the oxygen atom. Because of the great difference in electronegativity between oxygen and hydrogen the basically covalent bond has a large degree of ionic character. The electron pairs are not shared equally between the hydrogen and oxygen atoms but are more concentrated on the oxygen atom, leaving the hydrogen atoms in the molecules with, on the average, a net positive charge. The lines from the oxygen atom to the two hydrogens atoms in a single water molecule make an angle of 105°. When water crystallizes it forms a geometrical arrangement that allows each positively charged hydrogen atom of one molecule to lie close to the negatively charged oxygen atom of a neighboring molecule.

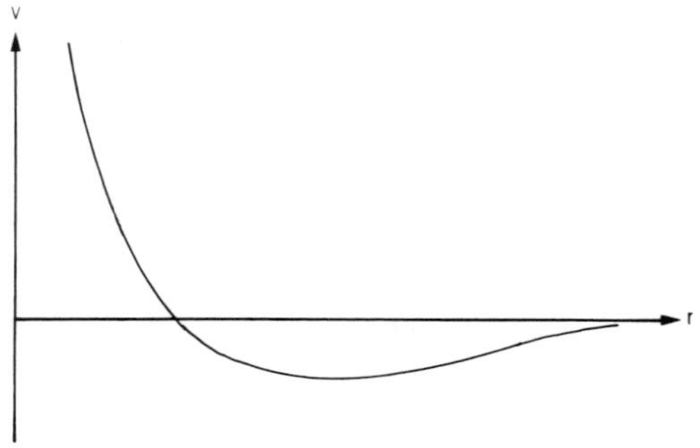

FIGURE 10.6 The interaction potential whose derivative gives the force between two inert gas atoms as a function of r, the distance between the atoms. It is strongly repulsive at short distances and weakly attractive at large distances.

Van der Waals: None of the above types of bonding can occur among inert gas molecules such as He, Ne, and Xe. As the name of the group suggests, these atoms form no stable chemical compounds. However, when two inert gas atoms are brought near one another there is a slight tendency for the electrons on the two atoms to coordinate their motion around their respective nuclei in such a way as to reduce the energy of the system. Thus there is a weak attractive force between two inert gas atoms. This force is called a Van der Waals force, after the name of the Dutch physicist who derived an equation to replace the ideal gas equation in cases in which the forces between gas molecules are significant. The weakly-attractive Van der Waals forces are sufficient to cause all inert gases to liquify at low temperatures and all but helium to crystalize at still lower temperatures. Helium will crystalize only under an external pressure of about thirty atmospheres. At atmospheric pressure it will remain liquid, even to absolute zero.

10.3. THE CRYSTAL LATTICE

Up to this point we have discussed crystals without ever having defined the term precisely. A perfect crystal is an arrangement of atoms or molecules that repeats itself periodically in all three dimensions. Another way of saying the same thing is that a crystal is a geometrical arrangement of atoms that has the following property. There exist three independent vectors **a**, **b**, and **c** (independent means that they do not all lie in a single plane) such that, if all the atoms in the crystal are simultaneously displaced an amount **a** (or **b**, or **c**) then every point in space will be occupied by the same kind of atom as it was before the displacement. It is clear that, according to this definition a perfect crystal would have to be infinite in all dimensions. Since real crystals have about 10^8 atoms in each direction one can expect that the properties of these imaginary perfect crystals will agree with those of real crystals except near the surface of the real crystal. To try to take surface effects into account would very much complicate the theory and we will therefore not do so. Our picture will usually be of two-dimensional rather than three-dimensional crystals. This helps to simplify the pictures but it does have the drawback that certain things that are planes in three-dimensional crystals are represented only by lines in two-dimensional crystals. The repeating unit in a crystal may be a single atom (in which case the crystal is called a *simple crystal*), it may be a small group of atoms, or a large and complex molecule. We will only consider simple crystals.

Any vector from the center of one atom to the center of any other atom is called a *lattice vector*. The collection of all the lattice vectors is the *crystal lattice*. The three smallest independent lattice vectors are called the *primitive lattice vectors*. We will denote them **a**, **b**, and **c**. Any other lattice vector can then be written in the form

$$\mathbf{V} = \mathrm{I}\mathbf{a} + \mathrm{J}\mathbf{b} + \mathrm{K}\mathbf{c} \tag{10.1}$$

where I, J, and K are integers.

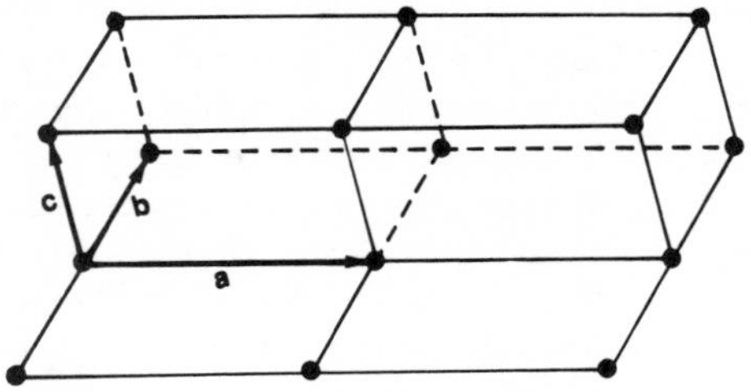

FIGURE 10.7 The primitive vectors, **a**, **b**, **c**, of a crystal lattice.

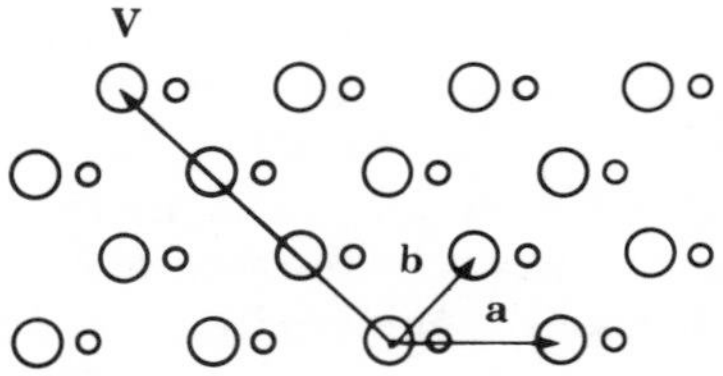

FIGURE 10.8
The lattice vector $V = -3\mathbf{a} + 3\mathbf{b}$.

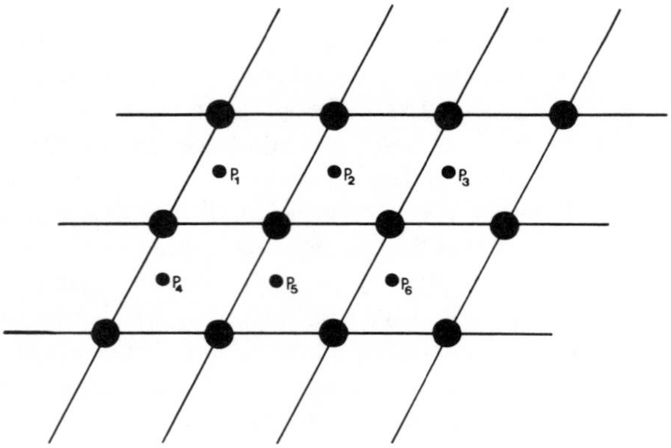

FIGURE 10.9 Any local property, such as the electron density in the crystal, is the same at each one of the points shown.

10.4. CRYSTAL SYMMETRY TYPES (BRAVAIS LATTICES)

The fundamental symmetry exhibited by all types of crystals is the symmetry with respect to translation by any lattice vector. If we consider any measurable physical property that depends upon position, such as $\rho(\mathbf{r})$, the electron density at position $\mathbf{r}$, or $\phi(\mathbf{r})$, the electrostatic potential at position $\mathbf{r}$, then that property is unaffected by any displacement by a lattice vector. That is:

$$\rho(\mathbf{r}) = \rho(\mathbf{r}+\mathbf{V})$$

and

$$\phi(\mathbf{r}) = \phi(\mathbf{r}+\mathbf{V})$$

where **r** has any value and **V** is any of the lattice vectors defined in Equation 10.1.

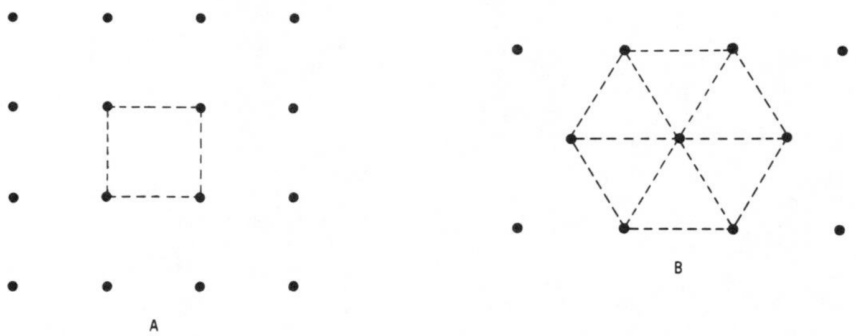

FIGURE 10.10 Rotational symmetries of two-dimensional lattices. A. The square lattice has a fourfold symmetry. B. The triangular lattice has a sixfold rotational symmetry.

In addition to this basic translational symmetry certain lattices exhibit other simple symmetries. Restricting ourselves to two-dimensional lattices, it is clear that the square and triangular lattices shown in Figure 10.10 are symmetric with respect to rotations of 90° and 60° respectively.

The square lattice is said to have a fourfold rotational symmetry, since the smallest rotation that exhibits the symmetry is a rotation by one fourth of a full revolution. The triangular lattice correspondingly has a sixfold rotational symmetry. A onefold rotational symmetry is no symmetry at all since a rotation by 360° brings one back to the original point in space. All two-dimensional crystal lattices exhibit twofold rotational symmetry. Some thought will convince the reader that the only other possible rotational symmetries for a two-dimensional lattice are fourfold, and sixfold symmetries.

QUESTION: How can we prove that no crystal lattice exists that exhibits fivefold symmetry?

ANSWER: Assume a crystal lattice exists that is symmetric with respect to a 72° rotation about a point P_0. (P_0 may or may not be a lattice point.) Choosing as a radius the distance between P_0 and the nearest lattice point construct a circle about P_0. By construction there can be no lattice points (except possibly P_0) inside the circle. By the rotational symmetry there must be at least five equally spaced points on

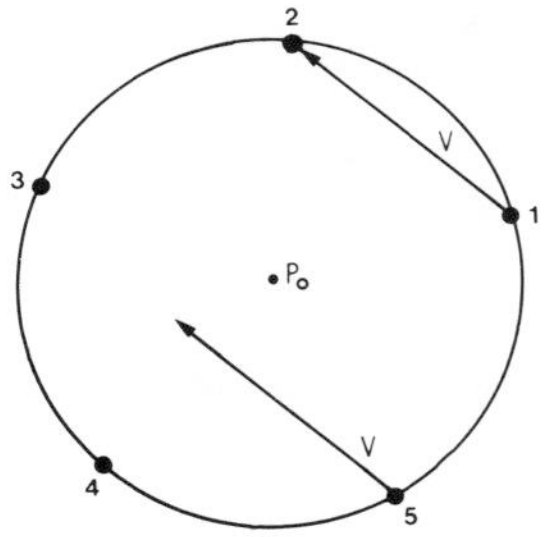

FIGURE 10.11

The proof that fivefold symmetry is impossible. See text for details.

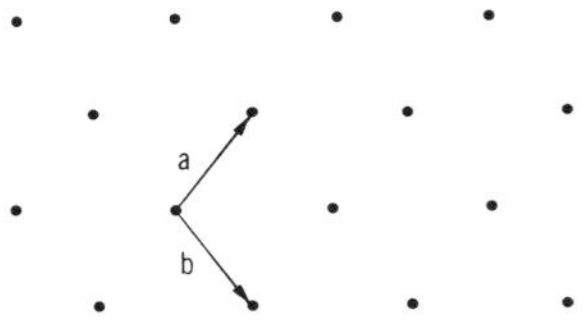

FIGURE 10.12

If $|\mathbf{a}| = |\mathbf{b}|$ then the lattice is symmetric with respect to reflection across a line bisecting the angle between **a** and **b**.

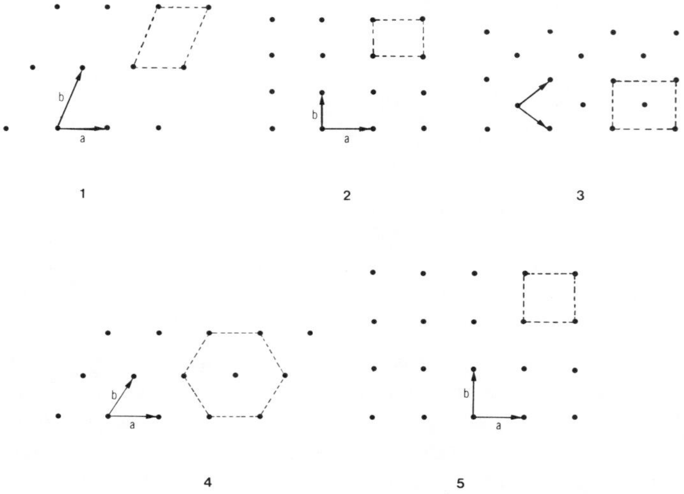

FIGURE 10.13

The five lattice types and the four symmetry classes in two dimensions.

Lattice	Symmetry Class Rotations	Reflections*
Oblique	2-fold	0
Rectangular Centered Rect.	2-fold	2
Hexagonal	6-fold	2
Square	4-fold	2

* 2 means that 2 independent reflection symmetries are exhibited.

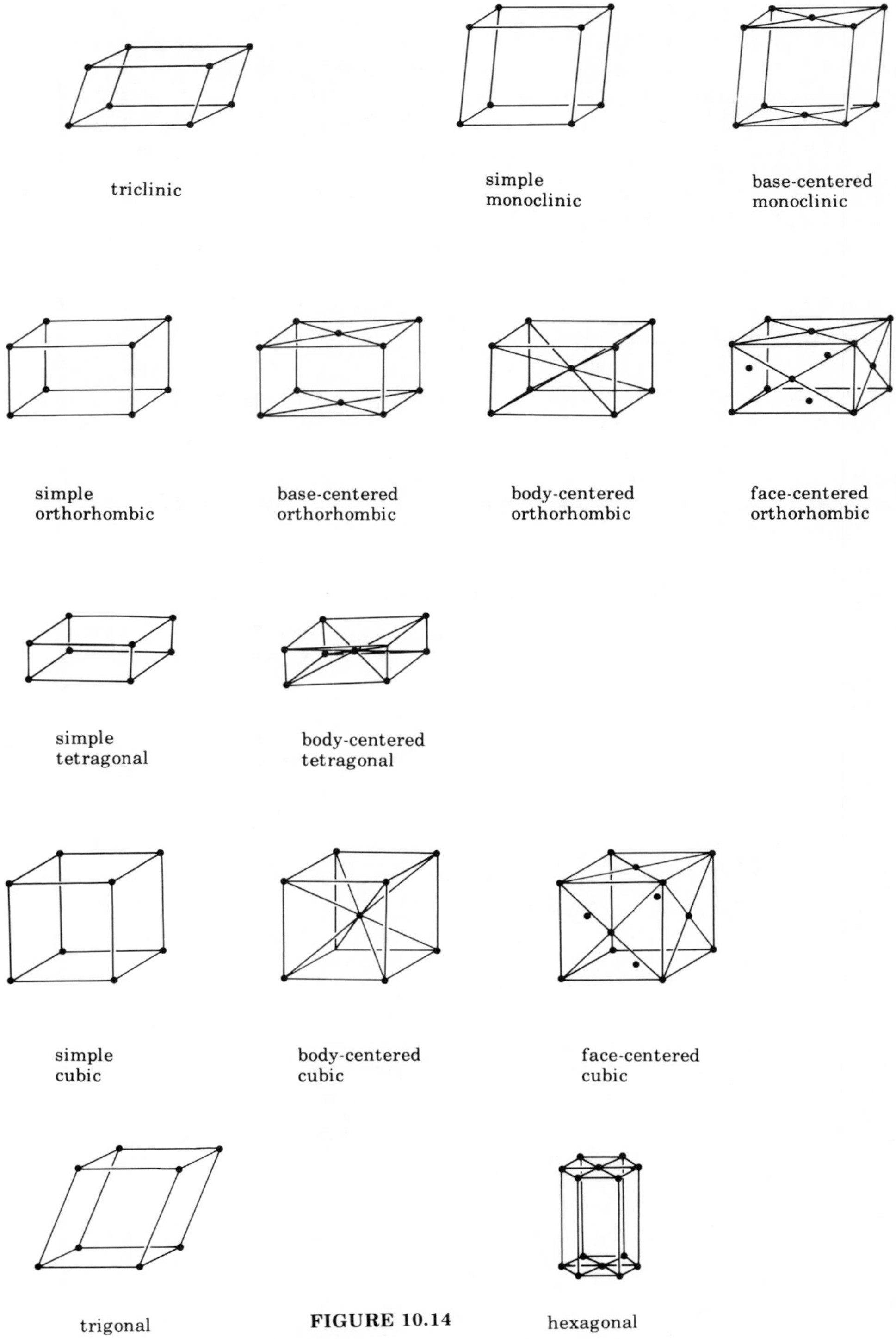
triclinic
simple monoclinic
base-centered monoclinic
simple orthorhombic
base-centered orthorhombic
body-centered orthorhombic
face-centered orthorhombic
simple tetragonal
body-centered tetragonal
simple cubic
body-centered cubic
face-centered cubic
trigonal
hexagonal

FIGURE 10.14

the circle. (Figure 10.11). We number them 1 to 5. The vector **V** from 1 to 2 must be a lattice vector. But that means that a displacement of **V** from point 5 would lead to another lattice point which would be inside the circle. This contradicts the fact that there are no points (other than P_0) inside the circle. By arguments like this one can prove that only twofold, fourfold, and sixfold symmetries are possible in two dimensions.

Another type of symmetry is mirror symmetry. For example, any two-dimensional lattice whose primitive lattice vectors are of equal length is symmetric with respect to reflection across the bisector of the angle between the two primitive vectors. (See Figure 10.12.)

If two lattices exhibit the same combination of rotational and mirror symmetries they are said to be in the same *symmetry class.* It can be shown that only four different symmetry classes exist for two-dimensional lattices. One of these symmetry classes contains two different types of lattices. The four symmetry classes are shown in Figure 10.13. In three dimensions things are appreciably more complicated. There are seven three-dimensional symmetry classes containing a total of fourteen different types of lattices. (Each type of lattice is called a Bravais lattice.) They are illustrated in Figure 10.14 and listed on Table 10.2.

TABLE 10.2
THE THREE-DIMENSIONAL BRAVAIS LATTICES

Symmetry System	Bravais Lattice	Characteristic Symmetries
Triclinic	Simple	None
Monoclinic	Simple Body-centered	One 2-fold rotation axis
Orthorhombic	Simple Base-centered Body-centered Face-centered	Three 2-fold rotation axes
Tetragonal	Simple Body-centered	One 4-fold rotation axis
Cubic	Simple Body-centered Face-centered	Four 3-fold rotation axes
Trigonal	Simple	One 3-fold rotation axis
Hexagonal	Simple	One 6-fold rotation axis

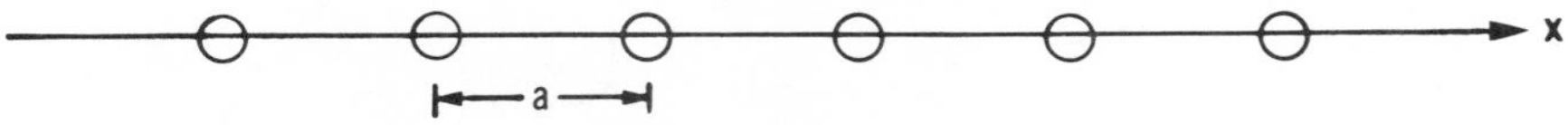

FIGURE 10.15 A one-dimensional 'crystal' is simply a set of particles in one dimension whose equilibrium positions are equally spaced.

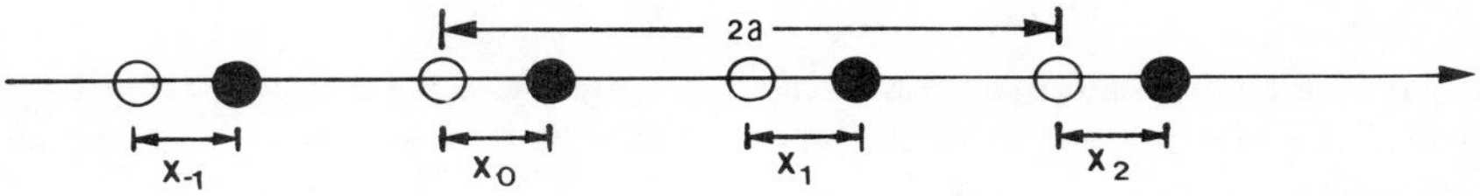

FIGURE 10.16 x_n gives the displacement from equilibrium of the nth particle. The open circles represent the fixed equilibriumn positions.

10.5. THE CLASSICAL THEORY OF LATTICE VIBRATIONS

In this section we will consider the motion of a crystal lattice using classical dynamics. We begin by considering a one-dimensional lattice, which is a collection of identical particles constrained to move on a line (the x axis). We assume that the equilibrium state of the system is one in which the particles are equally spaced with an interparticle spacing a. We are interested in describing the small amplitude vibrations of this system of particles. Therefore we will make the basic approximation that the displacement of any particle from its equilibrium position is small in comparison to the spacing between nearest neighbors. If we let X_n be the x coordinate of the nth atom and x_n be the displacement of the nth atom from its equilibrium position then

$$X_n = n\,a + x_n$$

where we have taken the equilibrium position of the zeroth atom as the origin of our coordinate system.

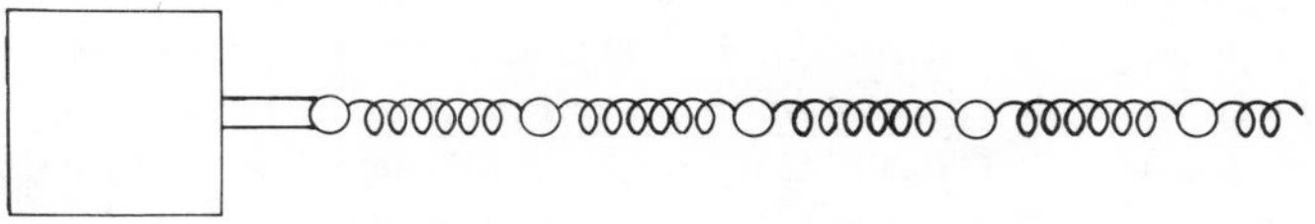

FIGURE 10.17 A simple model of a one-dimensional crystal plus a vibrator to produce compressional waves (the rod vibrates in and out, not up and down).

If the particles are all at their equilibrium positions, then, by the definition of equilibrium, the force on any particle is zero. However if the particles are displaced from those equilibrium positions they will exert forces on one another. A simple model system that exhibits these properties is a set of equal mass points connected by massless springs of equilibrium length a. In Figure 10.17 an infinite set of such particles and springs is shown. The first particle is attached to a device that can drive it with simple harmonic motion at any prescribed frequency. It is obvious that driving the first particle with harmonic motion will cause a traveling wave to propagate down the system to the right. Without carrying out a detailed mathematical analysis of this system we want to prove two important properties of the waves that can be propagated through this one-dimensional lattice. They are:

1. A wave can be propagated down the lattice only if its angular frequency lies within the range $0 < \omega \leqslant 2\sqrt{K/m}$ where K is the force constant of the connecting springs.
2. As the angular frequency of the wave is varied from $\omega = 0$ to $\omega = \omega_{max} = 2\sqrt{K/m}$ the wavelength of the wave varies from $\lambda = \infty$ to $\lambda = 2a$ where a is the lattice spacing.

First let us convince ourselves that there must be some maximum frequency for wave propagation. Suppose we cause the first particle to vibrate with some fixed amplitude, let us say an amplitude of .1a, and an extremely high frequency. When the first particle moves to the right it will compress the first spring and thereby apply a force on the second particle. The second particle will then also begin to move to the right. But if the frequency of the first particle's motion is very high the first particle will return to its equilibrium position and begin to apply a restoring force (to the left) on the second particle before that particle has had a chance to move very far. It is clear that for high enough frequency the second particle will have a smaller amplitude of vibration than the first. The spring, which has a fixed spring constant simply cannot apply sufficient force on the second particle to drive it to full amplitude at the prescribed high frequency. The amplitude of vibration of the second particle will be proportional to the amplitude of vibration of the first particle with a proportionality constant smaller than one. $A_2 = \gamma A_1$ $(\gamma < 1)$. The second particle acts as a source of driving force for the third. Again, because of the high frequency the amplitude will diminish by a factor of γ in going from the second to the third. The amplitude of vibration of the nth particle will be $\gamma^{n-1} A_1$. For large n the amplitude of vibration will be completely negligible. Thus only the particles near the driven one will vibrate and no wave will propagate through the lattice. We can obtain the same result in another way which will allow us to derive a definite formula for the cutoff frequency ω_{max} of this simple one-dimensional lattice. Suppose a wave of angular frequency ω is propagating through the crystal with no dimunition in amplitude. Then each particle will vibrate with angular frequency ω and amplitude A. (Although the paricles all have the same frequency and amplitude they have different phases since the wave reaches different particles at different times.) For fixed amplitude one obtains the maximum possible frequency when the restoring force is a maximum. It is easy to see that one would obtain the maximum restoring force if the particles on both sides of any given particle were vibrating exactly out of phase with that particle. (See Figure 10.18) In order to determine what that maximum frequency is we note that when the center particle in Figure 10.18 is displaced an amount x from its equilibrium position, each spring is changed in length by an amount 2x. Therefore the restoring force on the particle

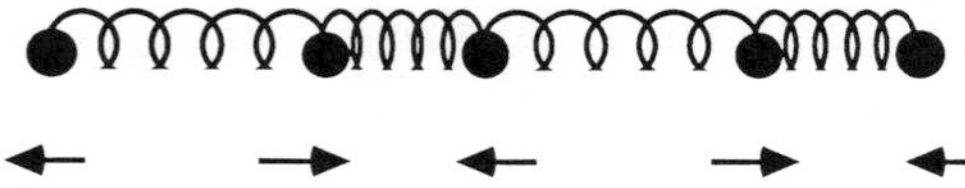

FIGURE 10.18 For fixed vibrational amplitude one gets the maximum restoring force on a particle when nearest neighbors vibrate exactly out of phase.

is equal to $-4Kx$. Thus the effective force constant (the ratio of force to displacement) is four times the force constant of one of the springs. This will yield a frequency of vibration equal to

$$\omega_{max} = \sqrt{4K/m} = 2\sqrt{K/m} \tag{10.2}$$

We now want to consider how the wavelength of the wave varies as the frequency is varied from zero to ω_{max}. In order to be able to easily visualize the wave we will draw diagrams in which the displacements of the particles are plotted in the vertical direction even though the actual displacements take place in the horizontal direction. Such a diagram is shown in Figure 10.19 for a wave whose wavelength is equal to 8a. One can see that, for very large wavelength, two neighboring particles will move almost in phase and therefore the spring between them will be compressed and expanded very little. Thus the restoring force is small which shows that the frequency of long wavelength waves must be small as one would intuitively expect. The maximum frequency occurs when neighboring particles move with opposite phases. As can be seen in Figure 10.20 this occurs when $\lambda = 2a$. Thus, as the frequency is varied from 0 to ω_{max} the wavelength ranges from $\lambda = \infty$ to $\lambda = 2a$.

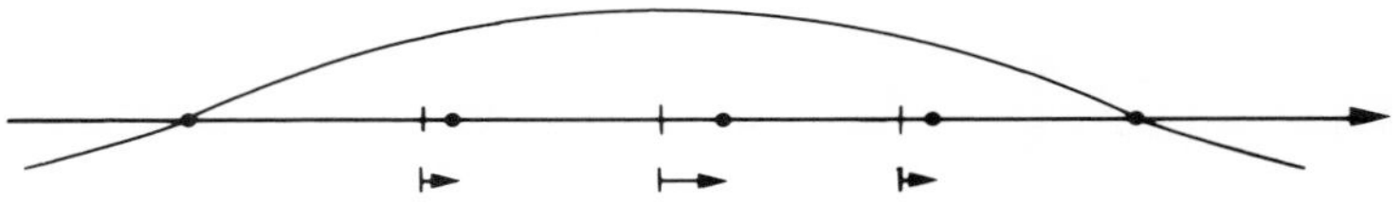

FIGURE 10.19 The relationship between the wavelength, $\lambda = 8a$, and the displacements of the particles.

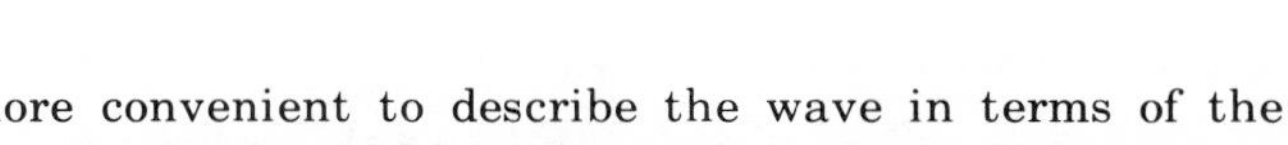

FIGURE 10.20 Nearest-neighbor particles are 180° out of phase when $\lambda = 2a$.

It is more convenient to describe the wave in terms of the wave vector k rather than the wavelength. k is positive for waves traveling to the right and negative for waves traveling to the left. The magnitude of k is given by $|k| = 2\pi/\lambda$. Considering the range of possible wavelengths we see that k has the range

$$-\frac{\pi}{a} \leqslant k \leqslant \frac{\pi}{a}$$

The full range of possible wave vectors of traveling waves in the one-dimensional lattice is called the *Brillouin zone*. It can be shown that, for the simple lattice model we have been considering, in which a given lattice particle interacts only with its two nearest neighbors, that the relationship between the frequency and wave vector of any wave is (ω_{max} is defined in Equation 10.2.)

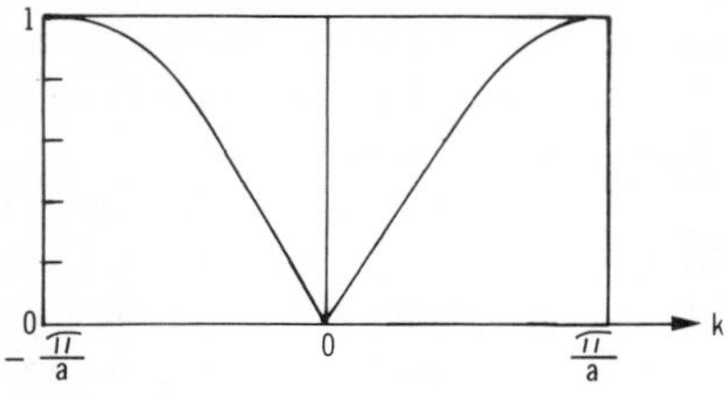

FIGURE 10.21

The relationship between ω/ω_{max} (plotted vertically) and k (plotted horizontally) for a one-dimensional crystal with nearest-neighbor forces. Since no propagating waves exist with wave vectors outside the Brillouin zone the function $\omega(k)$ is not defined there.

lattice particle interacts only with its two nearest neighbors, that the relationship between the frequency and wave vector of any wave is

$$\omega(k) = \omega_{max} \sin|k a/2| \tag{10.3}$$

QUESTION: Sound waves are compressional waves whose wavelengths are usually much larger than the crystal lattice spacing. What would be the velocity of sound waves in our simple one-dimensional lattice?

ANSWER: If $\lambda >> a$ then $ka/2 << 1$ and we can use the small angle approximation for the sine function in Equation 10.3. This gives

$$\omega = \omega_{max} |k a/2|$$

Using the relations $\omega_{max} = 2\sqrt{K/m}$, $|k| = 2\pi/\lambda$, $\nu = \omega/2\pi$, and $v_{sound} = \lambda\nu$, we get

$$v_{sound} = \lambda\nu = a\sqrt{K/m} \tag{10.4}$$

QUESTION: The speed of sound in a large single crystal of salt (NaCl) is 4730 m/s and the lattice spacing is 2.82 Å. Can we use this data to get a crude estimate of the effective spring constant K? (Since there are obviously no springs in salt the effective spring constant means the ratio of force to displacement for one of the lattice particles.)

ANSWER: Yes, we can use Equation 10.4 to obtain an order of magnitude estimate of K. For m we will use the geometric mean of the sodium and chlorine atomic masses.

$$m = (m_{Na} m_{Cl})^{1/2} = ((3.82 \times 10^{-26})(5.9 \times 10^{-26}))^{1/2} \text{ kg}$$
$$= 4.7 \times 10^{-26} \text{ kg}$$

From Equation 10.4 we obtain a formula for K.

$$K = m v^2/a^2 = (4.7 \times 10^{-26})(4730)^2/(2.82 \times 10^{-10})^2$$

which gives $K = 13$ N/m. Thus, typical force constants in real crystals are a few Newtons per meter.

10.6 LATTICE VIBRATIONS IN TWO AND THREE DIMENSIONAL CRYSTALS

In a two- or three-dimesional crystal the wave vector associated with a traveling vibrational wave is a two- or three-dimensional vector **k**, giving the direction of the propagation of the wave. For a two-dimensional rectangular lattice the range of possible wave vectors (i.e., the Brillouin zone) is a rather obvious generalization of what we found in the one-dimensional case. If the lattice spacing in the x direction is a and that in the y direction is b then the wave vector $\mathbf{k} = (k_x,k_y)$ can have any value in the range defined by

$$-\frac{\pi}{a} \leqslant k_x \leqslant \frac{\pi}{a}$$

$$-\frac{\pi}{b} \leqslant k_y \leqslant \frac{\pi}{b}$$

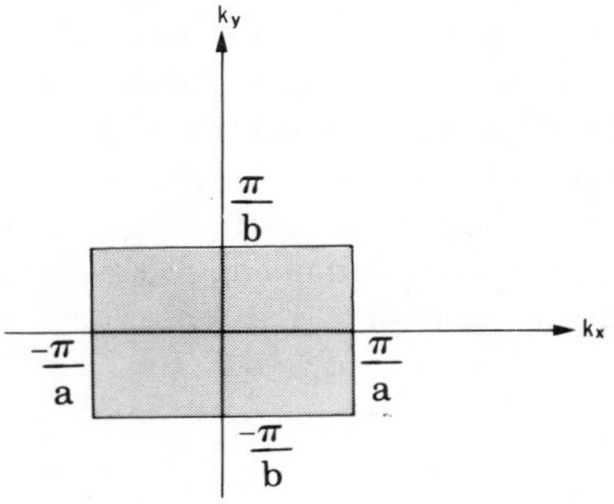

FIGURE 10.22

In a two-dimensional rectangular crystal the wave vectors of all lattice vibrational waves fall within the Brillouin zone shown.

For the other types of two-dimensional crystal lattices the Brillouin zone will not be a rectangle. It will always exhibit the same symmetries with respect to rotation and reflection as the crystal lattice. For example a triangular lattice with a sixfold symmetry axis leads to a hexagonal Brillouin zone which also has a sixfold symmetry axis. In three dimensions the Brillouin zone gives a three-dimensional range of allowed wave vectors. That is; the Brillouin zone is some three-dimensional region of wave vectors, $\mathbf{k}$, centered on $\mathbf{k} = 0$.

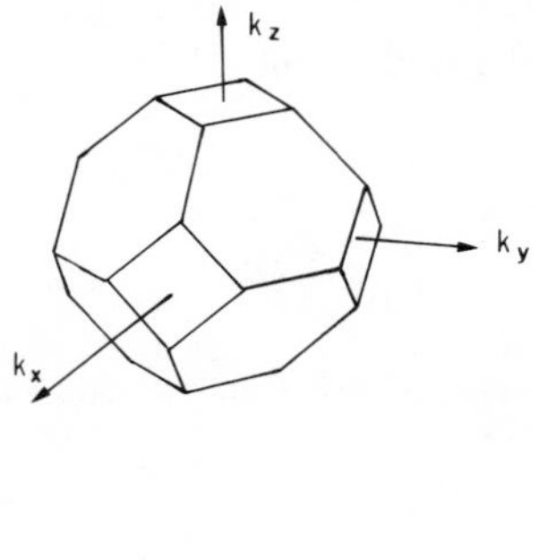

FIGURE 10.23

The Brillouin zone for the face-centered cubic lattice. All propagating lattice waves have wave vectors within this three-dimensional region in k-space.

10.7 POLARIZATION OF VIBRATIONAL WAVES

In two or three dimensions there is an additional complication in describing the lattice vibrational waves, namely their *polarization*. In Figure 10.24 we show a two-dimensional square lattice attached along one side to a mechanical vibrator. If the vibrator oscillates back and forth in the x direction it will create a

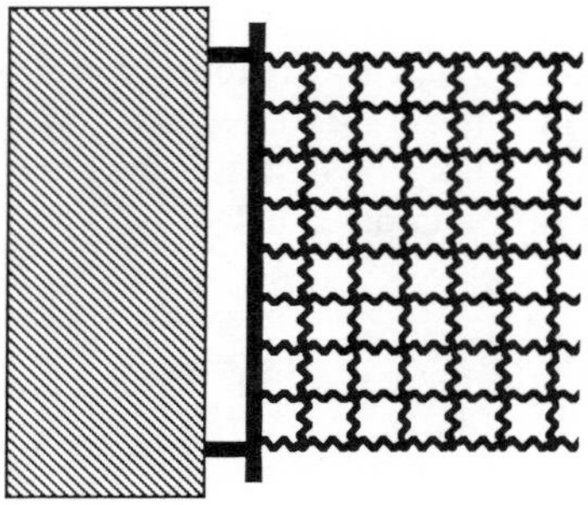

FIGURE 10.24

A device for producing lattice waves in a two-dimensional crystal. Either transversely-polarized or longitudinally-polarized waves can be produced.

traveling wave in the crystal. The wave will propagate in the x direction and the vibration of any lattice particle will also be in the x direction. However, if the vibrator oscillates in the y direction it will still create a traveling wave that propagates in the x direction. But, as this wave passes, the vibration of any lattice particle will take place in the y direction. We say that the first wave is polarized in the x direction and the second is polarized in the y direction. For any given wave vector (that is, any given wavelength and direction of propagation) there are two different possible directions of polarization of lattice vibrational waves in a two-dimensional crystal (and three in a three-dimensional one). The different possible polarization directions are perpendicular to one another.

10.8 THE DISPERSION RELATION

In a one-dimensional crystal the relationship between the frequency and wave vector of a vibrational wave is described by a function $\omega(k)$ defined for all k within the Brillouin zone. The function $\omega(k)$ is called the *dispersion relation* for the crystal. The dispersion relation for a crystal with nearest neighbor interactions is plotted in Figure 10.21. In a two-dimensional crystal there is a separate dispersion relation for each of the two possible polarizations, since waves with the same wave vector but different polarization have, in general, different frequencies. We will denote them as $\omega_1(\mathbf{k})$ and $\omega_2(\mathbf{k})$. Each of the functions, $\omega_1(\mathbf{k})$ and $\omega_2(\mathbf{k})$, may be pictured as a surface defined over the two-dimensional Brillouin zone of allowed wave vectors. (See Figure 10.25.) In real three-dimensional crystals there are three possible polarizations for waves with any given wave vector $\mathbf{k}$. Thus the relationship between frequency, polarization, and wave vector is given by three dispersion relations, $\omega_1(\mathbf{k})$,$\omega_2(\mathbf{k})$, and $\omega_3(\mathbf{k})$, each of which is defined throughout the three-dimensional Brillouin zone.

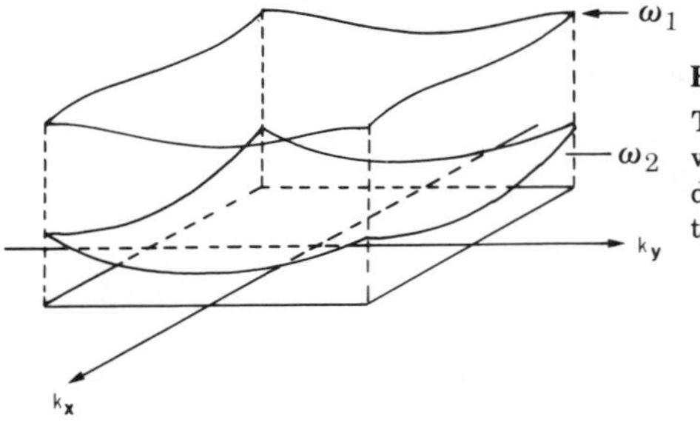

FIGURE 10.25

The two functions, $\omega_1(\mathbf{k})$ and $\omega_2(\mathbf{k})$, are each defined within the rectangular Brillouin zone. Thus they define two surfaces. The ω-axis in the vertical direction has been omitted for clarity.

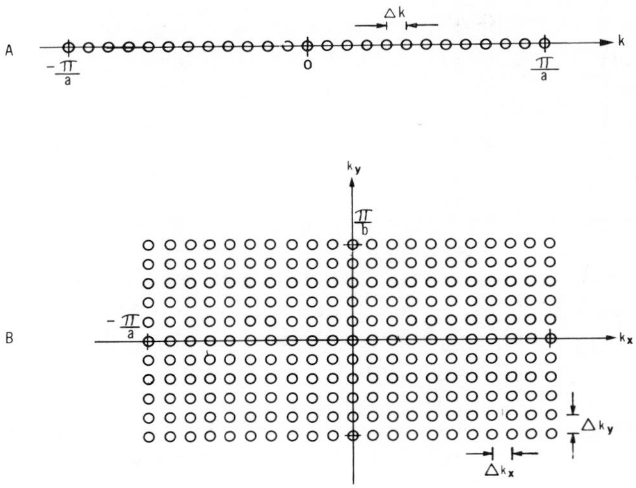

FIGURE 10.26 A. For a one-dimensional lattice containing a finite number, N, of particles there are N possible wave vectors of lattice vibrational waves. They are distributed throughout the Brillouin zone with a uniform spacing $\Delta k = 2\pi/Na = 2\pi/L$ where $L = Na$ is the total length of the crystal sample. (L is the one-dimensional "volume" of the crystal.)

B. For a two-dimensional rectangular lattice there are N possible wave vectors within the Brillouin zone. Each allowed wave vector "occupies" a rectangle of area

$$\Delta k_x \Delta k_y = (2\pi/L_x)(2\pi/L_y) = \frac{(2\pi)^2}{L_x L_y}$$

where L_x and L_y are sides of the crystal sample and therefore $L_x L_y$ is the two-dimensional "volume" of the crystal.

For a three-dimensional rectangular lattice each allowed wave vector occupies a parallelepiped of volume

$$\Delta k_x \Delta k_y \Delta k_z = \frac{(2\pi)^3}{L_x L_y L_z} = \frac{(2\pi)^3}{V}$$

10.9 QUANTIZED VIBRATIONAL WAVES*

A normal mode motion of any system is a motion in which all particles of the system vibrate harmonically about their equilibrium positions with a common frequency. Each one of the lattice vibrational waves is an independent normal mode of the lattice. The total energy of the lattice, when a number of vibrational waves are simultaneously passing through it, is equal to the sum of the energies associated with each one of the separate vibrational waves. When the crystal

Some of the following sections repeat analysis already done in the section of Chapter 9 on Quantum Field Theory. We repeat it here so that this chapter will be independent of Chapter 9.

lattice is analyzed according to quantum mechanics each one of the normal mode vibrations can be treated as an independent quantized harmonic oscillator. The energy associated with a particular lattice vibration cannot take an arbitrary value but must have one of the values appropriate to a quantum harmonic oscillator of its given frequency. Thus, for a one-dimensional crystal the energy associated with vibrational waves of wave vector k must have one of the values $(n + \frac{1}{2})\hbar\omega(k)$.

In our analysis we have considered crystal lattices that extend to infinity and therefore contain an infinite number of particles. We found that normal modes existed for any value of k within the Brillouin zone. If we had considered the more realistic case of a lattice containing a large but finite number of particles, N, our conclusions would have been modified in the following way. Instead of obtaining an infinite number of normal mode vibrations (one for each value of k) we would have obtained N closely but uniformly spaced allowed values of k within the Brillouin zone. (See Figure 10.26.) This is just a special case of a general property of vibratory systems. Any vibratory system has as many independent normal modes of vibration as it has coordinates to specify its configuration. A simple pendulum requires only one coordinate (the angle) to define its position and it obviously has only one normal mode of oscillation. A pair of coupled pendulums require two angles and correspondingly have two independent normal modes. The number of coordinates needed to completely define the configuration of a system is called the number of *degrees of freedom* of the system.

FIGURE 10.27

QUESTION: What about the system shown in Figure 10.27? It has two degrees of freedom but only one mode of vibration.

ANSWER: The one tricky point associated with the theorem that equates the number of normal modes with the number of coordinates is that a free translatory motion of the system must be counted as a normal mode with infinite period (and therefore zero frequency). That is, a system which can undergo uniform translation with no restoring force must be considered as a limiting case of a system whose center of mass is attracted to one point in space by an extremely weak restoring force. As the strength of the restoring force approached zero the period of the lowest frequency vibrational would go to infinity. If we count the infinite period translatory motion as a normal mode we then get as many normal modes as there are degrees of freedom.

As we have mentioned, the quantum theory predicts that each normal mode acts like a quantum mechanical harmonic oscillator. Thus the ground state of the system is the quantum state in which each one of the quantized vibrations is in its lowest possible energy state. The total energy of the lattice in its ground state is therefore equal to the sum of the ground state energies of the N normal modes.

That is

$$E_o = \sum_k \tfrac{1}{2}\hbar\omega(k) \qquad (10.5)$$

where the sum is taken over the N allowed wave vectors within the Brillouin zone. Any higher energy quantum state of the lattice is defined by specifying the quantum number of each of the vibrational modes. If, for each value of k, the vibrational mode with wave vector k has energy $(n_k + \frac{1}{2})\hbar\omega(k)$ where n_k is a nonnegative integer that gives the excitation level of that normal mode oscillation, then the total energy of the lattice is

$$E = \sum_k (n_k + \tfrac{1}{2})\hbar\omega(k) \qquad (10.6)$$

QUESTION: In aluminum the lattice vibrational frequencies run from $\nu = 0$ to about $\nu = 10^{13}$ Hz. ($\nu = \omega/2\pi$). Can we use this to obtain an estimate of the ground state vibrational energy of the aluminum lattice?

ANSWER: Yes. Consider a sample of aluminum containing N atoms. Since each atom has three coordinates, the system has 3N degrees of freedom. Thus the sum in Equation 10.5 will contain 3N terms. If we approximate the sum by setting all the values of $\omega(\mathbf{k})$ equal to $2\pi \times 10^{13}$ Hz we get

$$E_o = (3N)(\tfrac{1}{2})(10^{-34}\,\text{J-s})(2\pi \times 10^{13}\,\text{Hz})$$

or

$$\frac{E_o}{N} = 9.4 \times 10^{-21}\,\text{J} \approx 10^{-20}\,\text{J}$$

One way of interpreting this number is to ask at what temperature a purely classical lattice with the same normal mode distribution would have a thermal vibrational energy equal to E_o. That is, what classical temperature is the quantum mechanical zero point motion equivalent to? For a classical lattice at temperature T it can be shown that each normal mode has an average thermal energy equal to $k_B T$. Setting $3Nk_B T$ equal to our estimate of E_o we get $T = 424$ K. Since room temperature is about 300 K this shows that purely quantum mechanical energies are not at all negligible in comparison to thermal energies and one could therefore not treat the aluminum lattice by classical mechanics at room temperature.

10.10 THE DENSITY OF NORMAL MODES

The dispersion relations $\omega_1(\mathbf{k})$, $\omega_2(\mathbf{k})$, and $\omega_3(\mathbf{k})$ give detailed information on the relationship between the frequency of a vibrational wave and its wave vector. For determining thermodynamic properties of the crystal, such as the internal energy or specific heat, such detailed information is unnecessary. All that is needed is a knowledge of how many normal modes there are with angular frequencies within a specified frequency range. This information is given by a function $D(\omega)$ called the *normal mode distribution function* or more simply the density of normal modes. $D(\omega)$ is defined by the relation

$$D(\omega)\,d\omega = \begin{array}{l}\text{the number of normal modes with angular}\\ \text{frequencies within the range } \omega \text{ to } \omega + d\omega\end{array}$$

We will calculate $D(\omega)$ for two different cases, leaving some details of the calculation as exercises for the student.

$D(\omega)$ FOR A ONE-DIMENSIONAL LATTICE WITH NEAREST-NEIGHBOR FORCES

The dispersion relation for a one-dimensional lattice in which each particle interacts only with the two particles adjacent to it has been given in Equation 10.3. It is

$$\omega(k) = \omega_{max} \sin|ka/2|$$

The wave vector k is restricted to one of the N values that are uniformly distributed throughout the Brillouin zone, $-\frac{\pi}{a} < k < \frac{\pi}{a}$. Thus the number of normal modes with wave vectors lying in the interval dk is

$$dN = \frac{Na}{2\pi} dk$$

Since $\omega(-k) = \omega(k)$ we can restrict ourselves to positive wave vectors and, at the end, multiply our result for $D(\omega)$ by two to take account of the normal modes with negative k that we neglected. We can invert the equation for $\omega(k)$ to obtain

$$k(\omega) = \frac{2}{a} \sin^{-1}(\omega/\omega_{max})$$

which gives

$$dk = \frac{2}{a\sqrt{\omega_{max}^2 - \omega^2}} d\omega$$

It is now a simple step (see Problem 10.5) to obtain the density of normal modes.

$$D(\omega) = \begin{cases} \dfrac{2N}{\pi\sqrt{\omega_{max}^2 - \omega^2}} & \text{for} \quad \omega < \omega_{max} \\ 0 & \text{for} \quad \omega > \omega_{max} \end{cases}$$

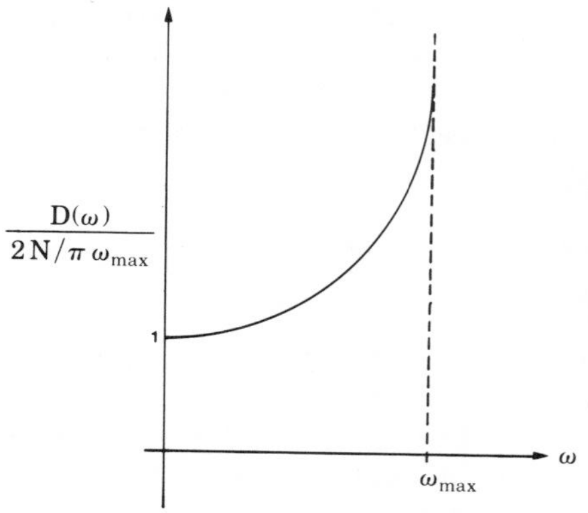

FIGURE 10.28

A graph of $D(\omega)$ for a one-dimensional crystal with nearest-neighbor forces. Although $D(\omega)$ becomes singular near $\omega = \omega_{max}$ the integral of $D(\omega)$ exists and is equal to N, the total number of normal nodes.

D(ω) FOR A THREE-DIMENSIONAL LATTICE IN THE DEBYE APPROXIMATION

The calculation of D(ω) using realistic models of crystal lattice structure and realistic interatomic forces is a very difficult and complicated business. Furthermore, one obtains D(ω) from such a calculation as a numerical function rather than as an explicit analytic function. That is, one obtains a graph of D(ω) rather than a formula. In any attempt to analyze physical effects that depend on the density of normal modes an explicit formula for D(ω) that is only reasonably accurate may be much more useful and convenient than a precise numerical function. The best compromise between simplicity and accuracy that has yet been found is a formula originally derived by Peter Debye. The formula, whose derivation we will discuss shortly, works quite well for a very wide variety of solids. The most important class of crystal structures for which the Debye formula is a *poor* approximation and hence should not be used are strongly anisotropic crystals. That is, crystals whose properties are very different in different directions. An example of a strongly anisotropic crystal is graphite, whose crystal structure is essentially a stack of sheets with strong covalent bonds connecting the atoms in any given sheet but only weak Van der Waals bonds between atoms in adjacent sheets.

To derive the Debye formula we first note that ordinary sound waves in solid are simply long wavelength vibrational waves. But sound waves are known to have the simple dispersion relation

$$\omega = v k$$

where v is the sound speed, which can be easily determined by experiment. Starting from this observation, a number of approximations lead to the Debye formula.

1. The sound wave dispersion relation, that is known to be correct for long wavelength vibrational waves, is used for *all* vibrational waves with a single value of v (in reality the wave velocity in a crystal depends on the frequency).

2. In solids, sound waves are polarized. For a given direction of propagation one longitudinal and two transverse polarizations are possible. The transversely polarized sound waves have a sound speed v_T that is different from the sound speed v_L of the longitudinally polarized sound waves. In the Debye approximation a single intermediate sound speed v is used for all polarizations. Because the sound speed appears in the final formula for D(ω) in the form v^{-3}, and because there are two transverse waves for each longitudinal wave, the intermediate speed v is determined by the relation

$$v^{-3} = \tfrac{1}{3} v_L^{-3} + \tfrac{2}{3} v_T^{-3}$$

3. The set of allowed wave vectors (the Brillouin zone) is taken to be a sphere of radius K in wave vector space. The value of K is chosen so that the total number of normal modes (counting the three polarizations) is equal to 3N as it should be. With these approximations one obtains the following formula for D(ω). (See Problem 10.6.)

$$D(\omega) = \begin{cases} \dfrac{3V}{2\pi^2 v^3}\,\omega^2 & \text{for} \quad \omega < \omega_{max} \\ 0 & \text{for} \quad \omega > \omega_{max} \end{cases}$$

where ω_{max} is given by

$$\omega_{max} = (6\pi^2 n)^{1/3} v$$

and n = N/V is the particle density in the crystal. We will make use of this formula later to calculate the heat capacity of a crystal.

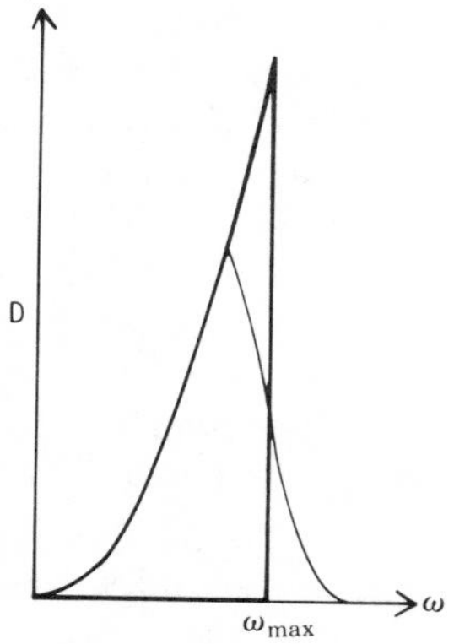

FIGURE 10.29

The Debye approximation for D(ω) compared with the exact function for a three-dimensional cubic crystal with nearest-neighbor forces. (Courtesy of C. Y. Fong.)

10.11 CRYSTAL MOMENTUM

The vibrational waves in a crystal carry momentum as well as energy. In order to see that this must be so, let us consider the process shown in Figure 10.30. The electron, moving parallel to a one-dimensional crystal, is assumed to interact weakly with the lattice particles by an electrostatic interaction. Because of the interaction the electron will create quantized vibrational waves in the lattice and will lose kinetic energy in doing so. If it loses kinetic energy it must slow down. Since the total momentum of an isolated system (the system being the electron *plus* the lattice) must be conserved, the momentum lost by the electron in slowing down must be carried away by the lattice waves the electron has created. If a vibrational wave of wave vector k is excited to a quantum level n_k then the momentum carried by that vibrational wave can (by a long and complicated argument) be shown to be

$$P_k = n_k \hbar k \tag{10.7}$$

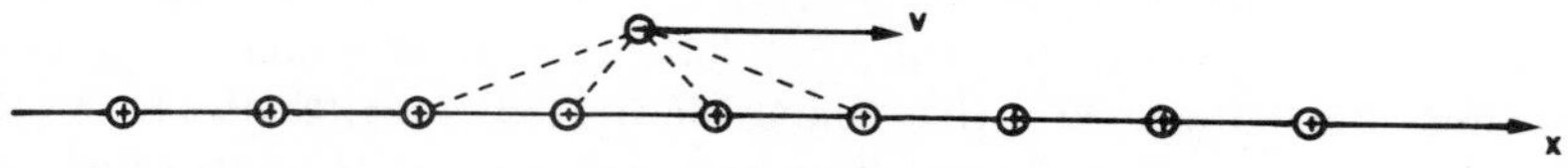

FIGURE 10.30 An electron, moving near a one-dimensional lattice will transfer energy and momentum to the lattice.

QUESTION: What happened to the $\frac{1}{2}$ that appears along with the n_k in the energy formula (Equation 10.6)? Do the zero point vibrational waves carry no momentum?

ANSWER: For any vibrational wave of wave vector k there is another one of wave vector $-k$. The zero point vibrational *energies* of these waves are both positive and make equal contributions to the ground state energy of the total lattice. On the other hand, the zero point *momenta* of the two waves are equal and opposite and therefore cancel. Thus we are free to entirely ignore any momentum associated with the ground state vibrational waves.

10.12 PHONONS

The ground state energy of a lattice could only be experimentally determined by measuring the work done in completely dismantling the lattice into its constituent particles*. For analyzing most of the processes that take place in solids it is convenient to ignore the ground state energy by treating E_0 as the zero of our energy scale. If we do that, we can express the energy of a crystal in terms of the quantum numbers n_k associated with each vibrational wave as

$$E = \sum_k n_k \hbar\omega(k) \qquad (10.8)$$

Using Equation 10.7 we can write the total momentum carried by all the vibrational waves as

$$P = \sum_k n_k \hbar k \qquad (10.9)$$

We now want to introduce another way of interpreting equations 10.8 and 10.9. At first sight the reinterpretation will probably seem foolishly contrived and artificial, but we will see that it has a fundamental significance that extends far beyond the field of solid state physics. Let us imagine a system of particles we will call *phonons.* The phonons are placed in a system in which the single-particle quantum states are assumed to have the following structure. Each quantum state is labelled by a quantum number k. The quantum state, ϕ_k, is an eigenstate of momentum with eigenvalue $p = \hbar k$, and an energy eigenstate of energy $E_k = \hbar\omega(k)$. The phonons are Bose-Einstein particles which means that any number of phonons can simultaneously occupy the same quantum state. An energy eigenstate of the whole system is defined by specifying how many phonons occupy each single-particle state ϕ_k. If, for each value of k, the quantum state ϕ_k contains n_k phonons (where n_k must naturally be a nonnegative integer) then the total energy and momentum of the complete system of phonons is given by formulas that are identical with equations 10.8 and 10.9. It would be understandable if the reader were very unimpressed by this strained analogy between the actual system of lattice vibrational waves and our fictitious system of phonons. However, a comparison of this analysis with that given in the chapter

---

Theoretically one could measure the ground state energy of a crystal by measuring its mass on a scale and then subtracting that number from the sum of the masses of the constituent particles and dividing by c^2. But the mass change due to chemical bonds is so small that it is entirely undetectable by present experimental methods.

on elementary particle physics, will make it clear that these hypothetical phonon particles bear the same relationship to the vibrational waves in a lattice that photon particles do to electromagnetic waves or that electrons do to electron field waves. According to the present view of quantum theory every type of particle is associated with the waves that can propagate in some corresponding field. This is not intended to convince the reader that he or she should picture phonons as little hard balls but rather that photons and electrons should not be pictured as little hard balls either. They are quantized waves and only get the characteristics of classical little hard balls when they are formed into localized wave packets.

In a three-dimensional crystal the quantum states of a phonon are described by a three-dimensional wave vector **k** and a polarization index, n = 1, 2, 3. The energy and momentum of a phonon in state (**k**,n) is

$$E = \hbar\omega_n(\mathbf{k})$$

and

$$\mathbf{p} = \hbar\mathbf{k}$$

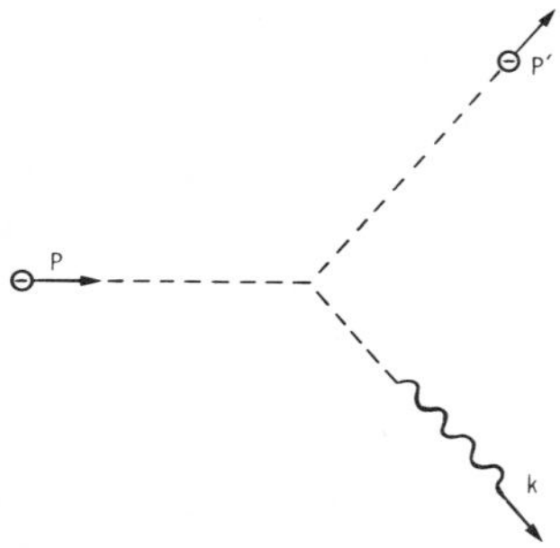

FIGURE 10.31
In real metals there exists an important process in which a conduction electron emits a phonon and thereby changes its energy and momentum.

10.13 THE PHONON GAS

In discussing the thermal properties of a crystal, such as its heat capacity and thermal conductivity, it is very convenient to picture the system of quantized lattice vibrations as a *phonon gas*. To a good approximation the phonons do not interact with one another, and therefore the phonon gas is almost an ideal gas. However there is one major difference between an ordinary gas and the phonon gas. The number of phonons is not a conserved quantity. Phonons can be created or destroyed in a number of ways. They can be generated or absorbed at the surface of the crystal due to the interaction of the crystal with its environment. That is, the particles in the surrounding air or the supporting table top can exchange energy with the crystal by producing or absorbing phonons. Also energy can be exchanged between the lattice vibrations and the electrons within the crystal. (we will discuss the possible quantum states of electrons within crystals in a later section.) This is of particular importance in metals where there is high density of mobile conduction electrons. The basic process is a transition in which an electron goes from a momentum state **p** to a momentum state **p**′ and simulataneously emits or absorbs a phonon.

10.14 LATTICE INTERNAL ENERGY AND HEAT CAPACITY

If the crystal is maintained in a constant temperature environment the total number of phonons and their distribution among the available phonon quantum states will reach some equilibrium value which depends only upon the temperature and the phonon despersion relation $\omega_n(\mathbf{k})$. In the chapter on quantum statistical mechanics we will show that in a crystal at temperature T the average number of phonons in the state $(\mathbf{k},n)$ is

$$\bar{N}(\mathbf{k},n) = \frac{1}{e^{E/k_B T} - 1} \tag{10.10}$$

where $$E = \hbar\omega_n(\mathbf{k})$$

The total energy of the phonon gas at temperature T can be obtained by multiplying the number of phonons in a particular quantum state by the energy of that quantum state and summing over all the quantum states in the Brillouin zone.

$$U_{ph}(T) = \sum_{\mathbf{k},n} \bar{N}(\mathbf{k},n)\,\hbar\omega_n(\mathbf{k}) \tag{10.11}$$

QUESTION: In Equation 10.3 the dispersion relation for a one-dimensional lattice with nearest-neighbor interactions is given as $\omega = \omega_{max} \sin|ka/2|$. What is the internal energy of such a lattice at 300 K (about room temperature) if the lattice spacing, a, is 1 Å, the maximum angular frequency, ω_{max}, is 10^{13} Hz, and the total number of particles is one mole? $(N = 6 \times 10^{23})$

ANSWER: The wave vector k must lie within the Brillouin zone $-\frac{\pi}{a} < k < \frac{\pi}{a}$. There are N normal mode wave vectors, uniformly spaced within the Brillouin zone. Therefore the number of allowed vectors within the interval from k to k+dk is equal to dk times N divided by the total length of the Brillouin zone.

$$\text{No. of normal modes in the interval dk} = \frac{Na}{2\pi}\,dk$$

The sum given in Equation 10.11 can be expressed as an integral over the Brillouin zone.

$$U_{ph}(T) = \frac{Na}{2\pi} \int_{-\frac{\pi}{a}}^{\frac{\pi}{a}} \bar{N}(k)\,\hbar\omega(k)\,dk$$

where $\bar{N}(k) = (e^{\hbar\omega(k)/k_B T} - 1)^{-1}$ and $\omega(k) = \omega_{max} \sin|ka/2|$.

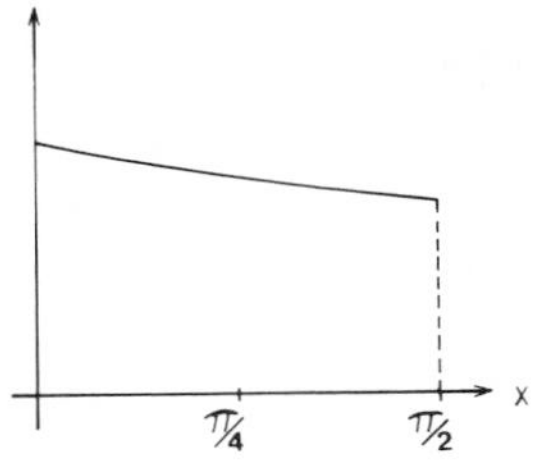

FIGURE 10.32
The function $\sin x/(e^{A \sin x} - 1)$ is quite smooth and is therefore easily integrated numerically.

Introducing the constant $A = \hbar\omega_{max}/k_B T = .26$ and the variable $x = ka/2$ we can write $U_{ph}(T)$ as

$$U_{ph}(T) = \frac{2N\hbar\omega_{max}}{\pi} \int_0^{\pi/2} \frac{\sin x\, dx}{e^{A\sin x} - 1}$$

The integral cannot be evaluated analytically. However it can easily be computed numerically using Simpson's rule. The result is

$$\int_0^{\pi/2} \frac{\sin x\, dx}{e^{A\sin x} - 1} = 5.56$$

This gives

$$U_{ph} = 2232\,J$$

For the same system a purely classical calculation of the thermal energy at temperature T would use the formula $U_{class} = N k_B T$. It would give $U_{class} = 2484\,J$ Thus, even though $k_B T \approx 4\hbar\omega_{max}$ the classical result is significantly different from the quantum mechanical one.

It is instructive to repeat the description of the calculation of the vibrational internal energy in the language of quantized normal mode vibrations. In doing so it becomes clear that the only real difference between the particle description, employing the concept of a phonon gas, and the wave description, in which one treats each normal mode as a quantized harmonic oscillator, is a difference in vocabulary. Both approaches lead to the same equations and to exactly the same predictions regarding physical measurements.

Each normal mode of vibration acts like an independent harmonic oscillator of angular frequency $\omega(\mathbf{k})$. At a temperature of absolute zero each of these harmonic oscillators will be found in its ground state. If we maintain the convention of measuring the internal energy from the ground state energy of the lattice the internal energy at $T = 0$ will be zero. At any positive temperature there will be a probability $P_n(T)$ of finding the oscillator in its nth energy state. The average energy of the oscillator (measured from its ground state energy) would be given by multiplying the energy of the nth quantum state (namely $n\hbar\omega$) by P_n and summing over all the possible quantum states. In Chapter 11 it is shown that the average energy for the normal mode of angular frequency ω is

$$\overline{E}_\omega = \sum_{n=0}^{\infty} P_n(T)\, n\hbar\omega = \frac{\hbar\omega}{e^{\hbar\omega/k_B T} - 1} \tag{10.12}$$

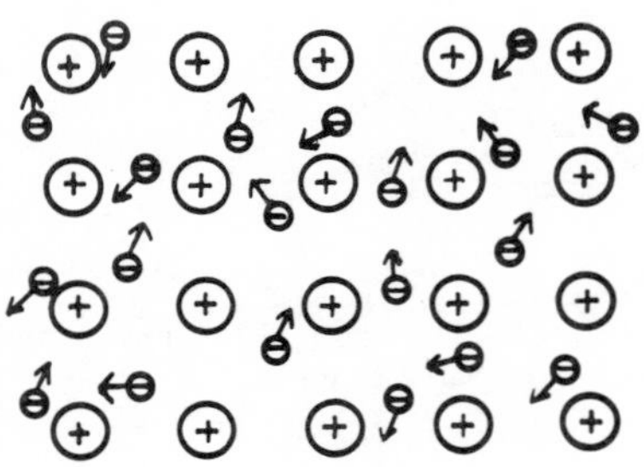

FIGURE 10.33

A metal can be pictured as a rigid lattice of positive ions permeated by a fluid of conduction electrons. Averaged over a large volume of space the total electric charge is zero.

If we take the sum of $\overline{E}_\omega$ over all the normal mode oscillators we obtain a form for the total energy that is identical to Equations 10.10 and 10.11. A more compact way of expressing the internal energy is to employ the normal mode distribution function. Since Equation 10.12 gives the average energy of an oscillator of angular frequency ω and the number of such oscillators in the frequency range $d\omega$ is $D(\omega)\,d\omega$ it is clear that the total energy of the assembly of oscillators at temperature T is

$$U(T) = \int_0^\infty D(\omega) \frac{\hbar\omega}{e^{\hbar\omega/k_B T} - 1} d\omega \tag{10.13}$$

10.15. ELECTRONS IN SOLIDS

Up to this point we have pictured a solid as a lattice of atoms without considering the question of whether the internal electronic structure of the atoms is affected by their being placed in the crystal. In some cases it is perfectly adequate to treat the atoms as if they were structureless particles with only three degrees of freedom each. It is particularly good approximation for the Van der Waals solids made up of noble gas atoms. The electronic structure of these weakly interacting atoms is affected very little by the presence of nearby atoms. However in considering conductors and semiconductors where electrical conductivity is an important property one cannot possibly avoid introducing the electronic degrees of freedom. One may then picture the solid as being composed of a collection of bare atomic nuclei surrounded by a sufficient number of electrons to make the whole assemblage electrically neutral. The nuclei are assumed to be fixed at the lattice sites. That is, while calculating the electronic states we ignore the zero point vibrations of the nuclei about their equilibrium positions.

Since any electron interacts electrostatically with all the nuclei and with every other electron it seems as if the problem of solving the Schrodinger equation for the electronic quantum states in such a system is hopelessly difficult. If the nuclei were located at random positions rather than at the regularly spaced lattice sites the problem of calculating the electronic states really would be completely intractable. However the lattice periodicity creates a certain translational symmetry in the system. Taking advantage of this symmetry it will be possible to give a qualitative description of the set of electronic energy eigenstates in a crystal. Although we will never look at the problem of

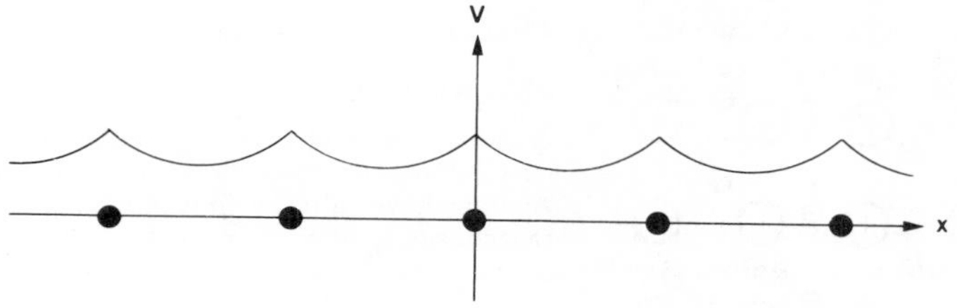

FIGURE 10.34 A potential, V(x), that is spatially periodic. $V(x) = V(x+a)$.

calculating the energy spectrum in detail it should be mentioned that, by making full use of the translational symmetry of the crystal, it is possible to reduce that problem from a hopelessly difficult one to a merely difficult but tractable one. We will begin by describing the solutions of the Schroedinger equation for a particle in a one-dimensional periodic potential. The mathematical proofs of the statements we will make are rather intricate and will therefore be omitted.

10.16. ENERGY BAND STRUCTURE IN ONE DIMENSIONAL CRYSTALS

We picture a one-dimensional crystal as a system of identical, positively charged nuclei distributed along a line with fixed equal spacing. A sufficient number of electrons are distributed throughout the one-dimensional system to make it electrically neutral. The electrons, whose spin variables we will temporarily ignore, are free to move along the line. The spacing between adjacent nuclei is a. An electron located at some position x within the system will, by virtue of its electrostatic interactions with the nuclei and all other electrons, have a potential energy, V(x), that depends on its location. Although, in the absence of any knowledge of the detailed distribution of the electrons within the system, we cannot predict the potential function V(x) in detail, there is one obvious characteristic that V(x) must have due to the basic periodicity of the system. That is that the potential energy of an electron at any location with respect to one nucleus must be the same as the potential energy of the electron at an equivalent location with respect to any other nucleus. Mathematically this is expressed by the equation

$$V(x) = V(x + n a) \tag{10.14}$$

where n is any integer. Such a potential is said to be periodic with spatial period a. The question we will answer in this section is; what is the basic structure of the energy spectrum of a particle in a periodic potential?

The energy eigenstates can be indexed by two quantum numbers, n and k. The quantum number n is a positive integer called the *band index.*

$$n = 1, 2, 3, \cdots$$

The quantum number k is related to a quantity p called the *crystal momentum* of the electron by a de Broglie relation

$$p = \hbar k$$

For a one-dimensional crystal with N lattice sites the wave vector k must be one of the N equally spaced allowed wave vectors within the first Brillouin zone. That is, it must lie within the interval

$$-\frac{\pi}{a} < k < \frac{\pi}{a}$$

Thus the crystal momentum of the electrons has the same set of allowed values as the momentum of a phonon in the same crystal.* For most purposes the crystal

Although the possible values of an electron's crystal momentum are the same as a phonon's momentum and both quantities have been written as $\hbar k$, it should be kept in mind that these are physically distinct quantities. One refers to electrons and the other to phonons. They have the same mathematical structure only because both the electron system and the atomic lattice are periodic systems with the same spatial period.

momentum can be treated as if it were simply the momentum of the electron in the crystal. It would take us too deeply into solid state theory to describe the subtle differences between the two concepts. For each value of n and k there is an energy eigenstate, $\phi_{nk}(x)$, that is a solution of the Schrodinger energy equation with an energy eigenvalue E_{nk}. If, for some particular band index n, we plot the value of E_{nk} as a function of k, we obtain a curve defined within the Brillouin zone.

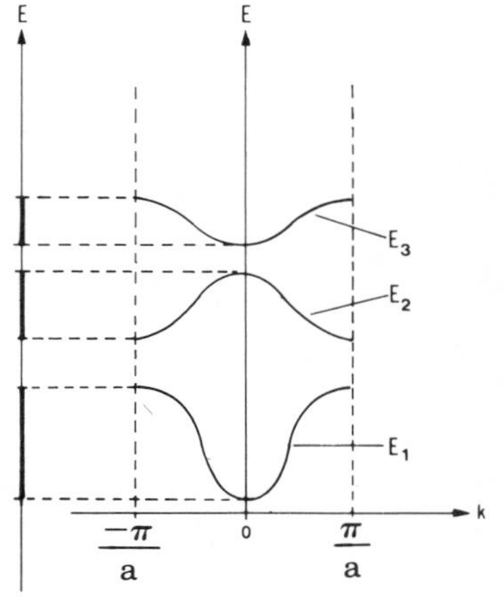

FIGURE 10.35

Electronic energy bands in a one-dimensional crystal. The first three energy bands, $E_1(k)$, $E_2(k)$, and $E_3(k)$ are plotted as functions of k (within the Brillouin zone) on the right. The range of possible energy values is shown by dark lines on the E-axis to the left. Between the allowed energy ranges are energy gaps that are ranges of energy in which no allowed energy values for electrons exist.

For given n the set of possible energy values that have n as a band index is called the nth *energy band.* It consists of N energy values that are so closely spaced as to effectively produce a finite continuous range of allowed energy values. The energy bands associated with adjacent band indices may overlap or they may be separated. If they are separated we say that the range of energies that separates the neighboring energy bands is a *forbidden zone* or an *energy gap* meaning simply that there do not exist any electronic quantum states with energies within the gap. The energy spectrum of electrons in a one-dimensional crystal therefore consists of a sequence of allowed energy bands separated by energy gaps. A given energy band will contain N energy states if it is the projection onto the energy axis of those states with a single band index. If the energy ranges of states with neighboring band indices overlap then the allowed energy band will contain 2N, 3N, or some other small integer times N discrete energy states. The most important thing to notice is that in all cases the number of quantum states within an allowed energy band is some simple multiple of the number of lattice sites in the crystal.

10.17. BAND STRUCTURE IN TWO AND THREE DIMENSIONS

When going from one to two or three dimensions the only way we need to modify what we have said about the energy spectrum is to make the wave vector, **k**, a two- or three-dimensional vector. **k** is restricted to the two- or three-dimensional Brillouin zone appropriate to the crystal structure. The energy eigenstates $\phi_{nk}(\mathbf{x})$ are functions in a two- or three-dimensional space. Their corresponding eigenvalues E_{nk} can, in the two-dimensional case, be pictured as a

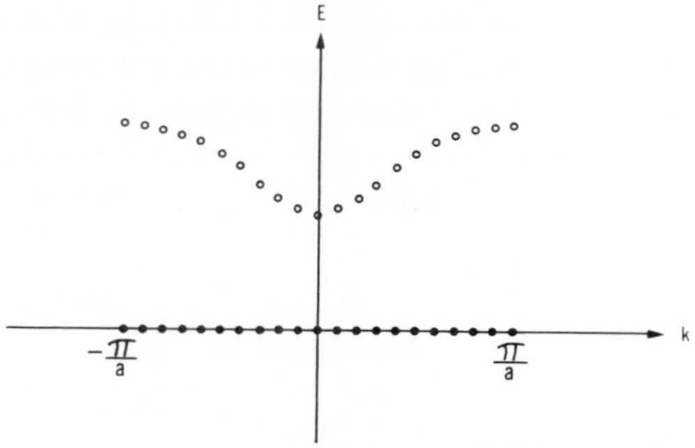

FIGURE 10.36 Each energy band actually contains N closely-spaced energy eigenstates.

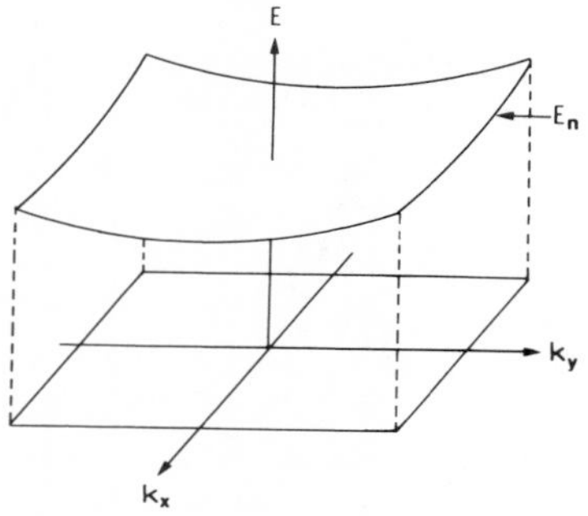

FIGURE 10.37

In a two-dimensional crystal, each energy band, $E_n(\mathbf{k})$, defines an energy surface over the Brillouin zone.

collection of energy surfaces, one surface for each value of n. Although the three-dimensional case is essentially the same it is not possible to give it a simple pictorial representation.

10.18. INSULATORS AND CONDUCTORS

At each lattice site in a crystal there is a nucleus of electric charge Ze. In order for the crystal to have zero net charge it must therefore contain Z electrons per lattice site or a total of ZN electrons. At zero temperature the crystal will be found in its ground state. That state is the lowest energy state consistent with the laws of quantum mechanics. Since electrons are fermions, one of the most important laws determining the ground state energy is the Pauli Exclusion Principle which states that no more than two electrons (with opposed spins) can occupy any one of the available electronic quantum states. The minimum energy is obtained by having the ZN electrons occupy the lowest $ZN/2$ states with 2 electrons per quantum state. Since the total number of electrons and the number of quantum states per allowed energy band are both integer multiples of the number of lattice sites, N, it is not uncommon to have exactly enough electrons to completely fill the states in a few allowed energy bands, leaving the quantum states of higher energy bands completely empty. For reasons we will give shortly, this situation yields a crystalline solid that is an electrical insulator rather than a conductor. An electrical conductor results when the borderline between the filled and empty quantum states occurs somewhere within an allowed energy range.

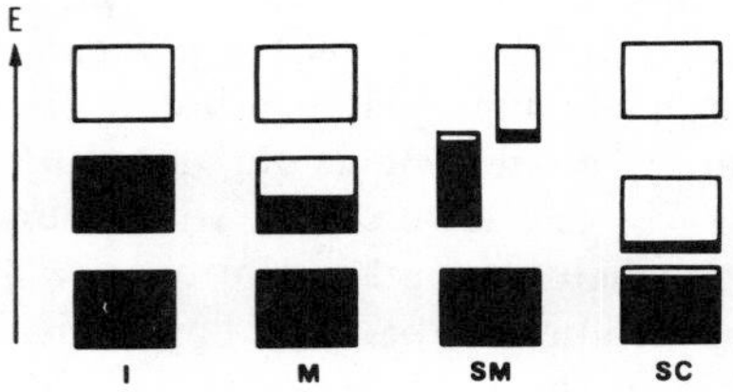

FIGURE 10.38
The filling of energy bands in: (I) an insulator, (M) a metal, (SM) a semimetal (for most purposes equivalent to a metal), (SC) a semiconductor.

To clearly understand why these two different patterns of quantum state occupation should lead to the two corresponding modes of electrical behavior let us consider the effect of placing a simple one-dimensional crystal in an electrical field. We first consider the case that we have said should correspond to a conductor. We assume that Z = 3 and that the two lowest energy bands, each of which contains N quantum states, are nonoverlapping. Without the electric field present the electrons would fill all the states of the lowest band (two electrons per state) and half the states of the second band. Within the second band the states would be occupied symmetrically about k = 0. That is, if state $\phi_{2,k}(x)$ were occupied then state $\phi_{2,-k}(x)$ would also be occupied. Now let us consider what happens to the system when an electric field is turned on. If we assume that the field points to the left then the negatively charged electrons will experience a force to the right.

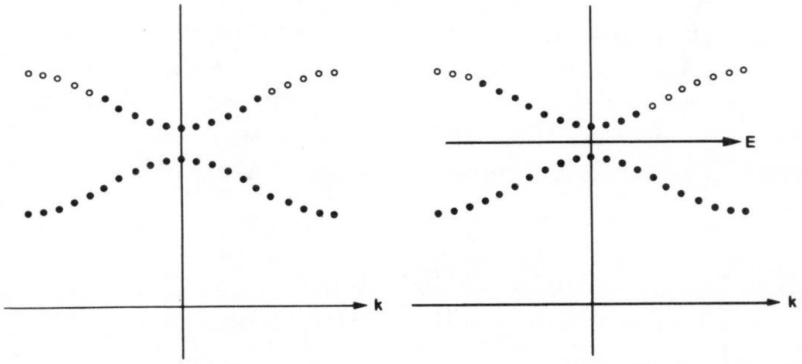

FIGURE 10.39 The effect of an electric field on the pattern of occupation in a one-dimensional metal.

In its ground state the net momentum and the average velocity of the electron system as a whole are zero. The electron system will begin to pick up energy and momentum from the electric field. From a quantum mechanical viewpoint this is accomplished by the electrons making transitions to states whose k vectors are to the right of the ones that they originally occupied. In other words the distrubution of occupied states within the second band will shift to the right so that those electrons will have a net momentum and give rise to a net electric current. Notice however that although an electric field is present the electrons in those states with n = 1 cannot make transitions to other k states because all k states in that band are completely filled. One could imagine an electron making a transition from a filled state in the lower band to an empty state in the upper band but for ordinary electric field strengths the energy gap between the two bands is simply too large to make such transitions energetically feasible. Thus the lower filled band plays no part in the electrical conductivity of the system. The upper band would be called the *conduction band* while the completely filled lower band is called a *valence band*.

From this discussion it is clear that if the number of electrons in a crystal is just sufficient to completely fill a few bands of quantum states so that there is no partially filled conduction band then the crystal will not be an electrical conductor. It will be an insulator.

10.19. SEMICONDUCTORS

In describing the pattern of filled and empty quantum states in an insulator we have made two important assumptions. The first assumption was that we are dealing with a perfect crystal, free from structural defects (such as misalignment of some crystal planes or vacant lattice sites) and containing no chemical impurities. The second assumption was that the crystal was in its quantum mechanical ground state. That second assumption would be strictly true only at an absolute temperature of zero. If either of these assumptions is not valid the material may exhibit an electrical conductivity that is intermediate between that of an insulator and that of a conductor. The crystal is then said to be a *semiconductor*. Technologically important semiconductors are usually produced by the addition of chemical impurities to an otherwise nearly perfect crystal. We will discuss those in greater detail later. First we will consider the effect of elevated temperatures on a crystal that, in its ground state, is an insulator. At any nonzero temperature some of the electrons will have sufficient energy to occupy

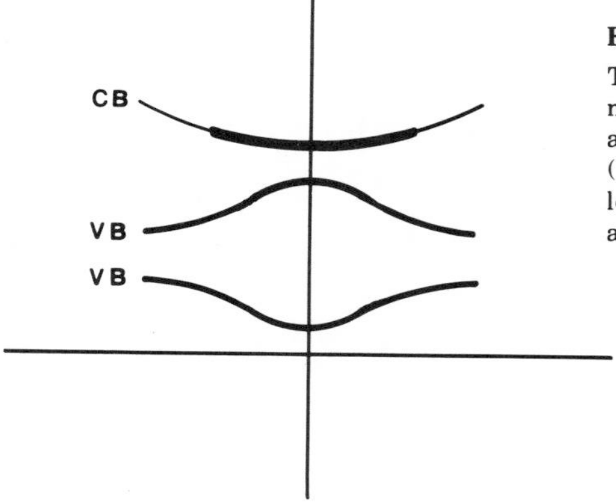

FIGURE 10.40

The filling of energy states in a metal in which the number of electrons is five times the number of atoms, N. The states of the two lower valence bands (VB) are each filled with two electrons. The N/2 lowest energy states of the conduction band (CB) are also filled.

states in the normally empty conduction band. The quantity that determines how many electrons are excited into the conduction band is the ratio of $k_B T$, the typical thermal energy of a particle, to E_g, the energy gap between the valence band and the conduction band. In the chapter on quantum statistical mechanics it will be shown that the number of electrons in the conduction band is proportional to $\exp(-E_g/2k_B T)$. Since the electrical conductivity is proportional to the number of electrons in the conduction band we would predict that K(T), the conductivity at temperature T, would satisfy the equation

$$K(T) = K_o e^{-E_g/2k_B T}$$

or

$$ln\, K(T) = ln\, K_o - E_g/2k_B T \qquad (10.15)$$

Figure 10.41 shows that this prediction is well satisfied for the crystal germanium.

In contrast to our theoretical models which are perfect crystals, any real crystal has a finite density of chemical impurities. As we will shortly show, those chemical impurities lead to a finite conductivity even at very low temperature. That contribution to the electrical conductivity due to chemical impurities is called *impurity conductivity* while that part of the conductivity due to thermally excited electrons is called *intrinsic conductivity*. It is only the intrinsic part of the conductivity that satisfies Equation 10.15. As the temperature is lowered the intrinsic conductivity goes to zero and therefore the impurity contribution to the conductivity becomes dominant, leading to large deviations from the predictions of Equation 10.15.

10.20. ELECTRICAL CONDUCTION BY HOLES

An empty electronic state in the normally filled valence band is referred to as a *hole*. If, within a certain volume ΔV of an insulating crystal, there are no electrons in the conduction band and no holes in the valence band then that region will be electrically neutral. The positive charges of the nuclei will exactly cancel the negative charges of the electrons in the valence band states. However, if within ΔV there are N_e electrons in the conduction band and N_h holes in the valence band then the net charge contained in ΔV will be

$$Q = -e N_e + e N_h$$

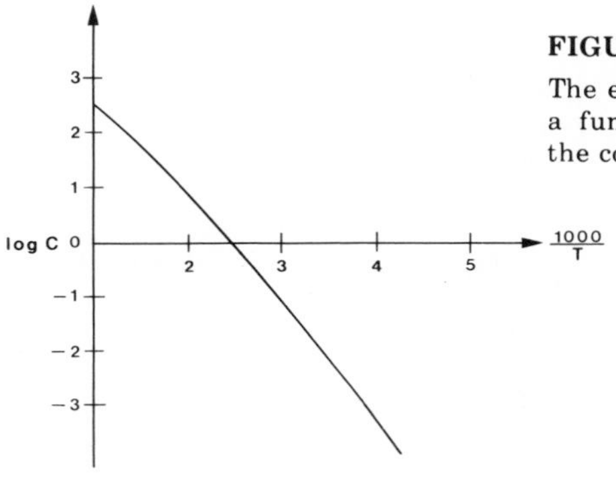

FIGURE 10.41

The electrical conductivity, C, of pure germanium as a function of x = 1000/T. The vertical axis gives the common logrithm of C in ohm $^{-1}$cm^{-1}.

Each hole in the valence band makes a contribution of $+e$ to the net charge in the region while each electron in the conduction band makes a contribution of $-e$. In an intrinsic semiconductor, in which the conduction electrons and holes are created by thermal excitation N_e must be equal to N_h since the excitation of an electron from the valence to the conduction band always produces one hole and one conduction electron. In a crystal containing impurities it is possible to have unequal members of conduction electrons and holes.

The presence of holes makes a contribution not only to the static electrical charge density, but if an electric field is imposed on the crystal the spatial location of the empty state (that is, the location of the hole) will drift in the direction of the electric field. Since the hole is positively charged its movement will constitute an electric current in the direction of **E**. This hole current is in addition to the current created by the moving electrons in the conduction band.

The crystal momentum and energy can also be expressed in terms of the density of holes and conduction electrons. If there is a hole in the valence band quantum state $\phi_{n,\mathbf{k}}$ (that is, if $\phi_{n,\mathbf{k}}$ is unoccupied) then the crystal momentum of the occuped state $\phi_{n,-\mathbf{k}}$ is not balanced by an electron in the state that is empty.

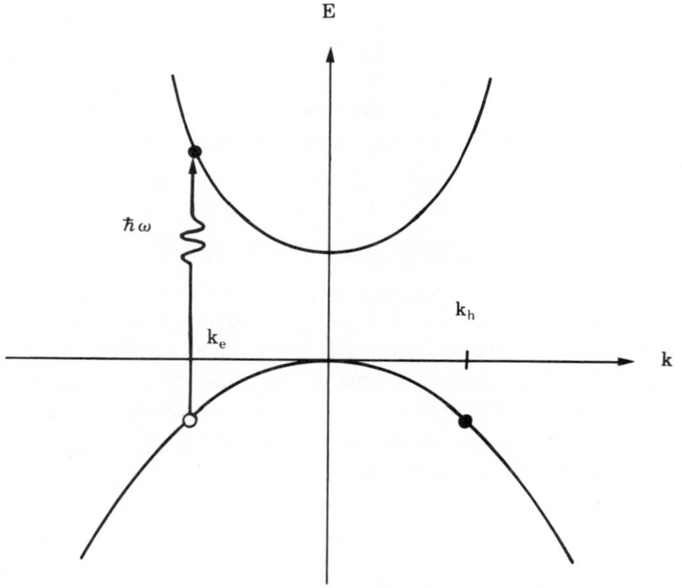

FIGURE 10.42 The movement of an electron from a valence band state to a conduction band state produces one conduction electron and one hole. The total valence band now contains an amount of momentum equal to the negative of the momentum of the empty state.

Thus the crystal will be left with a net momentum of $-\hbar\mathbf{k}$. If we want to assign that momentum to the hole then we must use the rule that an empty electronic state of wave number $\mathbf{k}$ corresponds to a hole of wave number $-\mathbf{k}$. If all this seems like a rather artificial construction its reasonableness might become much clearer if we consider the following mechanical analogue. We picture a fishtank with a single small bubble in it. We can predict the motion of the bubble in two different but equivalent ways. The first method we might use is to concentrate our attention on the water rather than the bubble. Using the equations of fluid dynamics we could calculate the flow pattern of the water. If we did our calculations correctly we would find that there was always an approximately spherical region that remained empty of water (i.e., the bubble). The location of that region as a function of time would give us the trajectory of the bubble. An alternative and clearly more reasonable way of determining the trajectory of the bubble is to first determine the general laws of "bubble dynamics". In comparison with the laws of particle dynamics the laws of bubble dynamics have certain peculiar features. If the fishtank is placed in a gravitational field $\mathbf{g}$ the bubble will begin the accelerate in a direction opposite to the direction of $\mathbf{g}$. Thus the effective gravitational mass of the bubble would be negative just as the effective electric charge of a hole is opposite to the electric charge of the real electrons. If we could find some way of applying a force $\mathbf{F}$ to the gas particles within the bubble without simultaneously applying forces to the water molecules surrounding it (admittedly a difficult feat) we could, by determining that ratio of $\mathbf{F}$ to the bubble acceleration, determine an effective dynamical mass for the bubble. That is, a parameter M to be used in the equation $\mathbf{F} = M\mathbf{a}$. We should not get any deeper into the interesting but not directly relevant topic of bubble dynamics. It was merely intended to clarify our objectives and our point of view as we consider the dynamics of holes in solids. In the very common situation where the density of holes is low it is much simpler to focus our attention on the few holes, treating them as particles with their peculiar set of dynamical laws, rather than to try to follow the dynamics of the billions of electrons in the filled valence states. Treated as a particle, the hole will have some effective dynamical mass m_h^* that will generally be different from the effective dynamical mass m_e^* of the conduction electrons. In order to see how m_h^* and m_e^* might be defined let us consider the energy band diagram in a typical one-dimensional insulator. (See figure 10.44.) We have assumed that there are two filled valence bands and that the upper one is separated from the lowest conduction band by an energy gap E_g. The diagram shows the case in which a single electron has been excited from a valence band state to a conduction band state. Thus, in the particle and hole language we would say that there is one electron in a state with energy E_e and momentum p_e and one hole with energy E_h and momentum p_h. We have already discussed the reason for drawing the hole momentum axis (the p_h axis) in a

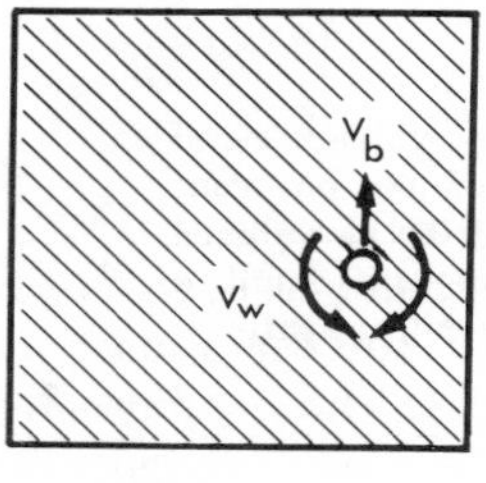

FIGURE 10.43

One can either focus on the flow pattern of the water, described by a vector field $\mathbf{v}_w(\mathbf{r})$, or else consider only the velocity of the bubble, described by the single vector $\mathbf{v}_b$. It is obviously simpler to do the latter.

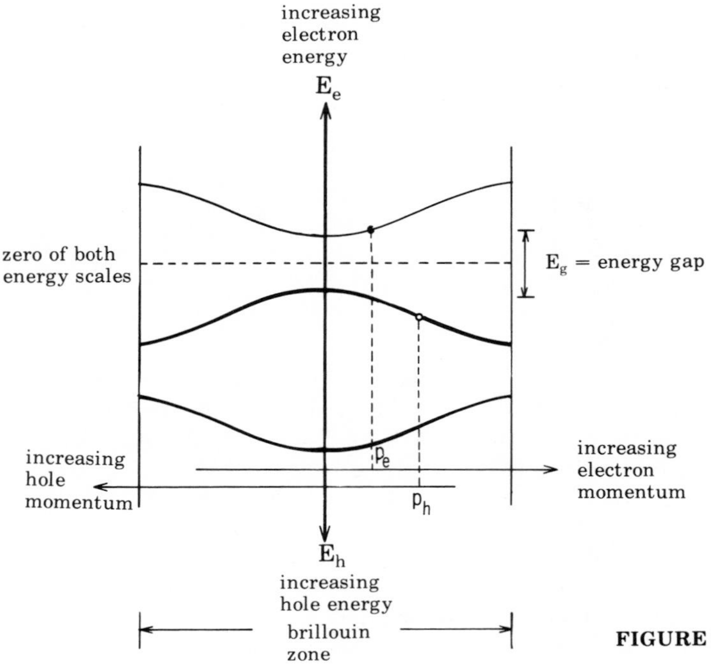

FIGURE 10.44

The momentum and energy axes of holes point oppositely to those of electrons in an energy band diagram. This is natural, since it takes more energy to create a hole that is lower down on the diagram, and therefore a lowdown hole has more energy.

direction opposite to the electron momentum axis. In order to see why the two energy scales must also be drawn in opposite directions we must ask ourselves what would be the total energy of the crystal in the one electron-one hole configuration shown on the diagram.

Let us call the ground state energy of the crystal E_{ground}. (E_{ground} is the energy of the crystal when all the valence band states are filled and all the conduction band states are empty.) In the diagram we have followed the traditional practice of choosing the arbitrary zero of the energy scale at the midpoint of the energy gap between the valence band and the conduction band. In order to bring the crystal from its ground state to the one electron-one hole state shown we would have to supply an amount of energy to the electron being excited equal to the sum of the positive quantities E_h and E_e shown in the figure. Thus the total energy of the state shown would be

$$E = E_{ground} + E_e + E_h$$

This energy is smallest if the occuped state in the conduction band is near the bottom of the band and the empty state in the valence band is near the top of that band. At ordinary temperatures this is usually the case for almost all the thermally excited electrons and holes. Near the bottom of the conduction band and the top of the valence band both function $E_e(p_e)$ and $E_h(p_h)$ can be approximated by parabolas. That is, we can write

$$E_e(p_e) = \tfrac{1}{2} E_g + \frac{1}{2 m_e^*} p_e^2 \qquad (10.16a)$$

and

$$E_h(p_h) = \tfrac{1}{2} E_g + \frac{1}{2 m_h^*} p_h^2 \qquad (10.16b)$$

where E_g is the gap energy. The quantities m_e^* and m_h^* are called the effective masses of electrons and holes respectively. Since they are determined by the details of the energy band structure there is no reason to expect m_e^* to agree with the mass of a free electron nor to expect m_e^* to be equal to m_h^*. When considering the important question of how electrons and holes in a crystal will react to weak electromagnetic fields one is concerned almost entirely with calculating the probability that an electron will make a transition from one state to another near the bottom of the conduction band or that a hole will make a transition between two states near the top of the valence band. In both cases one can ignore the constant term in the formula for E_e and E_h because the same constant appears in both the initial and the final state and therefore does not affect the energy change in the transitions. When the constant terms can be neglected the energy spectrum of the electrons and holes in the crystal is the same as the spectrum of a system containing two varieties of free particles of masses m_e^* and m_h^*.

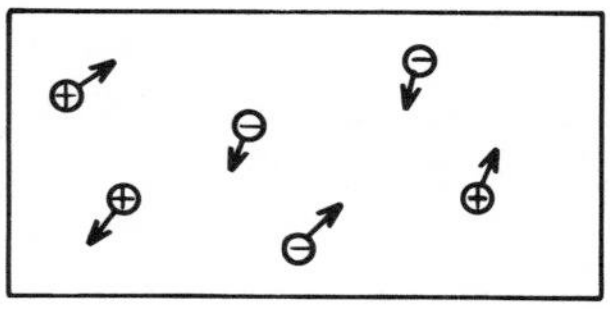

FIGURE 10.45

The combination of occupied states in the conduction band and empty states in the valence band leads to effects similar to a dilute gas of positive holes of mass m_h^* and negative electrons of mass m_e^*.

10.21. P TYPE AND N TYPE SEMICONDUCTORS

The semiconductors that are widely used in electronic circuitry do not rely on thermal excitation to produce the mobile electrons and holes needed for electrical conductivity. Instead, by incorporating small amounts of specific chemical impurities into the crystal structure as substitutions for the ordinary lattice atoms it is possible to produce controlled amounts of either mobile electrons or holes in the crystal. Semiconductors produced in this way are called *doped semiconductors*. They come in two varieties, called p-type (for positive type) and n-type (for negative type). In a p-type semiconductor, the mobile charges that are responsible for electrical conductivity in the material are positively charged holes. This is not to say that a p-type semiconductor has a net positive charge. Compensating the electrical charge of the holes are an equal number of negatively charged impurity atoms that are fixed in position and therefore make no contribution to the electrical conductivity of the crystal. In an n-type semiconductor the situation is exactly reversed. The impurities are in the form of positively charged ions that are immobile substitutional replacements for some of the usual atoms of the crystal. There is an equal density of mobile electrons within electron states of the conduction band. In the n-type semiconductor it is these conduction band electrons that account for the crystal's electrical conductivity.

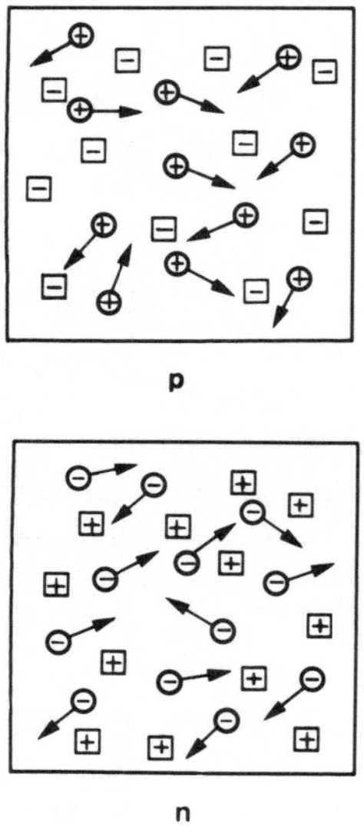

FIGURE 10.46

p-type semiconductors contain fixed negative ions and mobile positive holes. n-type semiconductors contain fixed positive ions and mobile negative electrons.

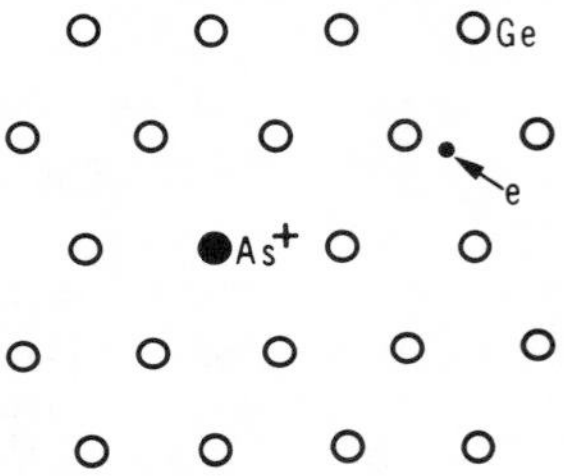

FIGURE 10.47

An arsenic ion in a germanium crystal. The extra electron can move throughout the crystal.

10.22. DOPED GERMANIUM SEMICONDUCTORS

Crystalline germanium (Ge) doped with arsenic (As) is a typical and technologically important example of an n-type semiconductor. The germanium atom has four electrons in its outermost shell. Those four electrons have the configuration $4s^2 4p^2$ which makes germanium a chemical analogue of carbon. (Carbon has the outer shell configuration $2s^2 2p^2$.) Pure germanium crystallizes with the same lattice geometry as diamond. In the diamond lattice each germanium atom has four nearest neighbors that are situated, with respect to the atom, at the vertices of a regular tetrahedron. (See Figure 10.2.) Each of the four outer electrons participate in a covalent bond with one of the neighboring atoms. If one replaces the germanium atom we are considering by an arsenic atom the geometry of the lattice will change very little. However the arsenic atom, which is directly to the right of germanium in the periodic table, has five electrons in its outer shell rather than four. Thus one of the electrons will be left without a bonding partner since all the outer electrons of all the germanium atoms are already involved in covalent bonds. The fact that covalent bonds are strong bonds means that electrons that are in covalent bonding states are in low energy states in comparison to electrons not involved in covalent bonds. Thus the unbonded electron associated with the arsenic atom will be in a much higher energy state than the other electrons in the crystal. It will therefore take much less energy to excite that electron into one of the empty conduction band states. The conduction band states are spread over the whole crystal, and so once the electron is in one of the conduction band states it will not remain in the vicinity of the impurity atom. That is, moving the extra electron into the conduction band is the same as ionizing the impurity atom. We can get a rough estimate of how much energy is needed to ionize the arsenic atom by calculating the binding energy of an electron of mass m_e^* (the effective mass of a conduction electron in the crystal) that is attracted to the positively charged arsenic ion by a Coulomb force $F = e^2/4\pi\epsilon r^2$ where ϵ is the dielectric constant of germanium. For this we need to use the hydrogen ground state energy formula with m_e replaced by m_e^* and $k = 1/4\pi\epsilon_0$ replaced by $1/4\pi\epsilon$. In germanium $\epsilon = 15.8\epsilon_0$ and $m_e^* \approx .1\,m_e$. This gives a binding energy of

$$E = \frac{m_e^* e^4}{2\hbar^2(4\pi\epsilon)^2}$$

$$= \frac{(.1)}{(15.8)^2}\,\frac{m_e\,e^4}{2\hbar^2(4\pi\epsilon_0)^2}$$

$$= \frac{(.1)}{(15.8)^2}(13.6\,\text{eV}) = .005\,\text{eV}$$

which corresponds to less than one percent of the usual energy gap between valence band states and conduction band states in germanium. (in Ge the energy gap is about .7 eV.) Thus the arsenic impurities contribute their electrons to the conduction band very easily. If we set the estimated binding energy, .005 eV equal to $k_B T$ in order to obtain an estimate of the temperature at which the arsenic impurities will be thermally ionized we get an ionization temperature of 63.3 K or $-210\,^\circ$C. At room temperature all of the unpaired electrons from the arsenic atoms can be assumed to be in conduction band states and therefore free to contribute to the electrical conductivity of the crystal. Impurities, such as arsenic, that supply electrons to the conduction band are called *donor impurities.*

As we have already mentioned, arsenic lies directly to the right of germanium in the periodic table of the elements. The element lying directly to the left of germanium is gallium (Ga). If gallium atoms are introduced into an otherwise pure germanium crystal they do not contribute electrons to the conduction band. Instead they absorb electrons from the filled valence band leaving mobile holes which account for the conductivity of Ga-doped Ge. For this reason the gallium atoms are called *acceptor impurities,* and the resulting semiconductor is of the p-type. The outer shell of gallium contains three electrons with the configuration $4s^2 4p$. Therefore, when a gallium atom replaces a germanium atom in the crystal, there must be, in the vicinity of the gallium atom, an unpaired electron on one of the germanium atoms. It takes very little

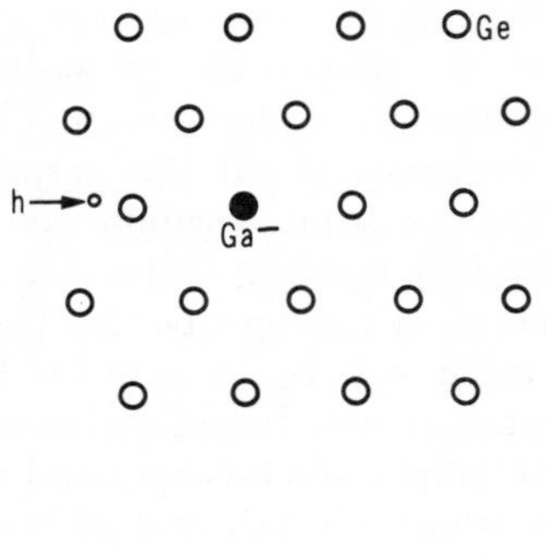

FIGURE 10.48
A gallium ion in a germanium crystal. The mobile hole can move throughout the crystal.

energy (about .01 eV) to transfer this unpaired electron to the gallium atom creating a positive hole and a negatively charged gallium ion. A calculation similar to the one we have done above can show that the positive hole is only weakly bound to the negative ion and at room temperature the holes act as mobile charges.

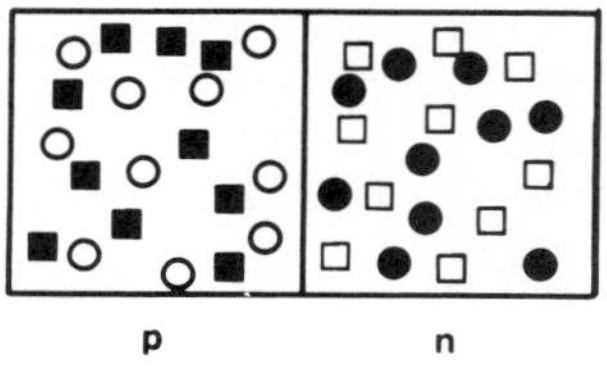

FIGURE 10.49
A pn-junction.

10.23. P N JUNCTIONS

Very interesting and useful things happen when a piece of p type and a piece of n type semiconductor are placed in contact to form a *pn junction* (also called a *pn diode*). Such a junction is shown in Figure 10.49. On the left is a sample of p type material which contains mobile positive holes (represented by white dots) and fixed negative impurity ions (the black squares). To the right of the junction the material is of n type, containing negative mobile charges (the black circles) and positive fixed ions (the white squares). (In general, white means +, black means −, square means fixed, and round means mobile.)

The most important new process that takes place near a junction is *electron-hole annihilation.* If a conduction electron and a hole find themselves in the same locality it is possible for the electron to make a transition from the conduction band state to the empty valence band state that is represented by the hole. The net result is that both the mobile electron and the hole disappear. Since the electron is making a transition from a state above the energy gap to a state below it, an amount of energy somewhat larger than the gap energy E_g is released in the process. This energy can be released either as one or more phonons that would contribute to the internal thermal energy of the crystal or it can be released as a photon. It is the second possibility that is responsible for light emission by the *light emitting diodes* (LED's) that were commonly used in calculator displays (they have now mostly been replaced by liquid crystal displays). Because of the large energy release involved, the electron-hole annihilation process is irreversible at ordinary temperatures.

In most real pn diodes both halves are parts of a single crystal sample and the junction is only the dividing surface between that portion of the crystal in which the impurities are donor atoms and that portion containing acceptor impurities. Therefore the junction presents no physical barrier to the movement of the mobile charges on either side of it. Electrons will tend to diffuse from the n side to the p side and holes in the other direction. But, because of the irreversible annihilation process, in the region near the junction where diffusion might be expected to create a mixture of holes and electrons the two types of mobile charges combine and what is prduced is a *depletion layer* which is narrow region on both sides of the junction that contains very few mobile charges of either type.

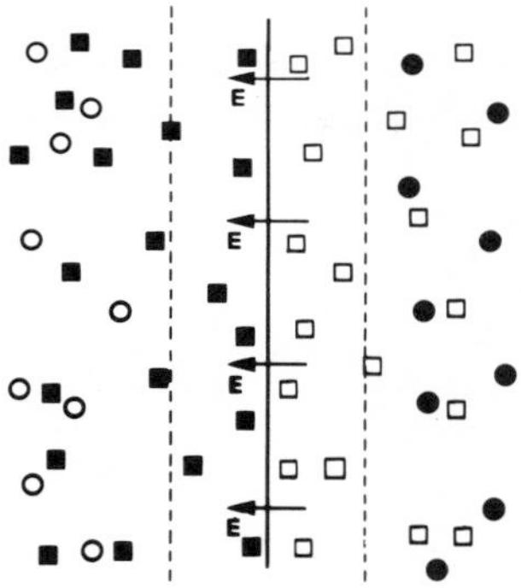

FIGURE 10.50

Due to pair annihilation, the region near the interface is depleted of mobile charge carriers. The uncompensated impurity ions therefore create an electric field normal to the interface.

Since nothing has caused the immobile impurity ions to vanish within the depletion layer those ions will cause that portion of the depletion layer within the p-type material to be negatively charged and that portion on the n side to be positively charged. An electric field will then point from the n side to the p side. At equilibrium that field will be sufficiently strong to prevent further diffusion of holes or electrons across the barrier. (See Figure 10.50.) Because of the electric field between them the n region and the p region will be at different electric potentials with a potential difference that is typically on the order of one volt.

If a battery of sufficient emf is connected to the two ends of the diode with the positive terminal on the p side (This is called *forward biasing* of the junction.) the electrons and holes can be driven against the field in the depletion layer and create a constant current through the diode. The current is carried by holes on the p side and by electrons on the n side. In the neighborhood of the junction the holes flowing in from one side and the electrons flowing in from the other side annihilate, but new electrons constantly enter the material from the battery wire on the n side, and new holes are created at the battery connection on the p side by having electrons leave the crystal and enter the battery wire. Thus the current is sustained as long as the battery is kept connected in the forward bias position. If the battery terminals are reversed quite a different thing happens. Then holes and electrons are drawn away from the junction on both sides and the size of the depletion layer is increased until the voltage drop across the depletion layer is equal to the battery voltage. At that point no further current flows. Thus the pn junction is a device that allows current to flow in one direction but not in the other. For that reason it is useful in rectifier circuits where one wants to convert alternating current to direct current.

10.24. TRANSISTORS

We will close this chapter by describing one of the many ways in which doped semiconductors can be used in devices that amplify or control electrical signals. All such devices as a group are loosely referred to as transistors. The particular device we will consider is the *field effect transistor* or FET. The essentials of an FET are shown in Figure 10.51. A single piece of doped semiconductor is shaped so as to have a narrow flat channel in the center. In the case shown the semiconductor is a silicon semiconductor of p-type but with an obvious modification of the circuit the same result could be obtained with an n-type

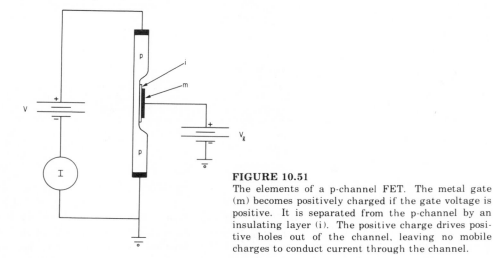

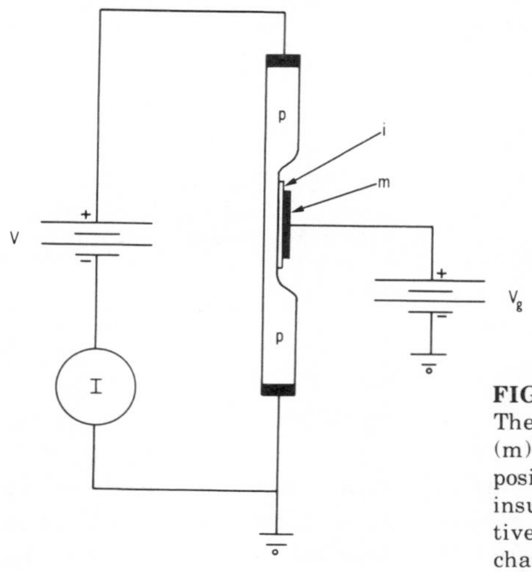

FIGURE 10.51
The elements of a p-channel FET. The metal gate (m) becomes positively charged if the gate voltage is positive. It is separated from the p-channel by an insulating layer (i). The positive charge drives positive holes out of the channel, leaving no mobile charges to conduct current through the channel.

semiconductor. The control voltage V_g is called the gate voltage. It is applied to a flat metal contact (the gate) that is separated from the semiconducting material by an extremely thin but unbroken layer of insulator. The insulating material is usually a surface layer of silicon oxide, produced by exposing the doped silicon semiconductor to an oxygen atmosphere at an appropriate temperature. If V_g is zero then the semiconductor contains a uniform density of mobile holes and will conduct a current, I, in the left-hand circuit that is proportional to the voltage V. If V_g is made positive the gate becomes positively charged. The positive charge on the gate will drive the mobile positive holes out of the channel, producing a negatively charged depletion region within the channel. Since the negative charges within the channel are immobile ions there will be no mobile charge carriers within the channel and the current I will be cut off. To see how amplification is achieved with this device let us assume that V_g is a fluctuating voltage. The power input at the gate will be proportional to the current in the gate circuit. Because of the existence of the insulating layer it is obvious that the current in the gate circuit will ordinarily be small. Thus the input power in the gate circuit is smaller than the output power which is proportional to VI. An increase in signal *power* is the defining characteristic of an amplifier.

SUMMARY

The atoms or molecules in an *amorphus* solid are not arranged in a geometrical lattice. Those in a *crystalline* solid are.

Solids come in five *bonding types.*

BOND TYPE	EXAMPLE	MELTING POINT
Covalent	Diamond	4100 K
Ionic	NaCl	1074 K
Metallic	Na	371 K
Hydrogen	H_2O	273 K
Van der Waals	Ne	24.3 K

A *crystal lattice* is a collection of vectors of the form

$$\mathbf{V} = I\mathbf{a} + J\mathbf{b} + K\mathbf{c} \quad (I, J, K \text{ integers})$$

where **a**, **b**, and **c** are the *primitive lattice vectors.*

A *simple crystal* has a single atom at each lattice site. A *molecular crystal* has a group of atoms at each site.

All crystals have *translational symmetry* in the three lattice directions. Some crystals also have *rotational* or *reflection* symmetries.

A rotational symmetry is called *n-fold* if the lattice is unchanged after a rotation by 360/n degrees.

For two-dimensional crystals, only 2-fold, 4-fold, and 6-fold rotational symmetries are possible.

The set of all possible crystal lattices are grouped into *symmetry classes* according to their rotational and reflection symmetries.

In each symmetry class there may be a few different crystal structures (or *bravais lattices*) that exhibit the particular combination of symmetries.

In the classical theory of lattice vibrations one studies the small amplitude oscillations of a crystal using classical mechanics.

For an infinite one-dimensional crystal with a lattice spacing, a, waves can be propagated with any wave vector in the *Brillouin zone*

$$-\frac{\pi}{a} < k < \frac{\pi}{a}$$

For a finite one-dimensional crystal with N particles, the possible values of k are N *equally spaced* points within the Brillouin zone.

As the wavelength of the wave varies from $\lambda = \infty$ to $\lambda = 2a$ the angular frequency varies from $\omega = 0$ to $\omega = \omega_{max}$.

For a one-dimensional crystal with nearest-neighbor Hooke's law forces $\omega_{max} = 2\sqrt{K/m}$ where K is the force constant and m is the mass of the atom.

In two- or three-dimensional crystals the Brillouin zone (the set of possible wave vectors for traveling waves in the crystal) is a two- or three-dimensional region in k-space surrounding $\mathbf{k} = 0$.

For the two-dimensional rectangular lattice with lattice vectors $\mathbf{a} = a\hat{\mathbf{x}}$ and $\mathbf{b} = b\hat{\mathbf{y}}$ it is the region defined by

$$-\frac{\pi}{a} < k_x < \frac{\pi}{a} \quad \text{and} \quad -\frac{\pi}{b} < k_y < \frac{\pi}{b}$$

For a two-dimensional lattice with an n-fold rotational symmetry it is an n-sided regular polygon.

For a one-dimensional crystal the dispersion relation is a function, $\omega(k)$, giving the angular frequency of a traveling wave of wave vector k.

For three-dimensional crystals the waves of a given wave vector can have one of three mutually perpendicular *polarizations.* The polarization direction is the direction that an atom actually vibrates as the wave passes.

Waves with different polarization but the same wave vector may have different frequencies. Thus there are three separate dispersion relations, $\omega_1(\mathbf{k})$, $\omega_2(\mathbf{k})$, and $\omega_3(\mathbf{k})$, for the three possible polarizations.

According to quantum mechanics, each normal mode acts like a quantized harmonic oscillator of angular frequency $\omega(k)$. There are N different normal modes for a one-dimensional crystal (3N for a three-dimensional one).

The ground state energy of the crystal is

$$E_o = \sum_k \tfrac{1}{2}\hbar\omega(k)$$

The complete energy spectrum is given by

$$E = E_o + \sum_k n_k \hbar\omega(k)$$

where, for each of the N allowed k within the Brillouin zone, $n_k = 0, 1, 2, \cdots$.

The *density of normal modes* is a function $D(\omega)$ defined by

$$D(\omega)\,d\omega = \begin{array}{c}\text{the number of normal modes with angular}\\ \text{frequencies between } \omega \text{ and } \omega + d\omega\end{array}$$

For the one-dimensional lattice with nearest-neighbor forces

$$\omega(k) = \omega_{max} \sin|ka/2|$$

and

$$D(\omega) = \begin{cases} \dfrac{2N}{\pi\sqrt{\omega_{max}^2 - \omega^2}} & \text{for} \quad \omega < \omega_{max} \\ 0 & \text{for} \quad \omega > \omega_{max} \end{cases}$$

The Debye formula for a three-dimensional crystal is

$$D(\omega) = \begin{cases} \dfrac{3V}{2\pi^2 v^3}\omega^2 & \text{for} \quad \omega < \omega_{max} \\ 0 & \text{for} \quad \omega > \omega_{max} \end{cases}$$

where $\omega_{max} = (6\pi^2 n)^{1/3} v$, n is the particle density and v is the sound speed in the crystal.

The Debye theory is based on three approximations.

1. The wave speed is assumed to be independent of the wavelength.

2. Both longitudinal and transverse waves are assumed to propagate with the same speed v, where

$$v^{-3} = \tfrac{1}{3} v_L^{-3} + \tfrac{2}{3} v_T^{-3}$$

3. The Brillouin zone is a sphere in wave vector space, with radius K chosen so that the number of allowed wave vectors within the sphere is 3N.

If a quantized vibrational wave of wave vector k is excited to the quantum level n_k the wave carries an amount of momentum

$$P_k = n_k \hbar k$$

The formulas for the energy and momentum of a set of quantized waves is the same as those for the energy and momentum of a set of Bose-Einstein particles called phonons. In the particle interpretation n_k is the number of phonons with momentum $p = \hbar k$.

In a three-dimensional crystal the quantum states of a phonon are described by a three-dimensional wave vector **k** and a polarization index, n = 1, 2, 3. The energy and momentum of a phonon in state (**k**,n) is

$$E = \hbar \omega_n(\mathbf{k})$$

and

$$\mathbf{p} = \hbar \mathbf{k}$$

For a crystal at temperature T the average number of phonons in the state of wave vector **k** and polarization n (n = 1, 2, 3) is

$$\overline{N}(\mathbf{k},n) = \frac{1}{e^{E/k_BT} - 1}$$

where $E = \hbar \omega_n(\mathbf{k})$.

$\overline{N}(\mathbf{k},n)$ can also be interpreted as the average excitation level of the normal mode (**k**,n) at temperature T.

The thermal vibrational energy is given in terms of the density of normal modes by

$$U(T) = \int D(\omega) \frac{\hbar \omega}{e^{\hbar \omega / k_BT} - 1} d\omega$$

The energy eigenstates of an electron in a periodic potential are defined by a *band index*, n = 1, 2, · · · , and a *wave vector*, **k**, where **k** must be in the Brillouin zone. The wave vector, **k**, is related to the momentum of the electron by the de Broglie relation, $\mathbf{p} = \hbar \mathbf{k}$.

The set of N very closely spaced energy values obtained by varying **k** for fixed n is called the nth *energy band*. If neighboring energy bands do not everlap then they are separated by an *energy gap*, which is a range of energy within which there are no allowed electronic energy states.

Each eigenstate can accept two electrons of opposite spin.

If the set of electrons in a given solid fills the energy states just up to an energy gap then the solid is an insulator.

If the set of electrons leave an energy band partially filled and partially empty then the solid is an electrical conductor because it is very easy for the electrons to make transitions to nearby energy states of different momentum.

The partially filled band in a conductor is called the *conduction band.* The completely filled bands are called *valence bands.*

In an insulator the conduction band is empty at zero temperature. At temperature T a small density of electrons are thermally excited into the conduction band, leaving holes in the valence band. The *intrinsic conductivity*, due to the thermal electrons and holes, is given by

$$K(T) = K_o e^{-E_g/2k_BT}$$

where E_g is the width of the energy gap.

By adding impurities to a crystal that would otherwise be an insulator an *impurity semiconductor* (or *doped semiconductor*) can be produced.

An *n-type* semiconductor contains *donor impurities* that are easily ionized to contribute mobile electrons to the conduction band.

A *p-type* semiconductor contains *acceptor impurities* that grab electrons from the regular crystal atoms to produce mobile positive holes.

The energy of electrons and holes of momentum **p** can be approximated by the formulas

$$E_e = \tfrac{1}{2} E_g + \frac{p^2}{2m_e^*} \quad \text{and} \quad E_h = \tfrac{1}{2} E_g + \frac{p^2}{2m_h^*}$$

where m_e^* and m_h^* are the *effective masses* of the electron and holes.

A *pn junction* is a junction between a p-type semiconductor and an n-type semiconductor. At the junction a *depletion layer* is produced in which there are no mobile charges. This is due to *electron-hole annihilation.*

PROBLEMS

10.1 The mass density of a NaCl crystal is $2.14\times 10^3 kg/m^3$. What is the distance between neighboring atoms in the crystal?

10.2* (a) Consider 2N electrons packed, two to a box, in N one-dimensional boxes, each of length d. Write a formula for the ground state energy of the system.

(b) Consider 2N electrons in a single one-dimensional box of length Nd. Taking account of the exclusion principle, write a formula for the ground state energy of the system.

Note: $1^2+2^2+\cdots+K^2 = \frac{1}{3}K^3+\frac{1}{2}K^2+\frac{1}{6}K$.

(c) For $N >> 1$ which of the above systems gives the lowest energy per particle? To what type of chemical bonding in solids could this calculation be related?

10.3* Prove that the mass density of a three-dimensional simple crystal is given by the formula $\rho = m/\mathbf{a}\cdot\mathbf{b}\times\mathbf{c}$, where m is the mass of one atom. What is the equivalent formula for a two-dimensional crystal?

10.4* Calculate the ground-state vibrational energy per particle for a one-dimensional crystal with nearest-neighbor forces if the force constant $K = 10\,N/m$, the mass $m = 20\,u$, and the lattice spacing $a = 1\,Å$. How does it compare with the work needed to move a particle .1 Å from its equilibrium position with all the other particles held fixed at their equilibrium positions?

10.5* Consider an electron in a cubic crystal with $a = b = c = 2\,Å$.

a. What is the maximum allowed value of $|\mathbf{k}|$?

b. Setting $m_e v_{max} = \hbar k_{max}$ what is v_{max}?

10.6 Complete the derivation given in Section 10.10 for the density of normal modes for the one-dimensional lattice with nearest-neighbor forces.

10.7** Consider a crystal of rectangular structure. Each allowed wave vector for vibrational waves occupies a parallelepiped in k-space of volume* $\Delta k_x \Delta k_y \Delta k_z = (2\pi)^3/V$, where V is the volume of the crystal sample. In the Debye approximation the Brillouin zone is taken to be a sphere in k-space of radius K.

a. What must be the value of K in order for the number of normal modes to be equal to 3N?

b. Show that $\omega_{max} = (6\pi^2 n)^{1/3} v$.

c. How many normal modes are within the spherical shell in k-space defined by $k \leqslant |\mathbf{k}| \leqslant k + dk$?

d. How many normal modes have angular frequencies within the range ω to $\omega + d\omega$?

*See the caption to Figure 10.26.

10.8* For an infinite sinusoidal wave the speed with which the wave fronts progress is call the *phase velocity* of the wave and is given by $v_{ph} = \omega/k$. In Chapter 4 (Equation 4.29 in particular) we saw that an isolated wave packet of electron waves does not travel with the phase velocity, but travels with a velocity (called the *group velocity*) that is given by $v_{gr} = d\omega/dk$. *This formula for the velocity of a wave packet is not restricted to electron waves (The Schroedinger equation was not actually used in deriving it). For the one-dimensional crystal with nearest-neighbor forces:*

a. Calculate v_{gr} *and* v_{ph} *as functions of* λ.

b. Show that $v_{gr} \rightarrow v_{ph}$ *as* $\lambda \rightarrow \infty$.

c. Determine what happens to v_{gr} *and* v_{ph} *as* $\lambda \rightarrow 2a$.

10.9** For crystals that are not strongly anisotropic the Debye formula for $D(\omega)$ is quite accurate at low frequencies. (It is based upon the sound wave dispersion relation which is valid for long wavelength.) Use this fact and Equation 10.13 to obtain a formula for $U(T)/N$ in the limit of very small T. It should give $U(T)/N$ as a function of v and T. Note: As $T \rightarrow 0$ $e^{\hbar\omega/k_BT} - 1 \rightarrow e^{\hbar\omega/k_BT}$.

10.10** Use Equation 10.13 and the fact that $\int_0^\infty D(\omega)\,d\omega = 3N$ to show that $U(T)$ approaches its classical value of $3Nk_BT$ at high temperatures. That is

$$U(T) \rightarrow 3Nk_BT \quad \text{as } T \rightarrow \infty$$

Note: $e^x \approx 1 + x$ for $x << 1$.

10.11* The classical value of the vibrational energy of a solid can be used if $k_BT >> \hbar\omega_{max}$. For copper, $v_L = 4.56\,km/s$ and $v_T = 2.25\,km/s$. The density of copper is $8.93 \times 10^3\,kg/m^3$. At what temperature is $k_BT = \hbar\omega_{max}$? That temperature is called the *Debye temperature* of copper.

10.12** Write a formula for the quantum mechanical ground state vibrational energy of a crystal using the Debye approximation. The formula should involve the speed of sound in the crystal, the particle density in the crystal, and fundamental constants.

10.13* Consider a resistor made of a one centimeter cube of pure germanium. What is the relationship between the temperature coefficient of resistance of that resistor and the slope of the curve shown in Figure 10.41?

CHAPTER ELEVEN

QUANTUM STATISTICAL MECHANICS

11.1. WHAT IS QUANTUM STATISTICAL MECHANICS?

Quantum statistical mechanics is the theoretical bridge between the microscopic world and the macroscopic world. The microscopic world is the world we have been studying since we took up the problems of quantum theory. It is a world of electrons and protons, of atoms and molecules in discrete states absorbing and emitting photons. The macroscopic world is the world of bulk matter. It is a world of stones and trees and water; all the things we can see and feel. The microscopic structure of these things is usually completely hidden from our view. The laws that determine the behavior of macroscopic things are mostly the laws of classical mechanics and the laws of thermodynamics. We have already, in some degree, shown how the laws of classical mechanics are a natural consequence of the fundamental laws of quantum mechanics. But what about the laws of thermodynamics? In particular, where does the rather strange law of increase of entropy come from? What *is* this entropy that must always increase? An atom does not have anything we call entropy. Neither does an aggregate of three or four atoms forming a molecule. When we put together 10^{23} atoms to form a chunk of lead, from where could a completely new property arise? Particularly one that is of prime importance in determining the time evolution of this macroscopic system. Does not the Schroedinger equation govern the time evolution of any aggregate of lead atoms? If it does, then any other law must be a direct consequence of the Schroedinger equation, because we cannot impose two mutually inconsistent laws on the time evolution of one system. These questions are all the subject of quantum statistical mechanics. The goals of quantum statistical mechanics can be separated into two fairly distinct parts. One is of importance to the conceptual foundations of thermodynamics while the other is of great practical value in the applications of thermodynamics to particular systems. The first goal is to show that the laws of thermodynamics follow logically from applying the laws of quantum mechanics to systems containing very many particles. The second goal is to develop useful formulas for calculating thermodynamic quantities such as the specific heat, the pressure, or the internal energy of a thermodynamic system by using our knowledge of the microscopic structure of the substance under consideration. Quantum statistical mechanics is a large and complex subject. In the small space we have available to devote to it we can only discuss a few questions and even those questions we will not try to answer with complete mathematical precision. With regards to the foundations of thermodynamics we will look primarily at one question. What is the entropy?

11.2. ENTROPY

According to the Second Law of Thermodynamics, the entropy of an isolated system that is not initially in its thermodynamic equilibrium state will spontaneously increase. The increase will continue until the equilibrium state is reached. That equilibrium state is the state of maximum possible entropy. In this section we will show how the thermodynamic law of entropy increase can be used to calculate the equilibrium state if we know the entropy functions of all components of a system. In the next section we will try to explain the connection between that law and the laws of quantum mechanics.

For a single sample of one substance (a bowl of water, a piece of ice, etc.) the entropy of the sample can be considered as a function of the total internal energy of the sample, E, the number of particles in the sample, N, and the volume occupied by the sample, V.

$$S = S(E, N, V)$$

For simplicity we are assuming that the sample contains only one type of particle. Otherwise the entropy is a function of the number of particles of each kind. If a system, such as that shown in Figure 11.1, is made up of a number of distinguishable parts or *components* then the entropy of the complete system is the sum of the entropies of its components.

$$S = \sum_{i=1}^{k} S_i(E_i, N_i, V_i) \qquad (k = \text{number of components})$$

The essential physical property of the entropy can be described as follows. If we take two or more systems that were initially isolated from one another and we allow them to interact so that they can exchange one or more of the quantities E, N, and V, then those quantities will change until a new equilibrium state is achieved. That new equilibrium state is the state that maximizes the total entropy with respect to all those variables that are allowed to change. A few examples will make this principle clear.

We consider two gases, initially kept in separated containers, that have energies E_1^o and E_2^o. (See Figure 11.2.) The two containers are brought into contact so that heat energy (microscopic kinetic energy) can flow from one gas to the other. What will be the values of E_1 and E_2 when thermal equilibrium is reached? Since energy is a conserved quantity we know that

$$E_1 + E_2 = E_1^o + E_2^o \equiv E^o$$

or $E_2 = E^o - E_1$. The entropy of the combined system, at equilibrium, is

$$S = S_1(E_1, N_1, V_1) + S_2(E^o - E_1, N_2, V_2)$$

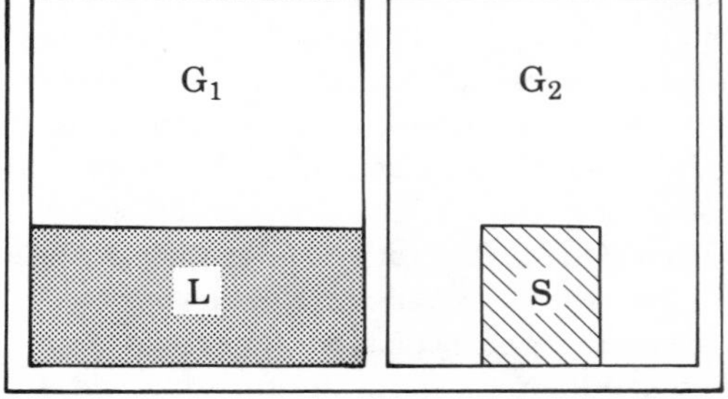

FIGURE 11.1

A system of four components (one liquid, two gases, and a solid) separated by a heat-conducting partition. For this system $S = S_{G_1} + S_{G_2} + S_L + S_S$

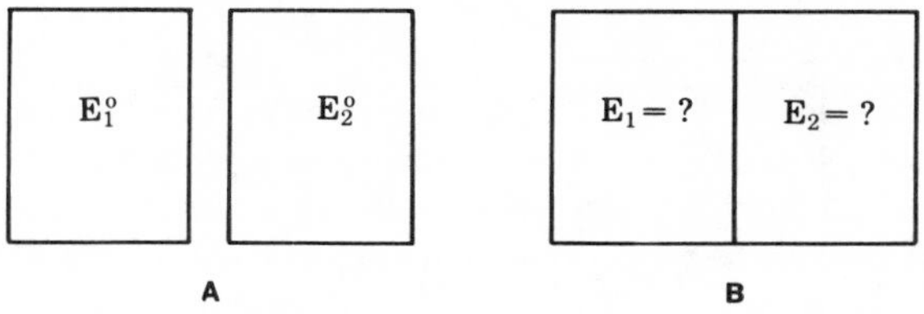

FIGURE 11.2 A. Two gas samples have initial volues of energy, particle number, and volume equal to (E_1^o, N_1, V_1) and (E_2^o, N_2, V_2).
B. They are allowed to transfer energy until they come into thermal equilibrium, at which time their energies are E_1 and E_2.

Since, in this problem, N_1, V_1, N_2, and V_2 are fixed we will no longer explicitly show those variables. The only free variable in this expression is E_1. The maximum value of S is given by that value of E_1 for which

$$\frac{\partial S}{\partial E_1} = \frac{\partial}{\partial E_1}(S_1(E_1) + S_2(E_o - E_1)) = \frac{\partial S_1(E_1)}{\partial E_1} - \frac{\partial S_2(E_2)}{\partial E_2} = 0$$

with $E_2 = E^o - E_1$. Thus the condition of thermal equilibrium is that

$$\frac{\partial S_1}{\partial E_1} = \frac{\partial S_2}{\partial E_2}$$

But we know that the condition of thermal equilibrium between two objects is that their temperatures be equal. Therefore the derivative $\frac{\partial S}{\partial E}$ for any substance must be related somehow to the temperature of the substance. When the usual units are used for entropy and temperature the relation between the two can be shown to be*

$$\frac{\partial S(E)}{\partial E} = \frac{1}{T(E)} \tag{11.1}$$

where T is the Kelvin temperature. Using this relation the condition for thermal equilibrium can be expressed in the usual way, namely

$$T_1(E_1) = T_2(E_2)$$

Our second example is the system shown in Figure 11.3. Two samples of gas with initial parameters (E_1^o, V_1^o, N_1^o) and (E_2^o, V_2^o, N_2^o) are separated by a thermally conducting piston that is pinned so as to prevent volume changes. The pin is then removed and the piston moves and finally settles at a new equilibrium position. Since the piston is thermally conducting both volume and energy can be exchanged, subject to the constraints

$$E_1 + E_2 = E_1^o + E_2^o \equiv E^o$$

and $V_1 + V_2 = V_1^o + V_2^o \equiv V^o$. The equilibrium values of E_1 and V_1 are those values for which the entropy,

$$S = S_1(E_1, V_1, N_1^o) + S_2(E^o - E_1, V^o - V_1, N_2^o)$$

is a maximum with respect to changes in E_1 and V_1. The conditions of a

---

*From the thermodynamic relation $dS = \frac{dQ}{T} = \frac{dE}{T} + \frac{p\,dV}{T}$ we can see that $\frac{\partial S}{\partial E} = \frac{1}{T}$ and $\frac{\partial S}{\partial V} = \frac{p}{T}$.

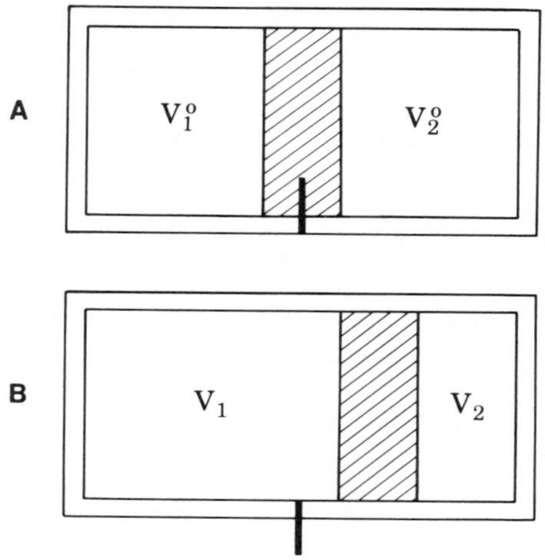

FIGURE 11.3

A. Two samples of gas, separated by a heat-conducting cylinder that is pinned in a nonequilibrium position.
B. After the pin is pulled the system comes into equilibrium by exchanging energy and volume.

maximum are that

$$\frac{\partial S}{\partial E_1} = 0 \text{ and } \frac{\partial S}{\partial V_1} = 0$$

The first equation gives $\frac{\partial S_1}{\partial E_1} = \frac{\partial S_2}{\partial E_2}$, which, according to Equation 11.1, is equivalent to $T_1 = T_2$. The second equation gives

$$\frac{\partial S_1}{\partial V_1} = \frac{\partial S_2}{\partial V_2}$$

This can be related to the condition that the mechanical pressures be equal by using the thermodynamic relation

$$\frac{\partial S}{\partial V} = \frac{p}{T} \tag{11.2}$$

Since $T_1 = T_2$ this gives the expected relation for mechanical equilibrium

$$p_1 = p_2$$

These examples illustrate how one can solve problems involving thermodynamic equilibrium *if one knows the entropy functions* $S_i(E_i, N_i, V_i)$ *of all components of the system.* There are two ways of getting those entropy functions. (a) Equations 11.1 and 11.2 can be used to determine the entropy functions experimentally. (b) The entropy can be calculated from a model of the microscopic structure of the substance, using theoretical relations that we will derive in this chapter. The advantage of the second procedure is that if the calculated entropy function gives correct predictions of experimental results then this tends to confirm one's picture of the microscopic structure of the substance. The disadvantage of the method is that the calculation can be carried out only for materials with simple microscopic structures.

11.3. THE APPROACH TO EQUILIBRIUM OF A QUANTUM SYSTEM

We will now look at what quantum mechanics has to say about the approach to equilibrium of very complex systems. By very complex we mean that the systems contain extremely large numbers of particles. At first sight any attempt to look at the time dependence of a system of 10^{23} particles using the Schroedinger equation seems foolishly futile. We can only solve the Schroedinger equation with reasonable accuracy for systems with quite small numbers of particles. There is no way we can hope to determine the eigenfunctions of 10^{23} interacting particles confined to a container. We will see that it is the very complexity of the system that will allow us to make some progress. For small systems one needs to know the detailed solutions of the Schroedinger equation in order to predict anything about the time evolution of the system. We will see that for very complex systems the details of the wave functions are not important. The quantity that is of most importance is the number of energy eigenfunctions that exist within a given energy interval.

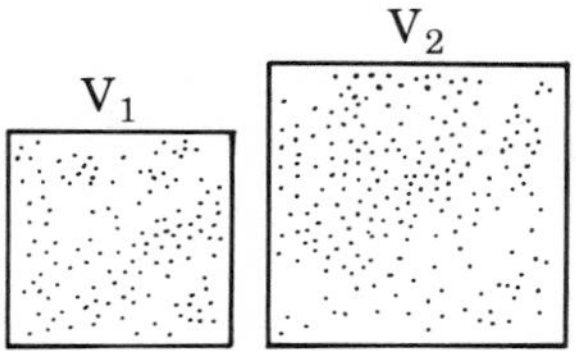

FIGURE 11.4
A system composed of two noninteracting parts.

The first question we will answer is *what are the energy eigenfunctions of a system composed of two noninteracting parts.* We picture a system such as that shown in Figure 11.4. Volume V_1 contains N_1 particles described by coordinates $x_1, \cdots, x_{N_1}$ and volume V_2 contains N_2 particles described by coordinates $y_1, \cdots, y_{N_2}$. The energy operator for the system is

$$\hat{E}_{12} = \hat{E}_1 + \hat{E}_2$$

where $\hat{E}_1$ only involves the coordinates $x_1, \cdots, x_{N_1}$ and $\hat{E}_2$ only involves the coordinates $y_1, \cdots, y_{N_2}$.

Let $\psi_1(x_1, \cdots, x_{N_1})$ be an energy eigenfunction (of energy E_1) for system 1 alone and $\psi_2(y_1, \cdots, y_{N_2})$ be an energy eigenfunction for system 2 alone. Then the wave function

$$\psi_{12}(x_1, \cdots, x_{N_1}, y_1, \cdots, y_{N_2}) = \psi_1(x_1, \cdots)\psi_2(y_1, \cdots)$$

is an eigenfunction for the combined system with an energy $E_{12} = E_1 + E_2$. We can prove this by first noticing that

$$\hat{E}_1 \psi_{12} = (\hat{E}_1 \psi_1)\psi_2 = E_1 \psi_{12}$$

and

$$\hat{E}_2 \psi_{12} = (\psi_1)(\hat{E}_2 \psi_2) = E_2 \psi_{12}$$

Therefore

$$(\hat{E}_1 + \hat{E}_2)\psi_{12} = \hat{E}_1 \psi_{12} + \hat{E}_2 \psi_{12} = (E_1 + E_2)\psi_{12}$$

We will not need to specify the structure of the quantum states in any more detail than this. We now want to look at the essential mechanism governing the approach to equilibrium of a complex system. Suppose we have a system composed of two initially separated components with energies E_1^o and E_2^o. (See

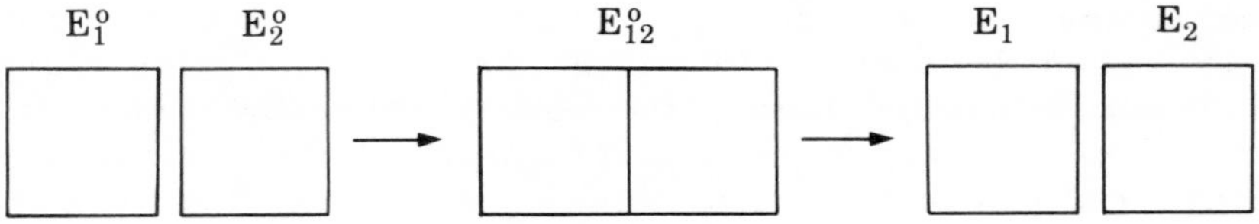

FIGURE 11.5 An initially separated system is allowed to exchange energy until it comes into equilibrium. How is the energy shared at equilibrium?

Figure 11.5.) The components are brought into contact for a long time and then separated again. We want to develop some criterion for calculating the most likely value of the energy of component 1 after the two components are separated. We know that the total energy must be conserved so that, if E_1 and E_2 are the energies of the two components after separation, then

$$E_1 + E_2 = E^o \equiv E_1^o + E_2^o$$

Let us suppose that, for some energy A, there are a total of 10 quantum states of the combined (but not interacting) system for which $E_1 = A$ and $E_2 = E^o - A$. Let us also suppose that, for some other energy, B, there are 10^8 quantum states with $E_1 = B$ and $E_2 = E^o - B$.

$$E_1 = A, \quad E_2 = E^o - A \qquad (10 \text{ states})$$

$$E_1 = B, \quad E_2 = E^o - B \qquad (10^8 \text{ states})$$

While the combined system is separated it stays in whatever state it happens to be in. But while the two parts are interacting the system can make transitions from any state to any other state of the same total energy. It is clear that, unless there is something that prevents or strongly suppresses transitions into the B states it is very much more likely that when the components are separated the system will be caught in one of the 10^8 B states rather than one of the 10 A states. Of course, all this depends upon our conjecture about the ratio of the number of B states to the number of A states. What makes this discussion relevant to real systems is the following fact which we will state without proof.

For a system composed of two macroscopic components and for any total energy E^o, the number of quantum states with energies E_1 and $E_2 = E^o - E_1$ for the two components is a function of E_1 with a very sharp peak. Therefore, if $E_1 = A$ is the value of E_1 at the peak then the number of quantum states of the combined system with energies $E_1 = A$ and $E_2 = E^o - A$ is much larger than the number of quantum states with any other combination of energies that adds up to E^o.

QUESTION: Consider a system, like that shown in Figure 11.5, composed of two identical components. Suppose each component is a helium gas with $N = 6.02 \times 10^{23}$ atoms. That is, each component contains one mole of gas. The volume of each container is $1\,m^3$, and the total energy of the system is 8000 J. (This would give the "reasonable" temperature of 47° C.)

(a) How many quantum states of the whole system are there for which

$$3999.99 \leqslant E_1 \leqslant 4000.01$$

(b) How many quantum states are there for which E_1 is anything outside the above range?

(c) What is the ratio of the number obtained in (a) to the number obtained in (b)?

ANSWER: It is possible to calculate both numbers to very high accuracy. The calculation is long and complicated and so we will just quote the results. The number of quantum states asked for in question (a) is

$$K_a = \left[\frac{V^{2N}(m/2\pi)^{3N}}{\hbar^{6N}(N!(\frac{3}{2}N)!)^2} (\frac{3N}{2})^2 E^{3N} \right] \sqrt{\pi}$$

where $V = 1\,m^3$, m = mass of a helium atom, $N = 6.02 \times 10^{23}$, and $E = 4000$ J.

The number requested in question (b) is

$$K_b = \left[\frac{V^{2N}(m/2\pi)^{3N}}{\hbar^{6N}(N!(\frac{3}{2}N)!)^2} (\frac{3N}{2})^2 E^{3N} \right] \frac{e^{-\frac{3}{2}N\epsilon^2}}{\epsilon\sqrt{3N/2}}$$

where $\epsilon = \frac{\Delta E}{E} = \frac{0.01\,J}{4000\,J} = 2.5 \times 10^{-6}$. The ratio of K_a to K_b is

$$\frac{K_a}{K_b} = \epsilon\sqrt{3\pi N/2}\, e^{\frac{3}{2}N\epsilon^2}$$

$$\approx e^{5.6 \times 10^{12}}$$

$$\approx (282)^{(10)^{12}}$$

When we say that there are many more states with the most probable values of E_1 and E_2 we really mean *very* many more.

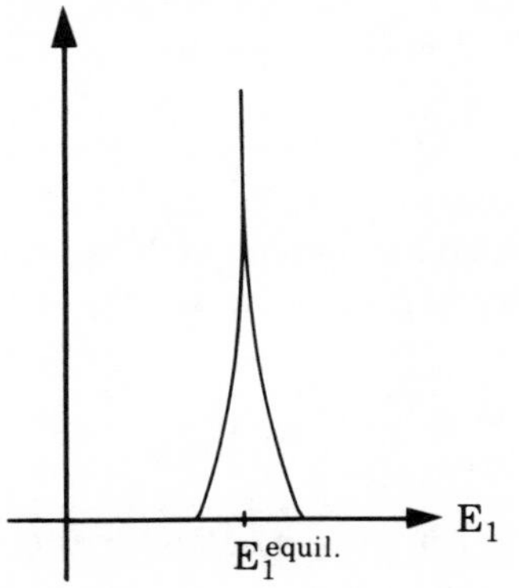

FIGURE 11.6

The total number of quantum states of a two-component system, $\mathbf{N}_{12}$, as a function of the energy of one component. There are many more quantum states of the complete system with $E_1 \approx E_1^{equil.}$ then there are with any other value of E_1.

Since the energy eigenvalues are discrete the number of quantum states with E_1 exactly equal to some number A is zero for most values of A. Therefore it is much more convenient to ask how many quantum states exist for which E_1 lies within some narrow interval ΔE centered at A. The density of energy eigenvalues is so large for a macroscopic system that, even with an extremely small value of ΔE the density function that has been averaged over the interval ΔE becomes a smooth function of A. It is that function that is plotted in Figure 11.6.

What has been stated regarding energy exchange between two macroscopic components is also true for particle exchange or volume exchange when these are allowed. It is therefore reasonable to postulate the following principle.

> If two macroscopic components are allowed to exchange energy, particles, or volume, these quantities will distribute themselves so as to correspond to the largest possible number of combined quantum states. That is, if $\mathbf{N}_{12}(E_1, N_1, V_1, E_2, N_2, V_2)$ is the number of quantum states of the complete system for which the two components have the parameters (E_1, N_1, V_1) and (E_2, N_2, V_2) respectively , then the quantum state of the system after it has interacted for a long time is overwhelmingly likely to have the values of the parameters that maximize $\mathbf{N}_{12}$ subject to the constraints imposed by the interaction.

It seems as if we have discovered the quantum mechanical meaning of the entropy. Here is a function of the thermodynamic parameters that is a maximum for the equilibrium values of those parameters. The only other property the entropy needs to have is additivity. That is, the entropy of the combined system should be the sum of the entropies of the components. Let us see if this is so for $\mathbf{N}_{12}$. Let $\mathbf{N}_1(E_1, N_1, V_1)$ be the number of quantum states of the first component alone with those values for the parameters. Let $\mathbf{N}_2(E_2, N_2, V_2)$ be the corresponding thing for the second component. By choosing any one of the $\mathbf{N}_1$ quantum states of the first component and multiplying it by any one of the $\mathbf{N}_2$ quantum states of the second component we can produce an energy eigenstate of the combined system with parameters $(E_1, N_1, V_1, E_2, N_2, V_2)$. There are $\mathbf{N}_1\mathbf{N}_2$ ways of doing this. Therefore

$$\mathbf{N}_{12}(E_1, N_1, V_1, E_2, N_2, V_2) = \mathbf{N}_1(E_1, N_1, V_1)\mathbf{N}_2(E_2, N_2, V_2)$$

Bad luck. The number of quantum states is not additive. It is multiplicative. However, the following trick will save the day. If $\mathbf{N}_{12}$ is a maximum at some value of the thermodynamic parameters then $ln(\mathbf{N}_{12})$ is also a maximum at the same value. But

$$ln(\mathbf{N}_{12}) = ln\,\mathbf{N}_1\mathbf{N}_2 = ln\,\mathbf{N}_1 + ln\,\mathbf{N}_2$$

Therefore the function

$$S_{12} = k\, ln\,\mathbf{N}_{12}$$

has all the properties of the entropy for any positive constant k. If we measure entropy in the conventional units, then k is Boltzmann's constant. For a system composed of K components

$$S(E_1, N_1, V_1, \cdots, E_k, N_k, V_k) = k_B \sum_{j=1}^{K} ln\,\mathbf{N}_j(E_j, N_j, V_j) \qquad (11.3)$$

where $\mathbf{N}_j$ is the number of quantum states of the jth component with the values (E_j, N_j, V_j) for the energy, particle number, and volume. In deriving this identification of the entropy we have used a certain statistical assumption that we should now state more exactly because we will have occasion to use it again in the next section.

> The probability that, after a long period of interaction, a composite system will be found with the values $(E_1, N_1, V_1, \cdots, E_k, N_k, V_k)$ for its thermodynamic parameters is proportional to the number of quantum states of the complete system for which the components have the indicated values of E, N, and V.

This postulate can only be expected to be satisfied in practice if transitions from the smaller number of quantum states with the initial, nonequilibrium values of the thermodynamic parameters to the very much larger number of quantum states with equilibrium values of those parameters are not prevented or strongly suppressed. There are many systems in which the transition rate is so small that the approach to equilibrium does not occur in any practical sense. One of the most common examples is glass. Glass is an extremely viscous supercooled liquid. The equilibrium state of a system with all the thermodynamic parameters of a piece of glass is a crystal structure. But the transition from the amorphous state to the crystalline state is so slow that even ancient samples of glass that have been protected from the elements show no signs of movement toward equilibrium.

11.4. THE GIBBS DISTRIBUTION

In the last section we always assumed that the complete thermodynamic system was isolated from its surroundings. It is only in that situation that the energy conservation principle can be used. It is a common experimental practice not to isolate the thermodynamic system of interest but rather to have it in thermal contact with a much larger system (called the reservoir) that, due to its large heat capacity, can exchange small amounts of energy with the system with negligible change in the reservoir temperature. The reservoir acts to stabilize the temperature of the thermodynamic system. Since the small system is not isolated we cannot predict its energy with absolute certainty. It will have some probability distribution with respect to its energy. We now want to show that that probability distribution is a simple function that depends only on the temperature of the reservoir.

We picture an isolated system made up of two components, a very large component called *the reservoir*, and a much smaller component that we call *the small system*. It has a total energy, E^o. The question we will answer is "What is

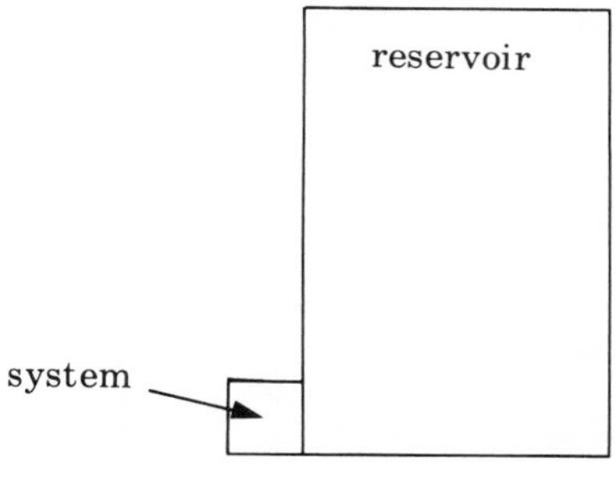

FIGURE 11.7
A small system in thermal contact with a large reservoir.

the probability that the small system will be in a particular quantum state, ψ_n, that has an energy E_n?"

The possible wave functions for the combined system that satisfy the conditions of the question are product functions of ψ_n and any reservoir quantum state with energy $E^o - E_n$. The number of those wave functions is equal to the number of reservoir wave functions with energy $E^o - E_n$. That is $\mathbf{N}_R(E^o - E_n)$. (We omit writing the variables N_R and V_R since they are fixed throughout the discussion.) Thus, by the postulate given in the last section, the probability of finding the small system in the state ψ_n is

$$P_n = C_o \mathbf{N}_R(E^o - E_n)$$

But the reservoir entropy is related to $\mathbf{N}_R$ by

$$S_R(E^o - E_n) = k_B \, ln \, \mathbf{N}_R(E^o - E_n)$$

or

$$\mathbf{N}_R(E^o - E_n) = e^{S_R(E^o - E_n)/k_B}$$

Since E_n is much less than E^o we can expand the entropy function about E^o and, using Equation 11.1, obtain

$$S_R(E^o - E_n) = S_R(E^o) - \frac{E_n}{T}$$

where T is the reservoir temperature. If we combine the normalization constant, C_o, with the quantity $e^{S_R(E^o)/k_B}$, that is also independent of E_n, we can write the probability as

$$P_n = C e^{-E_n/k_B T} \tag{11.4}$$

The probability function given in Equation 11.4 is called the *Gibbs distribution.*

QUESTION: In a dilute helium gas at room temperature what is the ratio of the number of atoms in the first excited state to the number of atoms in the ground state?

ANSWER: We can treat a single atom as our small system and all the other atoms as the reservoir. The probability that the atom will be found in its ground state is

$$P_o = C e^{-E_o/k_B T}$$

where E_o is its ground state energy. A similar result holds for any excited state. The ground state of helium is not degenerate but the first excited state is triply degenerate. (There are three different eigenfunctions with the same eigenvalue.) If we add the probabilities of those three degenerate states, then the desired ratio is

$$R = \frac{3 P_1}{P_o} = 3 e^{-\epsilon_1/k_B T} \tag{11.5}$$

where $\epsilon_1 = E_1 - E_o$ is the *excitation energy* of the first excited state. Notice that, in looking at the ratio of two probabilities the normalization constant drops out and the energy difference replaces the energy in the formula. Room temperature is about 300 K and $\epsilon_1 = 19.77\,\text{eV} = 3.168 \times 10^{-18}$ J. Therefore

$$R = 3e^{-765}$$

This number is so small that there is no real possibility that even one atom in the Goodyear blimp is in an excited state.

QUESTION: The probability, $P_n = Ce^{-E_n/k_BT}$, has a maximum for $E_n = E_o$, the ground state energy. Is E_o the most probable value of the energy?

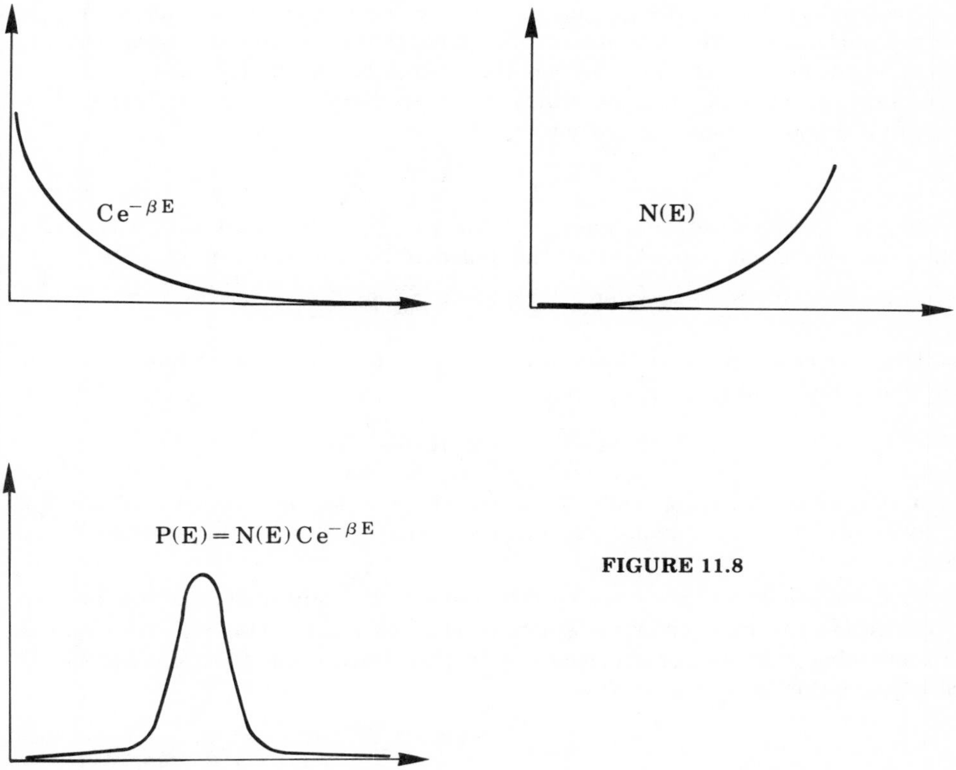

FIGURE 11.8

ANSWER: No. P_n is the probability that the system will be found in a particular quantum state. If we let

$$P(E)\,dE \equiv \begin{array}{c}\text{the probability that the system will be}\\ \text{found with energy between E and E+dE}\end{array}$$

and

$$N(E)\,dE \equiv \begin{array}{c}\text{the number of quantum states}\\ \text{whose energies lie between E and E+dE}\end{array}$$

then

$$P(E)\,dE = Ce^{-E/k_BT}\,N(E)\,dE$$

The function Ce^{-E/k_BT} decreases with increasing E, but the function N(E) rapidly increases with E. The product has a maximum that, for a macroscopic system, is extremely sharp.

11.5. THE HEAT CAPACITY OF A QUANTUM MECHANICAL SYSTEM

At room temperature helium atoms are all in their ground state because the excitation energy of helium is so much larger than k_BT. This lack of population of the excited states is not exhibited by other systems that have much lower excitation energies. Nor is it true for helium when the temperature is much higher than room temperature. When k_BT is comparable to the excitation energies of the molecules (or atoms) in a gas then a significant amount of the total internal energy of the gas may be in the form of the excitation energy of the gas molecules. We will assume for now that the kinetic energy of the translational motion of the gas molecules can be calculated using classical mechanics. For a sample of N molecules at temperature T the classical calculation gives a total translational kinetic energy of

$$E_{tran} = \tfrac{3}{2} N k_B T$$

To compute the total internal energy of the gas, U(T), we must add to E_{tran} the sum of the excitation energies of all the molecules. Let us call this $E_{ex}(T)$.

$$U(T) = \tfrac{3}{2} N k_B T + E_{ex}(T)$$

The constant volume heat capacity of the gas is defined by the thermodynamic relation

$$C_V \equiv \frac{dU}{dT} = \tfrac{3}{2} k_B N + \frac{dE_{ex}}{dT}$$

which shows that the heat capacity contribution due to the internal excitation of the molecules is simply added to the heat capacity due to their translational motion.

To calculate $E_{ex}(T)$ we have to determine how many molecules are in each excited state. The number of molecules in the nth excited state, N_n, is equal to the probability that any one molecule is in that state times the total number of molecules in the gas.

$$N_n = N P_n = N C e^{-E_n/k_BT}$$

The total excitation energy is the sum, over all possible excited states, of N_n times the excitation energy of the nth state, namely $\epsilon_n = E_n - E_0$. If we define a quantity, Q, by the relation $Q^{-1} = C e^{-E_0/k_BT}$ then $P_n = Q^{-1} e^{-\epsilon_n/k_BT}$ and E_{ex} can be written as

$$E_{ex} = N Q^{-1} \sum_n \epsilon_n e^{-\epsilon_n/k_BT} \qquad (11.6)$$

We only count the excitation energy rather than the total energy because we want our zero energy state to be a state of no translational kinetic energy with all the molecules in their internal ground states.

The first problem we will face in any attempt to apply Equation 11.6 is the calculation of the normalization constant Q. The condition that the Gibbs distribution be normalized is that

$$\sum_n P_n = 1$$

This easily leads to a formula for Q.

$$Q = \sum_{n} e^{-\epsilon_n/k_BT} \tag{11.7}$$

We will illustrate how the normalization constant can be evaluated by carrying out the evaluation for some important cases.

11.6. VIBRATIONAL SPECIFIC HEATS

The first example is the harmonic oscillator energy spectrum. The results of this calculation will be used to compute the internal energy associated with the vibrational states of a gas of diatomic molecules and the heat capacity due to lattice vibrations in a solid. For a harmonic oscillator

$$\epsilon_n = (n+\tfrac{1}{2})\hbar\omega_0 - \tfrac{1}{2}\hbar\omega_0 = n\hbar\omega_0$$

Therefore

$$Q = \sum_{n=0}^{\infty} e^{-n\hbar\omega_0/k_BT}$$

If we let $x = e^{-\hbar\omega_0/k_BT}$ then

$$e^{-n\hbar\omega_0/k_BT} = x^n$$

and

$$Q = \sum_{n=0}^{\infty} x^n$$

This is the well-known geometric series whose sum is $1/(1-x)$.

$$Q = \frac{1}{1-e^{-\hbar\omega_0/k_BT}}$$

Now that we have the normalization constant we can evaluate E_{ex} using Equation 11.6. This can most easily be done by employing the following device, which makes no use of the detailed form of the harmonic oscillator spectrum. We define the inverse temperature β, by

$$\beta = \frac{1}{k_BT}$$

Then

$$Q = \sum_{n} e^{-\beta\epsilon_n} \qquad \text{and} \qquad E_{ex} = NQ^{-1}\sum \epsilon_n e^{-\beta\epsilon_n}$$

We notice that

$$E_{ex} = -NQ^{-1}\frac{dQ}{d\beta} \tag{11.8}$$

For the harmonic oscillator spectrum this formula gives the total vibrational energy as

$$E_{vib} = N\frac{\hbar\omega_0}{e^{\beta\hbar\omega_0}-1} \tag{11.9}$$

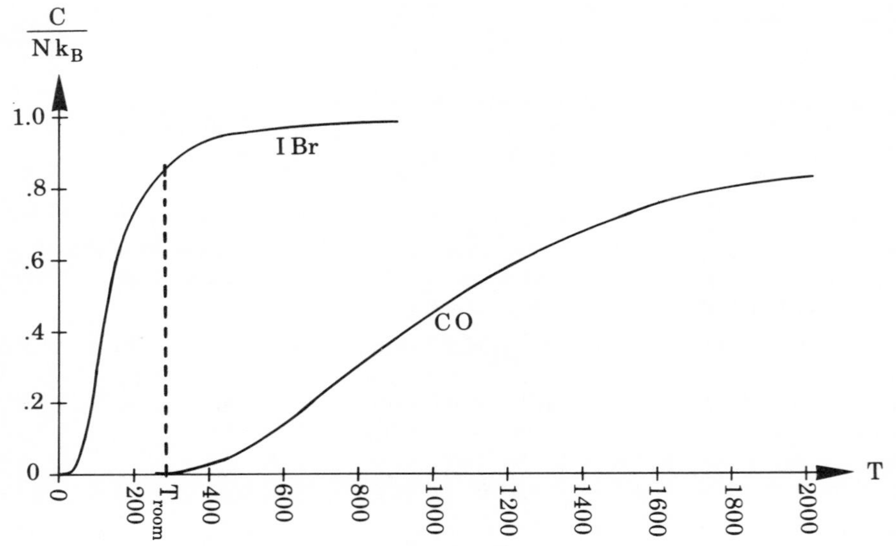

FIGURE 11.9 The vibrational contribution to the heat capacity of I Br and C O.

The contribution of the vibrational energy to the heat capacity per particle is plotted in Figure 11.9 for two common diatomic gases. One of them (IBr) has a fairly low vibrational frequency, while the other (CO) is a more tightly bound molecule with a rather high vibrational frequency. From the plot it is clear that at room temperature the vibrational degrees of freedom make a major contribution to the heat capacity of iodine bromide but at the same temperature the carbon monoxide molecules are almost all in their vibrational ground states.

Before making any comparison of the theoretical calculations with the total heat capacities of real diatomic gases it is important that the contribution of rotational degrees of freedom be added to the translational and vibrational heat capacities. This is discussed in more detail in a later section.

11.7. THE HEAT CAPACITY OF A SOLID

Equation 11.9 can be applied to a simple model, due to Einstein, of the vibrational energy of a solid. In the early days of the quantum theory Einstein's model was successful in explaining the mysterious vanishing of the heat capacity of a solid at low temperature. One assumes that each atom in the solid finds itself in a potential that can be approximated by a three-dimensional harmonic oscillator potential. We will assume that the potential is isotropic although only a trivial modification is required if the vibrational frequencies in the three different directions are not equal. Since the vibrations of a three-dimensional oscillator in the three perpendicular coordinate directions proceed entirely independently of one another a single three-dimensional oscillator is equivalent to three one-dimensional oscillators. Thus a solid, made up of N such oscillators would have the same vibrational energy as 3N one-dimensional oscillators. The internal energy at temperature T could therefore be predicted by Equation 11.9 with N replaced by 3N.

$$E = 3N \frac{\hbar\omega_0}{e^{\beta\hbar\omega_0} - 1} \qquad (11.10)$$

The heat capacity is given, in this model by the formula

$$C = \frac{dE}{dT} = 3Nk_B \frac{x^2 e^x}{(e^x - 1)^2}$$

where $x = \hbar\omega_0/k_BT$. The theory contains the single parameter ω_0 that must be chosen so as to give the best fit to the experimentally measured heat capacity of a particular solid. If T is much larger than $\hbar\omega_0/k_B$ then $x << 1$ and $(e^x - 1)^2 \approx x^2$. In that limit

$$C \approx 3Nk_B \qquad (T >> \hbar\omega_0/k_B)$$

This is the result of a classical calculation using the same model. The classical model had been worked out before Einstein's analysis and it was just the deviations, at lower temperatures, from the classical predictions that were unexplained. If $T << \hbar\omega_0/k_B$ then $x >> 1$ and $x^2 e^x/(e^x - 1)^2 \approx x^2 e^{-x}$, giving

$$C \approx \frac{3N\hbar^2\omega_0^2}{k_BT^2} e^{-\hbar\omega_0/k_BT} \qquad (T << \hbar\omega_o/k_B)$$

For small T the exponential dominates the T^{-2} factor and C goes to zero.

QUESTION: At 300 K the heat capacity of one mole of diamond is found to be 6 J/K. Using the Einstein model, what would its heat capacity be at 1000 K?

ANSWER: We are given that

$$C = 3Nk_B \frac{x^2 e^x}{(e^x - 1)^2} = 6 \text{ J/K}$$

where $x = \hbar\omega_0/k_BT$. Using the fact that $N_0 k_B \equiv R = 8.3$ J/K we get

$$\frac{x^2 e^x}{(e^x - 1)^2} = .72$$

This transcendental equation for x can be solved graphically by plotting the function on the left and seeing for what value of x it is equal to .72. When this is done we find that x = 2.0. Since T = 300 K this gives

$$\omega_0 = \frac{(2.0)(1.38 \times 10^{-23})(300)}{(1.05 \times 10^{-34})} = 7.9 \times 10^{13} \text{ Hz}$$

At T = 1000 K the value of x will be

$$x = \frac{(1.05 \times 10^{-34})(7.9 \times 10^{13})}{(1.38 \times 10^{-23})(1000)} = .60$$

and at that temperature

$$C = 3R \frac{x^2 e^x}{(e^x - 1)^2} = 24.2 \text{ J/K}$$

Although the Einstein model gives a dramatic improvement over the classical model in predicting the thermodynamic behavior of solids at low temperatures it does not really give good quantitative agreement with experiment, particularly in the very interesting region near absolute zero where the model predicts that C should go to zero exponentially as T goes to zero ($C \propto e^{-\alpha/T}/T^2$ as $T \rightarrow 0$) but the experiments show that C goes to zero as a cubic function of T. ($C \propto T^3$ as $T \rightarrow 0$) The reason why the Einstein model gives a poor representation of the heat capacity at low temperatures is revealed by the following argument. Suppose a system has a number of different normal modes of vibration with different frequencies. Whenever $T << \hbar\omega/k_B$ for any particular normal mode then that normal mode becomes "frozen" and does not contribute to the heat capacity. Thus, at very low temperatures, the important vibrational modes are those of lowest frequency. All others have been frozen. But Einstein's model gives a very poor representation of the physical nature of the lowest frequency vibrational modes of a solid. Because the atoms in a solid interact strongly with one another the low frequency vibrational modes are ones in which large numbers of atoms vibrate coherently. In fact they are simply sound waves with macroscopic wavelengths. In order to obtain accurate quantitative results for the heat capacity of a solid we must use a better representation of the actual vibrational normal modes of the system.

The proper way to carry out a quantum mechanical calculation of the heat capacity of a solid is to use the density of normal modes function, $D(\omega)$, introduced in Chapter 10. $D(\omega)\,d\omega$ gives the number of vibrational modes of the crystal lattice that have angular frequencies within the range ω to $\omega + d\omega$. It is clear from Equation 11.9 that the average thermal energy of an oscillator of angular frequency ω is

$$\overline{E}_\omega = \frac{\hbar\omega}{e^{\beta\hbar\omega} - 1} \qquad (11.11)$$

Therefore the average thermal energy of the complete set of lattice vibrations at temperature T is

$$U(T) = \int_0^\infty D(\omega)\,\frac{\hbar\omega\, d\omega}{e^{\beta\hbar\omega} - 1} \qquad (11.12)$$

This is Equation 10.13 of Chapter 10. When this equation is used, along with more accurate approximations for $D(\omega)$, the theoretical values of the vibrational heat capacities of solids agree very well with experimental measurements at all temperatures.

11.8. DERIVATION OF PLANCK'S FORMULA

Equation 11.11 for the average thermal energy of an oscillator of angular frequency ω is identical to the equation given at the bottom of page 71 for the average energy of a normal mode of the electromagnetic field within a cavity. This shows that the normal mode oscillations of the electromagnetic field can be treated as a system of independent quantized harmonic oscillators. Combining Equation 11.11 with the formula for the number of normal modes per frequency interval (also given at the bottom of page 71), namely $D(\nu) = 8\pi\nu^2/c^3$, we obtain the Planck formula for the electromagnetic radiation energy in a cavity at temperature T (recall that $\hbar\omega = h\nu$).

$$\frac{E}{V} = \frac{8\pi h}{c^3} \int_0^\infty \frac{\nu^3 d\nu}{e^{\beta h\nu} - 1} \qquad (11.13)$$

QUESTION: Why does the light given off by a hot object change its color as the object cools?

ANSWER: The perceived color of an object depends on the distribution of the radiant energy as a function of frequency. If we assume that the object radiates as a black body then we can use the Planck distribution to describe the energy radiated by the object. If, in Equation 11.13, we introduce the variable $x = h\nu/k_BT$ then the equation becomes

$$\frac{E}{V} = \frac{8\pi (k_BT)^4}{(ch)^3} \int_0^\infty \frac{x^3\,dx}{e^x - 1} \qquad (11.14)$$

At fixed T, the variable x is proportional to the frequency. Thus the function

$$\phi(x) = \frac{x^3}{e^x - 1}$$

describes how the radiant energy is distributed as a function of frequency. It has a maximum at the point x_m defined by

$$\phi'(x_m) = 0$$

The value of ν at which the maximum occurs is given by

$$\frac{h\nu_m}{k_BT} = x_m$$

or

$$\nu_m = (k_B x_m/h)\,T \qquad (11.15)$$

Thus the maximum in the energy distribution as a function of frequency occurs at a frequency that is proportional to the absolute temperature. (This fact is called the *Wien displacement law.*) The perceived color is approximately the color of light of frequency ν_m and it therefore moves through the spectrum from red to violet as T increases.

11.9. THE ROTATIONAL HEAT CAPACITY OF A DIATOMIC GAS

In Chapter 7 we saw that the energy levels of a diatomic molecule are

$$E_{nl} = (n + \tfrac{1}{2})\hbar\omega_0 + A\,l(l+1)$$

where n and l are the vibrational and rotational quantum numbers respectively, and $A = \hbar^2/2\mu r_0^2$ (See Equation 7.41). There is another quantum number that is needed to uniquely define the quantum state but that does not appear in the expression for the energy. It is m, the magnetic quantum number, that gives the magnitude of the z-component of angular momentum. For a given value of l there are $2l+1$ different quantum states with energy E_{nl}. Taking this degeneracy into account the normalization constant for the Gibbs distribution (See Equation 11.7) can be written as

$$Q = \sum_{n=0}^{\infty}\sum_{l=0}^{\infty}(2l+1)\,e^{-\beta(\hbar\omega_0 n + A\,l(l+1))}$$

Since the sum over n and l are independent and the exponential factors into $e^{-\beta\hbar\omega_o n} e^{-\beta A l(l+1)}$ we get

$$Q = \sum_{n=0}^{\infty} e^{-\beta\hbar\omega_o n} \sum_{l=0}^{\infty} (2l+1) e^{-\beta A l(l+1)}$$

$$= Q_{vib}\, Q_{rot}$$

where Q_{vib} is the normalization constant for a one dimensional oscillator. Using Equation 11.8 to evaluate the vibrational and rotational contributions to the internal energy gives

$$E_{ex} = -N Q^{-1} \frac{dQ}{d\beta} = -N Q_{vib}^{-1} \frac{dQ_{vib}}{d\beta} - N Q_{rot}^{-1} \frac{dQ_{rot}}{d\beta}$$

$$= E_{vib} + E_{rot}$$

Thus the vibrational and rotational terms separate and we need only evaluate E_{rot} and add it to our previously computed expression for E_{vib}.

$$Q_{rot} = \sum_{l=0}^{\infty} (2l+1) e^{-\beta A l(l+1)} \tag{11.16}$$

The infinite sum defining Q_{rot} can be computed exactly only in the high temperature limit. In that limit one can approximate the sum by an integral over l.

$$Q_{rot} = \int_{0}^{\infty} (2l+1) e^{-\beta A l(l+1)} dl$$

This integral is evaluated by letting $y = \beta A\, l(l+1)$. Then $dy = \beta A (2l+1) dl$ and

$$Q_{rot} = \frac{1}{\beta A} \int_{0}^{\infty} e^{-y} dy = \frac{1}{\beta A}$$

This gives

$$E_{rot} = N k_B T$$

As one would expect, the high temperature result agrees with the result of a calculation using classical mechanics for the rotational motion.

QUESTION: Above what temperature is the classical result a useful approximation?

ANSWER: The spacing between the first two energy levels is equal to A. If the typical thermal energy, namely $k_B T$, is much larger than the spacing between discrete quantum states then many terms in the sum make substantial contributions and we can use the integral approximation. Thus, a reasonable criterion for the use of the high temperature limit is that $k_B T >> A$ Since $A = \hbar^2/2\mu r_o^2$ the criterion we get for the case of the potassium chloride molecule is

$$T >> \frac{\hbar^2}{2 k_B \mu r_o^2} = \frac{(10^{-34})^2}{(2)(1.4\times 10^{-23})(3\times 10^{-26})(2.8\times 10^{-10})^2}$$

or $T >> .15\,K$. For most diatomic molecules the classical approximation is good at all except very low temperatures. The worst case is H_2 in which T must be larger than 100 K (− 173°C).

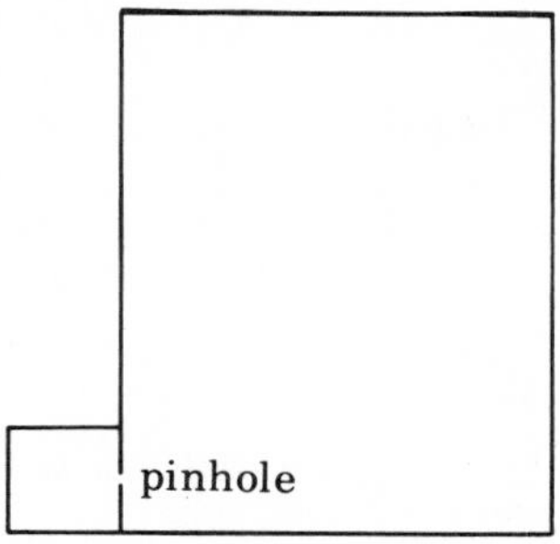

FIGURE 11.10

The small system can exchange both energy and particles with the reservoir.

11.10. THE GIBBS DISTRIBUTION FOR AN OPEN SYSTEM

In Section 11.4 we looked at a small system that could exchange energy with a large reservoir. We saw that the probability that the small system will be found in a quantum state of energy E_n is proportional to $e^{-\beta E_n}$ where $\beta = 1/k_B T$. We now consider what happens when the system can exchange both energy and particles with the reservoir. A system in that situation is called an *open system.* An example of such a system is shown in Figure 11.10. The reservoir and system are connected by a pinhole that allows particles to move from one to the other. The question we ask is the following. What is the probability that the system will contain N particles and be in a particular N-particle quantum state of energy E? The derivation of the probability can be carried out in a way that is very similar to the analysis given in Section 11.4. The details are left as an exercise. (See Problem 11.13.) The result is that the probability that the open system will be found in a quantum state ψ_n, that has energy E and particle number N, is equal to

$$P_n = C e^{-\alpha N - \beta E} \tag{11.17}$$

where $k_B \alpha = \dfrac{\partial S_R}{\partial N}$, $k_B \beta = \dfrac{1}{T} = \dfrac{\partial S_R}{\partial E}$, and $S_R(E_R, N_R)$ is the entropy of the reservoir. The parameter α is called the *affinity* of the reservoir.

QUESTION: What is the physical meaning of the "affinity", α, and why is it given that name?

ANSWER: We can best answer that question by looking at some physical process in which the affinity plays a dominant role. We will consider a system that consists of a liquid component and a gas component in contact within a closed container. We assume that the temperature and pressure in the liquid are equal to those in the gas. The total entropy of the system is the sum of the entropies of the components.

$$S = S_l(E_l, V_l, N_l) + S_g(E_g, V_g, N_g)$$

Let us assume that the two components are *not* in equilibrium and compute the rate of change of S as follows.

$$\frac{dS}{dt} = \frac{\partial S}{\partial E_l}\frac{dE_l}{dt} + \frac{\partial S}{\partial V_l}\frac{dV_l}{dt} + \frac{\partial S}{\partial N_l}\frac{dN_l}{dt} + \frac{\partial S}{\partial E_g}\frac{dE_g}{dt} + \frac{\partial S}{\partial V_g}\frac{dV_g}{dt} + \frac{\partial S}{\partial N_g}\frac{dN_g}{dt}$$

Conservation of energy, total volume, and particles implies that

$$\frac{dE_g}{dt} = -\frac{dE_l}{dt}, \quad \frac{dV_g}{dt} = -\frac{dV_l}{dt}, \quad \text{and} \quad \frac{dN_g}{dt} = -\frac{dN_l}{dt}$$

Equality of temperature and pressure imply that

$$\frac{\partial S}{\partial E_l} = \frac{\partial S}{\partial E_g} \quad \text{and} \quad \frac{\partial S}{\partial V_l} = \frac{\partial S}{\partial V_g}$$

Thus $\frac{dS}{dt}$ can be written in the simple form

$$\frac{dS}{dt} = (\alpha_l - \alpha_g)\frac{dN_l}{dt}$$

where $\alpha_l = \frac{\partial S_l}{\partial N_l}$ and $\alpha_g = \frac{\partial S_g}{\partial N_g}$. But, by the Second Law of Thermodynamics, the entropy of a closed system will increase until internal equilibrium is reached. Thus

$$(\alpha_l - \alpha_g)\frac{dN_l}{dt} > 0$$

which means that particles will condense on the liquid ($\frac{dN_l}{dt} > 0$) if $\alpha_l > \alpha_g$ and particles will evaporate from the liquid ($\frac{dN_l}{dt} < 0$) if $\alpha_l < \alpha_g$. In general, when two components can transfer any type of particle then that particle will spontaneously move to the component with the larger affinity.

11.11. THE IDEAL QUANTUM GAS

The simplest (but most unrealistic) example of an ideal quantum gas is a system of N identical noninteracting particles in a one-dimensional periodic box. The single-particle quantum states of such a system are the one-dimensional plane waves.

$$\phi_n(x) = L^{-1/2} e^{ik_n x}$$

where $k_n = (2\pi/L)n$, and the corresponding single-particle energies are $E_n = \hbar^2 k_n^2/2m$.

Because the particles are identical the overall state of the N-particle system can be defined by specifying how many particles are in each of the single-particle energy states. For truly identical particles there would be no physical meaning to specifying *which* particles were in a particular state. If N_n is the number of particles in state ϕ_n then the total energy of the system is given by

$$E = \sum_n N_n E_n$$

In a system in which there was absolutely no interaction between the particles, any distribution at all of the particles among the various energy states would remain fixed forever. For instance, if half the particles were moving to the right with one velocity while the other half were moving to the left with another they would remain in that peculiar state indefinitely. Such a situation is obviously unrealistic. In any real system there are always some small interactions between the particles that will cause an unrealistic velocity distribution such as the one just described to gradually be transformed into a more normal equilibrium distribution of the particles among the available quantum states. It is our aim to calculate the resulting equilibrium distribution. We can do so using the following argument.

We consider the collection of particles in the nth quantum state as a *small system* and the collection of particles in all the other states as a *reservoir*. The particles in the small system interact weakly with the particles in the reservoir. The small system is an open system in the sense that particles in the reservoir can make transitions into the nth quantum state and therefore become part of the small system and vice versa. This is just the situation we have considered in the last section. Therefore we can state that, at equilibrium, the probability that the small system (the nth quantum state) will contain exactly N_n particles is given by

$$P(N_n) = C e^{-\alpha N_n - \beta E}$$

But N_n particles in the nth energy level have an energy $N_n E_n$. Therefore

$$P(N_n) = C e^{-(\alpha + \beta E_n) N_n} \qquad (11.18)$$

11.12. THE IDEAL FERMI-DIRAC GAS

For a system of Fermi-Dirac particles the possible values of N_n, the number of particles in the nth quantum state, are 0 and 1. For this case the normalization constant, C, in Equation 11.18 is quite easy to evaluate.

$$\frac{1}{C} = \sum_{N_n} e^{-(\alpha + \beta E_n) N_n} = 1 + e^{-\alpha - \beta E_n} \qquad (11.19)$$

The average number of particles in the nth quantum state is given by

$$\overline{N}_n = \sum_{N_n} P(N_n) N_n = \frac{e^{-\alpha - \beta E_n}}{1 + e^{-\alpha - \beta E_n}} = \frac{1}{e^{\alpha + \beta E_n} + 1}$$

It should be clear that this analysis made no use of the detailed nature of the system other than the facts that the particles had only very weak interactions and that they were Fermi-Dirac particles. Thus, we can state that in any system of weakly-interacting fermions at equilibrium the average number of particles in a single-particle quantum state of energy E is

$$\overline{N}(E) = \frac{1}{e^{\alpha + \beta E} + 1} \qquad (11.20)$$

It is convenient, in analyzing this *Fermi-Dirac distribution function*, to introduce a parameter, called the *Fermi energy*, that is defined as that energy at which a single-particle state is equally likely to be found occupied or empty. That is

$$\overline{N}(E_F) = \frac{1}{e^{\alpha + \beta E_F} + 1} = \frac{1}{2}$$

For a given temperature the affinity can be written in terms of the Fermi energy by solving the above equation for α. One gets

$$\alpha = -\beta E_F$$

Using this equation and the relation $\beta = 1/k_B T$ we can rewrite Equation 11.20 in the form

$$\overline{N}(E) = \frac{1}{e^{(E - E_F)/k_B T} + 1} \qquad (11.21)$$

This function is plotted, for two different temperatures, in Figure 11.11.

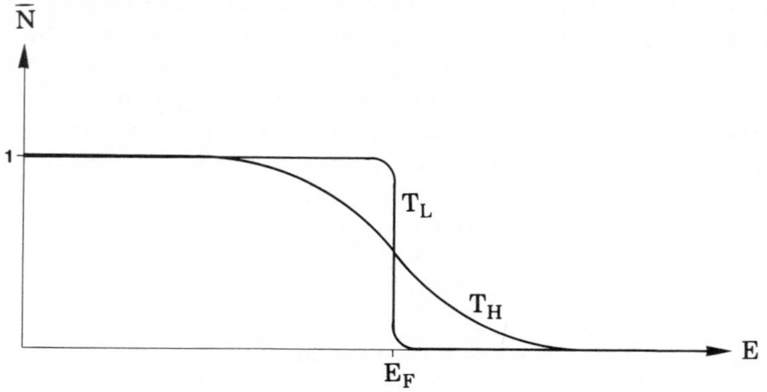

FIGURE 11.11 The Fermi-Dirac distribution function for two temperatures. $k_B T_H >> E_F$ and $k_B T_L << E_F$.

The limiting form of the Fermi-Dirac distribution function as $T \to 0$ is very easy to understand. All of the energy states up to the Fermi energy are filled with particles while all of the states above the Fermi energy are empty. It is obvious that this is the lowest possible energy state consistent with the Pauli restriction that no more than one fermion can occupy a given single-particle state. That is, as $T \to 0$ the system settles into its lowest possible energy state. The Fermi energy at $T = 0$ can therefore be determined by the condition that the total number of quantum states of energy less than E_F be equal to the number of particles in the system. It should be clear from this analysis that, in this section, we are treating states of opposite spin as independent single-particle states.

QUESTION: For a system of 10^{10} spin-½ fermions of mass m_e in a one-dimensional periodic box of length $L = 1\,\mathrm{m}$ what is the Fermi energy at $T = 0$?

ANSWER: The allowed values of the wave vector, k, are $k_n = 2\pi n/L$. These can be viewed as points of an infinite one-dimensional lattice of spacing $2\pi/L$ along the k-axis. For each allowed wave vector there are two possible spin states. The states of lowest energy are the states for which the magnitude of k_n is the smallest. Thus, at $T = 0$, the states with $-K_F \leqslant k_n \leqslant +K_F$ are each occupied by two particles of opposite spin. K_F is determined by the condition that the number of lattice points in the interval $(-K_F, +K_F)$ be equal to $N/2$.

$$\frac{2K_F}{(2\pi/L)} = \frac{N}{2}$$

This gives $K_F = \dfrac{\pi N}{2L}$ and

$$E_F = \frac{\hbar^2 K_F^2}{2m_e} = 1.5 \times 10^{-18}\,\mathrm{J} = 9.34\,\mathrm{eV}$$

For a three-dimensional gas of spin-½ fermions the states with energy less than E_F are the states for which

$$\frac{p^2}{2m} < E_F$$

If we plot these in a three-dimensional *momentum space* they form a sphere, called the *Fermi sphere.* At $T = 0$ any momentum state within the Fermi sphere is occupied by a pair of particles of opposite spin. Any momentum state outside the Fermi sphere is unoccupied. It can be shown by an argument similar to the one given in the question and answer above that a momentum state at the surface of the Fermi sphere has a wave vector of magnitude

$$K_F = (3\pi^2 n)^{1/3} \tag{11.22}$$

where $n = N/V$ is the particle density. Therefore, a three-dimensional fermion system has a Fermi energy

$$E_F = \frac{\hbar^2 K_F^2}{2m} = \frac{\hbar^2 (3\pi^2 n)^{2/3}}{2m} \tag{11.23}$$

QUESTION: Suppose we consider the system of conduction electrons in copper as a gas of spin-½ fermions. What is the Fermi energy of the system and how does it compare with a typical room temperature thermal energy, $k_B T_{room}$? The density of copper is $8.93 \times 10^3\,\mathrm{kg/m^3}$.

ANSWER: Using the density of copper and the mass of a copper atom ($m_{Cu} = 63.5\,u = 1.05 \times 10^{-25}\,\mathrm{kg}$) we find that the number of atoms per cubic meter in copper is 8.47×10^{28}. Copper is a divalent metal and therefore has two conduction electrons per atom. The number of conduction electrons per cubic meter is thus twice the number of atoms per cubic meter. This gives

$$E_F = \frac{(1.05 \times 10^{-34}\,\mathrm{J\text{-}s})^2 (3\pi^2\, 1.69 \times 10^{29}\,\mathrm{m^{-3}})^{2/3}}{(2)(9.1 \times 10^{-31}\,\mathrm{kg})}$$

$$= 1.78 \times 10^{-18}\,\mathrm{J} = 11.1\,\mathrm{eV}$$

Taking $T_{room} = 300\,\mathrm{K}$ we see that

$$\frac{E_F}{k_B T_{room}} = 429$$

which shows that at room temperature the thermal energy of the conduction electrons in copper is very small in comparison to the minimum kinetic energy allowed by the exclusion principle. The room temperature distribution of conduction electrons in any metal is very close to the discontinuous distribution that occurs at $T = 0$. In other words, in Figure 11.11 the curve marked T_L is the relevant one for the conduction electrons in a metal.

11.13. THE IDEAL BOSE-EINSTEIN GAS

We saw that, for any quantum system of weakly-interacting particles, the probability of finding a quantum state of energy E occupied by N particles is given by

$$P(N) = C e^{-(\alpha + \beta E)N}$$

For Fermi-Dirac particles the possible values of N are only 0 and 1. For Bose-Einstein particles there is no such restriction on the allowed values of the occupation number N. Thus, in calculating the normalization constant, C, (in order to derive a completely explicit formula for P(N)) we must sum over all nonnegative integers. That is

$$\frac{1}{C} = \sum_{N=0}^{\infty} e^{-(\alpha+\beta E)N}$$

This is essentially the same as the infinite sum we evaluated in Section 11.6. We define a quantity $Z = e^{-(\alpha+\beta E)}$ and notice that $e^{-(\alpha+\beta E)N} = Z^N$. Therefore

$$\frac{1}{C} = \sum_{0}^{\infty} Z^N = \frac{1}{1-Z}$$

This gives the following formula for the probability of finding exactly N Bose-Einstein particles in a state of energy E.

$$P(N) = (1-e^{-(\alpha+\beta E)})\, e^{-(\alpha+\beta E)N} \qquad (11.24)$$

It is left as a problem for the reader to prove that the average occupation number of a state of energy E in a Bose-Einstein system is given by the following *Bose-Einstein distribution function.*

$$\overline{N}(E) = \frac{1}{e^{\alpha+\beta E}-1} \qquad (11.25)$$

QUESTION: Looking at Equation 11.13 for the energy density of the "photon gas" in a cavity at temperature T it is clear that the integrand contains the Bose-Einstein distribution function with $\alpha = 0$. Why is α always zero for a photon gas?

ANSWER: Since photons can be created or destroyed there is no conservation law maintaining the total number of photons fixed. Therefore, according to the principles of thermodynamics, the photon number will adjust itself so as to maximize the entropy. The condition of maximum entropy is that

$$\frac{\partial S}{\partial N} = 0$$

But, $k_B \alpha = \dfrac{\partial S}{\partial N}$ and therefore $\alpha = 0$.

11.14. THE ABSORPTION AND EMISSION OF RADIATION

The power of thermodynamic arguments to give important information about fundamental physical processes is nicely illustrated by the following analysis, due to Einstein. We consider a cavity containing a dilute gas of atoms. The system is kept at temperature T. For simplicity, we assume that the atoms have only two energy levels, E_0 and E_1, with an excitation energy, $\epsilon = E_1 - E_0$. (The argument could be generalized to any number of energy levels.) Let

N_0 = the number of atoms in the ground state

N_1 = the number of atoms in the excited state

and

ρ = the density of photons of frequency $\nu = \epsilon/h$

The atoms will be constantly exchanging energy with the radiation field by absorbing and emitting photons of energy ϵ. We will make certain simple assumptions about the absorption and emission processes.

1. For any given atom, in its ground state, the probability per unit time of its absorbing a photon and making a transition to its excited state is proportional to ρ, the density of photons of the appropriate energy. With this assumption the rate at which atoms are making transitions from the ground state to the excited state is

$$R_{0\rightarrow 1} = A N_o \rho$$

The constant A is called the *absorption coefficient.*

2. For any given atom, in its excited state, the probability per unit time of its emitting a photon and making a transition to its ground state contains one term that is independent of the photon density and represents the *spontaneous emission* of a photon by the atom, and one term that is proportional to the photon density, ρ, and represents *stimulated emission.* That is

$$R_{1\rightarrow 0} = (B_{sp} + B_{st}) N_1 \rho$$

B_{sp} and B_{st} are called the *coefficients of spontaneous and stimulated emission* respectively. The coefficient of spontaneous emission seems very natural, but the necessity for including the B_{st} seems very questionable. Einstein showed that a classical analysis of the interaction between charged particles and radiation required that both coefficients be nonzero. We will simply show that it is impossible to set B_{st} equal to zero without producing a violation of the equations we have already derived for the equilibrium distributions of radiation and particles.

At equilibrium the number of atoms in the ground state must remain constant. Therefore

$$\frac{dN_o}{dt} = R_{1\rightarrow 0} - R_{0\rightarrow 1} = 0$$

which implies that

$$B_{sp} + B_{st}\,\rho = A\rho\,\frac{N_o}{N_1} \qquad (11.26)$$

But we have already shown that at equilibrium,

$$\frac{N_o}{N_1} = e^{\beta\epsilon} \text{ and } \rho = \frac{D(\nu)}{e^{\beta h\nu} - 1} = \frac{8\pi\nu^2}{c^3(e^{\beta h\nu} - 1)}$$

Using this, and the fact that $\epsilon = h\nu$, we can write Equation 11.26 in the form

$$(e^{\beta\epsilon} - 1)\frac{c^3 B_{sp}}{8\pi\nu^2} + B_{st} = A e^{\beta\epsilon}$$

The parameters A, B_{sp}, and B_{st} are the probabilities of fundamental microscopic processes. They would be expected to depend on the detailed nature of the quantum states of the atom and on the energy, ϵ, but they cannot depend on the temperature, T, which describes the distribution of the atoms and photons among their possible energy states and has nothing to do with the probability that a particular atom in a known state will or will not absorb a photon. Thus, the only way Equation 11.38 can be valid at all temperatures is for the terms multiplying $e^{\beta\epsilon}$ and the constant terms to be separately equal. From that we get that

$$A = B_{st} = \frac{c^3}{8\pi\nu^2} B_{sp} \qquad (11.40)$$

Only one of the three, independently defined, coefficients, A, B_{st}, and B_{sp} is truly independent. Given any one of them, the laws of quantum statistical mechanics will give the other two.

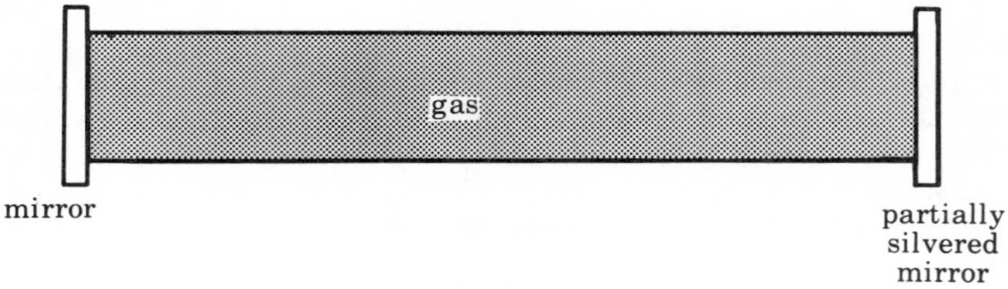

FIGURE 11.12 The elements of a gas laser.

11.15. THE LASER

Stimulated emission is an important element in the operation of a laser. The essentials of a gas laser are shown in Figure 11.12. Other types of lasers may use liquids or solids as the working substance. The gas is in a tube with mirrors at the ends. One of the mirrors is partially transmitting. By some means the molecules of the gas are caused to make transitions to a variety of excited states. This might be done by shining a strong light through the transparent sides of the tube, or it might be done by passing an electric current through the gas, or in some other way. It is vital to the operation of the device that the molecules have a low-lying excited state that is metastable. Being metastable means that its lifetime with respect to spontaneous emission is much larger than the lifetimes of the other states. We will assume that it is the first excited state. After a short time many of the molecules that have been excited make transitions into the metastable state. (See Figure 11.13.) Because that state is metastable the molecules do not quickly make transitions to the ground state and the population of that state in comparison to the population of the ground state increases far above the equilibrium value. If one of the metastable molecules emits a photon in a direction parallel to the tube then that photon is likely to be reflected a number of times by the mirrors before leaving through the front mirror. During its travel back and forth through the gas it will stimulate other molecules to emit photons. There is an aspect of stimulated emission that was not mentioned in the last section but is important for laser action. Namely, when one photon stimulates a molecule to emit another photon the second photon is emitted into exactly the same quantum state as is occupied by the first one. This is possible because photons are bosons and therefore any number of them can occupy a single state. In this way the occupation number of a single photon state can rise exponentially.

The more photons there are in the state already the stronger is the stimulated emission and therefore the faster do new photons enter the state. Of course photons are constantly leaking out the partially transmitting front mirror. This produces a powerful beam of extremely coherent light. Since the photons are all in the same quantum state they move in the same direction with the same phase. In order to keep the laser action going the mechanism for exciting molecules from the ground state must be kept in continuous operation so as to prevent depletion of the important metastable state.

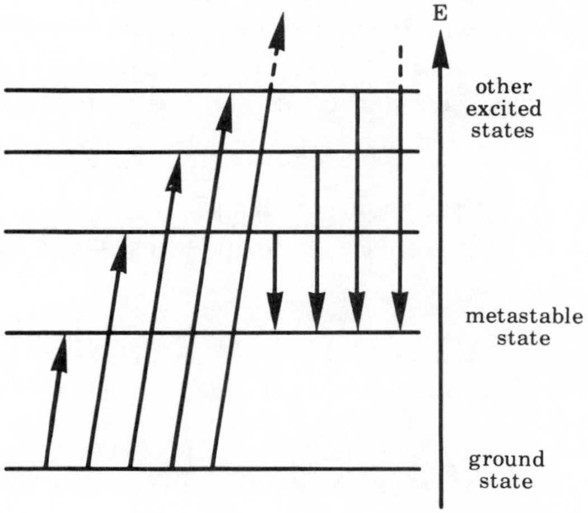

FIGURE 11.13 The energy level diagram of a material that is suitable for use in a laser.

SUMMARY

The entropy of a sample S(E, N, V), is related to **N**, the number of quantum states of that sample that have the values, E, N, and V, by

$$S = k_B \, ln \mathbf{N}$$

The fundamental assumption of quantum statistical mechanics is that the probability that, after a long period of interaction, a composite system will be found with the values $(E_1, N_1, V_1, \cdots, E_k, N_k, V_k)$ for its thermodynamic parameters is proportional to the number of quantum states of the complete system for which the components have the indicated values of E, N, and V.

If a system has been in contact with a reservoir at temperature T, the probability that the system will be found in a particular quantum state, ψ_n, is equal to

$$P_n = C e^{-\beta E_n}$$

where E_n is the energy of the state and $\beta = 1/k_B T$.

The total internal energy of a dilute gas at temperature T is equal to

$$U(T) = \tfrac{3}{2} N k_B T + E_{ex}(T)$$

where $E_{ex}(T)$ is the excitation energy of the gas particles. The excitation energy, E_{ex}, is given by the following sum, taken over all the quantum states of a single gas particle.

$$E_{ex} = N Q^{-1} \sum_n \epsilon_n e^{-\beta \epsilon_n}$$

where $\epsilon_n = E_n - E_o$ is the excitation energy of the nth quantum state and

$$Q = \sum_n e^{-\beta \epsilon_n}$$

The excitation energy can also be calculated by the formula

$$E_{ex} = -N Q^{-1} \frac{dQ}{d\beta}$$

For a single harmonic oscillator

$$Q = \sum_{n=0}^{\infty} e^{-\beta \hbar \omega_o n} = \frac{1}{1 - e^{-\beta \hbar \omega_o}}$$

In a system with a number of normal modes of vibration each vibrational mode makes a contribution to the internal energy at temperature T of

$$E = \frac{\hbar \omega_o}{e^{\beta \hbar \omega_o} - 1} \qquad \text{(Energy per vibration mode)}$$

The Einstein model of a solid assumes that each atom is located in a three-dimensional harmonic oscillator potential and that each atom vibrates independently. It leads to the following formula for the internal vibrational energy of a solid.

$$E_{vib} = 3N \frac{\hbar \omega_o}{e^{\beta \hbar \omega_o} - 1}$$

The exact quantum mechanical formula for the contribution of the vibrational modes to the heat capacity of a solid is given in terms of the density of normal modes function $D(\omega)$ by

$$E_{vib} = \int_0^\infty D(\omega)\frac{\hbar\omega\, d\omega}{e^{\beta\hbar\omega} - 1}$$

Treating the normal modes of the electromagnetic field within a cavity as quantized oscillators one can derive Planck's formula for the energy density

$$\frac{E}{V} = \frac{8\pi h}{c^3}\int_0^\infty \frac{\nu^3 d\nu}{e^{\beta h\nu} - 1}$$

The internal energy of a gas of diatomic molecules is a sum of translational, vibrational, and rotational terms.

$$U(T) = \tfrac{3}{2} N k_B T + E_{vib} + E_{rot}$$

where

$$E_{vib} = N\frac{\hbar\omega_0}{e^{\beta\hbar\omega_0} - 1}$$

and

$$E_{rot} = -N Q_{rot}^{-1}\frac{d Q_{rot}^{-1}}{d\beta}$$

with

$$Q_{rot} = \sum_{l=0}^{\infty}(2l+1)e^{-\beta A l(l+1)}$$

and $A = \hbar^2/2\mu r_0^2$. At room temperature, for all diatomic molecules, one can use the high temperature limit

$$E_{rot} = N k_B T$$

If a system can exchange energy and particles with a reservoir the probability of finding the system in a particular N-particle quantum state, ψ_n, that has energy E, is given by the Gibbs distribution.

$$P_n = C e^{-\alpha N - \beta E}$$

For a Fermi-Dirac system at temperature T the average number of particles in a single-particle quantum state of energy E is

$$\overline{N}(E) = \frac{1}{e^{(E - E_F)/k_B T} + 1}$$

For a Bose-Einstein system the average number of particles in a state of energy E is

$$\overline{N}(E) = \frac{1}{e^{\alpha + \beta E} - 1}$$

If a collection of atoms is exposed to radiation the rate at which atoms make transitions from a state with energy E_0 to a state with energy $E_1 > E_0$ is

$$R_{0\to1} = A N_0 \rho$$

The transition rate in the other direction is

$$R_{1\to 0} = B_{sp} N_1 + B_{st} N_1 \rho$$

A, B_{sp}, and B_{st} are called the absorption coefficient, the coefficient of spontaneous emission, and the coefficient of stimulated emission respectively. N_0 and N_1 are the numbers of atoms in the two states, and ρ is the density of photons of energy $\epsilon = E_1 - E_0$.

Statistical mechanical arguments give the following relationships among A, B_{sp}, and B_{st}.

$$A = B_{st} = \frac{c^3}{8\pi\nu^2} B_{sp}$$

The laser is a device in which an extremely large number of photons are created in a single quantum state by means of stimulated emission from a metastable energy level.

PROBLEMS

11.1 Suppose the entropy of one mole ($N = N_o$ = Avogadro's number) of a substance is found, by experiment, to be

$$S(E, N_o, V) = R\, ln(V E^2)$$

where $R = N_o k_B$. What is the entropy function $S(E, N, V)$ for an arbitrary amount of the substance?

11.2* The entropy function for a certain gas is

$$S = 3 N k_B\, ln(V E^{3/2} / N^{5/2})$$

Calculate the energy density, E/V, and the pressure, p, as functions of the temperature, T, and the density, $n = N/V$.

11.3** Consider a system composed of 3 quantized harmonic oscillators, each of angular frequency ω_o. Calculate the entropy, S, for each of the following energy values.

$$E = 2\hbar\omega_o,\ 3\hbar\omega_o,\ 4\hbar\omega_o,\ 5\hbar\omega_o$$

11.4 The first excited state of helium lies 19.77 eV above the ground state and has a degeneracy of three. At what temperature would the number of atoms in the first excited state be equal to the number in the ground state?

11.5** (a) Evaluate the normalization constant

$$Q = \sum_n e^{-\beta \epsilon_n}$$

for a particle in a one-dimensional periodic box of length L in the high temperature limit. (b) Use the result to calculate the average value of the energy of such a particle at temperature T.

11.6** Each of the N atoms of a certain solid has an angular momentum of $\hbar/2$ and a magnetic moment μ. The solid is placed in a magnetic field B. The magnetic energy of the ith atom is then μB or $-\mu B$ depending on whether its spin component in the direction of B is plus or minus $\hbar/2$. Ignoring all other forms of energy:

(a) Determine the normalization constant (for one atom).

$$Q = \sum_n e^{-\beta \epsilon_n}$$

(b) Using Equation 11.8 determine the magnetic energy of the solid as a function of T.

(c) Determine the contribution of the magnetic energy to the heat capacity.

(d) Taking $\mu = 1$ Bohr magneton (9.27×10^{-24} J/T) and B = 2 T draw a rough plot of C_{mag} as a function of T.

11.7 The heat capacity per mole of the rhombic phase of sulfur is 21.30 J/K-mole at T = 300 K. Using the Einstein theory, predict the heat capacity per mole at T = 100 K.

11.8** Consider a violin string of length L, linear mass density μ, and tension τ.
(a) Calculate the density of normal modes function, $D(\omega)$, of the system.
(b) Calculate the heat capacity due to the quantized normal modes of the string. Let any sum over normal modes include modes of arbitrarily small wavelength. Remember that the waves are transverse waves and thus there are two normal modes for each wavelength.

11.9 For what range of temperature would a black body produce its most intense radiation within the visible spectrum? ($\lambda = .4\mu m\ .7\mu m$)

11.10* The moment of inertia of the carbon monoxide (CO) molecule is $1.36 \times 10^{-46}\,kg\text{-}m^2$.
(a) Calculate and plot N_l/N_0 for $l = 0, 1, 2, 3, 4, 5, 6$ at $T = 15\,K$, where N_l is the number of CO molecules with angular momentum $L = \sqrt{l(l+1)}\,\hbar$.
(b) What is the most probable rotational energy of a molecule at that temperature and how does it compare with the energy $\epsilon = k_BT$.

11.11* At high temperatures the angular momentum quantum number, l, can be treated as a continuous variable. Doing so obtain a formula for the most likely value of l as a function of T for a gas of diatomic molecules of moment of inertia I.

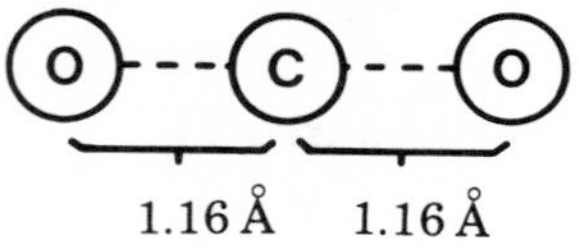

FIGURE P.11.12

11.12* CO_2 is a linear molecule with dimensions shown in Figure P.11.12
(a) Calculate its moment of inertia.
(b) At what temperature is $N_3/N_2 = \frac{1}{2}$? (N_l is the number of particles of angular momentum $L = \sqrt{l(l+1)}\,\hbar$.)

11.13** Derive Equation 11.17.

11.14 Show that, when $\overline{N}(E) << 1$, both the Fermi-Dirac and the Bose-Einstein distribution functions can be approximated by the *Maxwell-Boltzmann* distribution

$$\overline{N}(E) = \text{const}\cdot e^{-E/k_BT}$$

11.15* According to Equation 10.16, the energies of electrons and holes in solids can be approximated, near the energy gap, by the expressions

$$E_e = \tfrac{1}{2}E_g + \frac{p^2}{2m_e^*} \quad \text{and} \quad E_h = \tfrac{1}{2}E_h + \frac{p^2}{2m_h^*}$$

where E_g is the energy gap between the valence and conduction bands. At ordinary temperatures $k_BT << E_g$ and, for those electron and hole states that are likely to be filled, $p^2/2m^* << E_g$. Using these two approximations and the Fermi-Dirac distribution function, show that the density of particles and holes are both proportional to $\exp(-E_g/2k_BT)$. This is the justification for Equation 10.15.

11.16* Derive Equation 11.25 from Equation 11.24.

INDEX

Absolute space and time 22

Affinity, physical meaning of 367

Alpha-particle scattering 78

Alpha decay 227

Ammonia molecule 211

Angular momentum
  quantization of 139
  operators for 160
  quantum number 185

Antineutrino 227

Atom
  planetary 79
  many-electron 196

Atomic mass unit 221

Barrier penetration 133

Baryons, quark structure of 278

Bead on circular wire 138

Beta decay 227

Beta-stable nuclei
  curve of 234
  equation for 235

Blackbody radiation 71

Black holes 59

Bohr
  theory of hydrogen 83
  radius 84

Bose-Einstein particles 193

Bosons 193
  fundamental 276

Boundary conditions 116
  periodic 138

Box
  particle in 106
  periodic 138

Brillouin zone
  for lattice vibrations 313
  two- and three-dimensional 315

Bubble chamber 272

Carbon-14 dating 238

Cavity radiation 67

Cloud chamber 272

Chromodynamics, quantum 276

Classical physics 66
  relation to quantum 100

Colliding beam machines 267

Color
  charge 277
  field 277

Commutation of operators 169

Compton effect 74

Compton wavelength 76

Configuration, atomic 194

Confinement
  inertial 245
  magnetic 241
  quark 277

Contraction, length 11, 21

Counter
  Geiger 270
  multiwire 271
  proportional 271

Covalent bond
  in carbon compounds 207
  in solids 302

Crystal
  lattice 305
  -momentum of vibrational waves 322
  symmetry types 306

Curvature of space 58

Cyclotron 263
  frequency modulated 264

Davisson-Germer experiment 89

De Broglie
  waves 85
  relations 88

Debye approximation 321

Decay
  nuclear 226
  gamma 226
  beta 227
  alpha 227
  constant 229
  strong, EM, and weak 285

Degrees of freedom 318

Density of normal modes
  defined 319
  for electromagnetic field 71
  for one-dimensional lattice 320
  for three-dimensional lattice 321

Deuterium 240

Diatomic molecule
  potential for 200
  rotation 204
  spectrum 206
  vibration 201

Diffraction
  of electron waves 90, 97
  of light waves 97

Dispersion relation 316

Distribution function
  blackbody 68
  Planck 71
Doppler
  shift 28
  broadening 230
Eigenvalues
  operator 156
  measurements and 158
  energy 156
Einstein's theory of:
  atomic spectra 80
  general relativity 51
  heat capacity of solids 362
  photoelectric effect 73
  radiation emission and absorption 372
  special relativity 12
Electromagnetic
  interactions 281
  theory and relativity 4
Electron
  capture in beta decay 228
  scattering by protons 99
  spin 192
  states in solids 327
  velocity and wave equation 123
  waves 90, 97
Empirical mass formula 223
Energy
  and work done 47
  discrete, and standing waves 106
  operator 156
  operator in three dimensions 160
  relativistic 46
Energy bands
  in insulators and conductors 330
  in one dimension 328
  in two and three dimensions 329
Entropy
  in thermodynamics 350
  quantum interpretation of 356
Equivalence principle 48, 51
Ether, velocity of 6
Exclusion principle 193
Field effect transistor 341
Fermi-Dirac particles 193
Fermi distance unit 219
Fermion 193
  fundamental 275
Fine structure constant 84
Fizeau experiment 6, 25
Force
  nuclear 218
  strong 218, 284
Franck-Hertz experiment 81
Fundamental commutator theorem 170
Galileo 1
Gamma decay 226
Gaussian function 145
Geometrical principle 54
Geometry
  curvature of space 58
  spacetime 22
Gibbs distribution 357
  of open system 367
Grand unified theories 279
Hadrons, quark structure of 278
Half-life 229
Harmonic oscillator 141
  emission spectrum of 144
  energy spectrum of 145
  ground state wave function of 143
  normalized wave functions of 145
  quantum length for 142
  wave packet solution 146
Heat capacity
  quantum systems 360
  rotational 365
  solids 362
  vibrational 361
Heisenberg uncertainty relation 110
Helicity 227
Hermite polynomials 145
High energy physics 259
Hofstadter, R. 167
Holes
  electrical conduction by 333
  momentum and energy of 334
Hybrid state 208
Hydrogen
  Bohr theory of 83
  bonding in solids 304
  molecule 84
  molecular ion 199
  nonradial solutions 183
  quantum numbers of 187
  radial solutions 180
  spectrum 82, 182
  table of solutions 182, 187
Inertial confinement 245
Inertial frame 12
Insulators, band theory of 330
Ionic bonding in solids 302

Lattice
Bravais 306
defined 305
one-dimensional 311
vibrations 311
Metallic bonding 304
Length contraction 11, 21
Light waves and electron waves 97
Linear accelerator 266
Line width
Doppler broadening of 230
and Mossbauer effect 231
natural 230
nuclear 230
recoil broadening of 231
Lorentz-Fitzgerald contraction 11
Lorentz transformation 14
equations for 18
Lorentz transformer 26
Magnetic
confinement 241
field, motion of particle in 244
quantum number 185
spectrometer 167
Mass
and energy 44
formula, empirical 223
gravitational and inertial 50
number of nucleus 222
zero 48
Measurements
and eigenvalues 158
wave function after 163
Mercury's orbit 52
Mesons
quark structure of 279
and Yukawa theory 251
Metric, spacetime 55
Michelson-Morley experiment 8
Molecular
ion, hydrogen 199
potential 200
rotation 202, 204
spectrum 206
structure 198
vibration 201
Momentum
conservation 36
operator 157
relation to energy 41
relativistic 39
time component of 39
Mossbauer effect 231
Moving clock problem 53
Neutron
induced fission 250
-neutron force 219
number 222
quark structure of 278
star 221
Normalization of wave functions 109
Neutrino 227
Normal modes
defined 317
density of 319
of lattice 317
Nuclear
binding energies 222
decay modes 226
energy 241
fission 245, 251
fluid 220
force 218
fusion 239
linewidth 230
liquid drop model 221
potential 219
shells 237
size 221
stability 235
Nuclei
beta stable 234
electron scattering by 167
masses of 221
Nucleon
defined 218
-nucleon force 218
-nucleon scattering 218
Operator 156
angular momentum 160
commutation of 169
eigenvalues of 156
energy 156
momentum 157
position 159
Orthogonal quantum states 210
Periods, lengths of 197
Phonon
defined 323
gas 324
Photoelectric effect 73
Photon energy and momentum 76
Pion
and nuclear force 252
quark structure of 279
Planck distribution 71
derivation of 364
Planck's constant 71

Plasma
confinement 242
defined 241
instabilities 244
pn junction 340
Polarization of lattice waves 315
Position operator 159
eigenvalues of 159
Potential
localized attractive 117
localized repulsive 117
positively divergent 117
step 131
Pound-Rebka experiment 57
Probability
density 105, 107
density for hydrogen 183
of obtaining an eigenvalue 161
Particles
conservation of by wave equation 133
table of 49
zero mass 48
Pauli exclusion principle 193
Periodic table 194
Proton
decay 279
quark structure of 278
Quantization
of angular momentum 139
of electromagnetic energy 71
of energy 79
of lattice vibrations 317
Quantum field theory 290
Quantum number
angular momentum 185
magnetic 185
principal 185
Quantum state
definition 155
orthogonal 210
setting up 163
Radiation by atoms 122
Radiative decay, law of 228
Reduced mass
in diatomic molecule 203
in hydrogen atom 177
Reflection coefficient 132
Relativistic
dynamics 36
energy 46
momentum 39
particle in magnetic field 167
velocity 37
Relativity
Einstein's theory of 12
Galilean 1
general 51
principle of 1
special 12
Resonance absorption 232
Ritz combination principle 81
Rotation, molecular 202
Rotational heat capacity 365
SCF method 196
Scattering
electron-nucleon 99, 167
nucleon-nucleon 219
Schroedinger equation
and barrier penetration 133
derivation 102
interpretation of solution 104
and particle conservation 133
radial solutions for hydrogen 180
solution of 130
spurious solutions of 115
in three dimensions
with a potential 113
with step potential 131
Selection rules for molec. transitions 205
Self-consistent field approximation 196
Semiconductors 332
doped 338
holes in 333
p-type and n-type 337
Shells
atomic 196
nuclear 237
Solids
amorphous 301
bonding in 302
crystalline 301
heat capacity of 362
Spacetime
geometry of 22
metric 55
Spectrometer, magnetic 167
Spectrum
Einstein's interpretation of 80
harmonic oscillator 145
hydrogen 82
Spherical coordinates 184
Spin, electron 192
Stability, nuclear 235
Standard quantum experiment 165
Standing waves and discrete energies 106

States
setting up 163
stationary 107
and wave functions 163

Strong interactions 284

Symmetry
in crystals 306
energy 224

Synchrotron 265

Tetrahedral compounds 211

Time
dilation 20
component of velocity 38
component of momentum 39

Tracers, isotopes as 230

Transistors 341
FET 341

Transmission coefficient 132

Trouton-Noble experiment 7

Tunneling 119, 135

Twin paradox 43

Uncertainty principle 110

Van de Graaff generator 262

Van der Waals bonding in solids 305

Velocity
addition, relativistic 24, 26
phase and group 123
relativistic 37
time component of 38

Vibration, molecular 201

Vibrational-rotational spectrum 206

Vibrational waves
in one-dimensional lattice 311
quantized 317
polarization of 314
in two- and three-dim. lattice 314

Wave functions
normalization of 109
physical interpretation of 105
supplemental conditions on 116

Weak interactions
charged 283
neutral 281

Weightlessness 48

White dwarfs 60

Wien displacement law 365

Yukawa
potential 251
prediction of pion 252
theory of nuclear force 251

Zeeman effect 189

PERIODIC TABLE OF THE ELEMENTS

Group I	Group II	Transition Elements										Group III	Group IV	Group V	Group VI	Group VII	Group VIII
1 **H** 1.00797																	2 **He** 4.0026
3 **Li** 6.939	4 **Be** 9.0122											5 **B** 10.811	6 **C** 12.01115	7 **N** 14.0067	8 **O** 15.9994	9 **F** 18.9984	10 **Ne** 20.183
11 **Na** 22.9898	12 **Mg** 24.312											13 **Al** 26.9815	14 **Si** 28.086	15 **P** 30.9738	16 **S** 32.064	17 **Cl** 35.453	18 **Ar** 39.948
19 **K** 39.102	20 **Ca** 40.08	21 **Sc** 44.956	22 **Ti** 47.90	23 **V** 50.942	24 **Cr** 51.996	25 **Mn** 54.9380	26 **Fe** 55.847	27 **Co** 58.9332	28 **Ni** 58.71	29 **Cu** 63.54	30 **Zn** 65.37	31 **Ga** 69.72	32 **Ge** 72.59	33 **As** 74.9216	34 **Se** 78.96	35 **Br** 79.909	36 **Kr** 83.80
37 **Rb** 85.47	38 **Sr** 87.62	39 **Y** 88.905	40 **Zr** 91.22	41 **Nb** 92.906	42 **Mo** 95.94	43 **Tc** (99)	44 **Ru** 101.07	45 **Rh** 102.905	46 **Pd** 106.4	47 **Ag** 107.870	48 **Cd** 112.40	49 **In** 114.82	50 **Sn** 118.69	51 **Sb** 121.75	52 **Te** 127.60	53 **I** 126.9044	54 **Xe** 131.30
55 **Cs** 132.905	56 **Ba** 137.34	57 **La*** 138.91	72 **Hf** 178.49	73 **Ta** 180.948	74 **W** 183.85	75 **Re** 186.2	76 **Os** 190.2	77 **Ir** 192.2	78 **Pt** 195.09	79 **Au** 196.967	80 **Hg** 200.59	81 **Tl** 204.37	82 **Pb** 207.19	83 **Bi** 208.980	84 **Po** (210)	85 **At** (210)	86 **Rn** (222)
87 **Fr** (223)	88 **Ra** (226)	89 **Ac#** (227)															

*Lanthanide Series

58 **Ce** 140.12	59 **Pr** 140.907	60 **Nd** 144.24	61 **Pm** (147)	62 **Sm** 150.35	63 **Eu** 151.96	64 **Gd** 157.25	65 **Tb** 158.924	66 **Dy** 162.50	67 **Ho** 164.930	68 **Er** 167.26	69 **Tm** 168.934	70 **Yb** 173.04	71 **Lu** 174.97

#Actinide Series

90 **Th** 232.038	91 **Pa** (231)	92 **U** 238.03	93 **Np** (237)	94 **Pu** (242)	95 **Am** (243)	96 **Cm** (247)	97 **Bk** (247)	98 **Cf** (249)	99 **Es** (254)	100 **Fm** (253)	101 **Md** (256)	102 **No** (253)	103 **Lw** (257)

A number in parentheses represents the mass number of the most stable isotope of a radioactive element.

PHYSICAL CONSTANTS

Name	Symbol	Value	Units
Bohr radius	a_o	5.29177×10^{-11}	m
Speed of light	c	2.9979246×10^{8}	m/s
Electron charge	e	1.60218×10^{-19}	C
Gravitational constant	G	6.67×10^{-11}	$N\text{-}m^2/kg^2$
Planck's constant	h	6.62612×10^{-34}	J-s
	$\hbar$	1.05458×10^{-34}	J-s
Boltzmann's constant	k_B	1.3807×10^{-23}	J/K
Coulomb force constant	k	8.98755×10^{9}	$N\text{-}m^2/C^2$
Electron mass	m_e	9.1095×10^{-31}	kg
Proton mass	m_p	1.67265×10^{-27}	kg
Neutron mass	m_n	1.67495×10^{-27}	kg
Avogadro's number	N_A	6.0220×10^{23}	mol^{-1}

DEFINITE INTEGRALS

$$\int_0^\infty e^{-x} x^n \, dx = 1 \cdot 2 \cdot 3 \cdots n = n!$$

$$\int_0^\infty e^{-ax} x^n \, dx = n!/a^{n+1}$$

$$\int_{-\infty}^\infty e^{-ax^2} \, dx = \sqrt{\pi/a}$$

$$\int_{-\infty}^\infty e^{-ax^2} x^2 \, dx = \frac{\sqrt{\pi/a}}{2a}$$

$$\int_{-\infty}^\infty e^{-ax^2} x^{2n} \, dx = \frac{1 \cdot 3 \cdot 5 \cdots (2n-1)}{(2a)^n} \sqrt{\pi/a}$$